Stochastic Financial Models

CHAPMAN & HALL/CRC
Financial Mathematics Series

Aims and scope:
The field of financial mathematics forms an ever-expanding slice of the financial sector. This series aims to capture new developments and summarize what is known over the whole spectrum of this field. It will include a broad range of textbooks, reference works and handbooks that are meant to appeal to both academics and practitioners. The inclusion of numerical code and concrete real-world examples is highly encouraged.

Series Editors

M.A.H. Dempster
Centre for Financial Research
Judge Business School
University of Cambridge

Dilip B. Madan
Robert H. Smith School of Business
University of Maryland

Rama Cont
Center for Financial Engineering
Columbia University
New York

Published Titles

American-Style Derivatives; Valuation and Computation, *Jerome Detemple*
Analysis, Geometry, and Modeling in Finance: Advanced Methods in Option Pricing, *Pierre Henry-Labordère*
Credit Risk: Models, Derivatives, and Management, *Niklas Wagner*
Engineering BGM, *Alan Brace*
Financial Modelling with Jump Processes, *Rama Cont and Peter Tankov*
Interest Rate Modeling: Theory and Practice, *Lixin Wu*
An Introduction to Credit Risk Modeling, *Christian Bluhm, Ludger Overbeck, and Christoph Wagner*
Introduction to Stochastic Calculus Applied to Finance, Second Edition, *Damien Lamberton and Bernard Lapeyre*
Monte Carlo Methods and Models in Finance and Insurance, *Ralf Korn, Elke Korn, and Gerald Kroisandt*
Numerical Methods for Finance, *John A. D. Appleby, David C. Edelman, and John J. H. Miller*
Portfolio Optimization and Performance Analysis, *Jean-Luc Prigent*
Quantitative Fund Management, *M. A. H. Dempster, Georg Pflug, and Gautam Mitra*
Robust Libor Modelling and Pricing of Derivative Products, *John Schoenmakers*
Stochastic Financial Models, *Douglas Kennedy*
Structured Credit Portfolio Analysis, Baskets & CDOs, *Christian Bluhm and Ludger Overbeck*
Understanding Risk: The Theory and Practice of Financial Risk Management, *David Murphy*
Unravelling the Credit Crunch, *David Murphy*

Proposals for the series should be submitted to one of the series editors above or directly to:

CRC Press, Taylor & Francis Group
4th, Floor, Albert House
1-4 Singer Street
London EC2A 4BQ
UK

Chapman & Hall/CRC FINANCIAL MATHEMATICS SERIES

Stochastic Financial Models

Douglas Kennedy

Trinity College
Cambridge, UK

CRC Press
Taylor & Francis Group
Boca Raton London New York

CRC Press is an imprint of the
Taylor & Francis Group, an **informa** business
A CHAPMAN & HALL BOOK

CRC Press
Taylor & Francis Group
6000 Broken Sound Parkway NW, Suite 300
Boca Raton, FL 33487-2742

First issued in paperback 2018

ISBN-13: 978-1-4200-9345-2 (hbk)
ISBN-13: 978-1-138-38145-2 (pbk)

Library of Congress Cataloging-in-Publication Data

Kennedy, Douglas.
Stochastic financial models / Douglas Kennedy.
p. cm. -- (Chapman & Hall/CRC Financial mathematics series)
Includes bibliographical references and index.
ISBN 978-1-4200-9345-2 (hardcover : alk. paper)
1. Investments--Mathematical models. 2. Stochastic analysis. I. Title. II. Series.

HG4515.2.K46 2010
332.63'2042--dc22 2009044114

Visit the Taylor & Francis Web site at
http://www.taylorandfrancis.com

and the CRC Press Web site at
http://www.crcpress.com

Contents

Preface

The bulk of this work originated in lecture notes prepared for two courses that I introduced for students of mathematics at the University of Cambridge, dating back to 1992, and with which I was associated for fifteen years. The first of these was for final-year undergraduates and it covered the material in Chapter 1 and roughly half of each of Chapters 2, 4 and 5. The second was for first-year graduate students and it dealt with most of Chapters 2–6 at an accelerated pace; it did not have the former course as a prerequisite.

While students were expected to have a good prior knowledge of elementary probability theory and perhaps some acquaintance with Markov chains, no background in measure-theoretic probability was assumed for either course, although in each case students had the opportunity to take a course on that topic concurrently; for the more advanced option, an introductory course on stochastic calculus was also available to be taken at the same time. Apart from the intrinsic interest of presenting the material on mathematical finance, a major pedagogical motivation for introducing the courses was to stimulate students to learn more about probability, martingales and stochastic integration by exposing them to one of the most important and exciting areas of application of those topics.

The introduction to mathematical finance presented here is designed to slot in between those works that provide a survey of the field with a relatively light mathematical content and those books at the other end of the spectrum, which take no prisoners in their rigorous, formal approach to stochastic integration and probabilistic ideas. In many places in the book the slant is toward a classical applied mathematical approach with a concentration on calculations rather than necessarily seeking the greatest generality. To avoid breaking the flow of material, where concepts from measure-theoretic probability are required, for the most part they are not introduced in the main body of the text but have been gathered in the mathematical preliminaries in Appendix A; the reader is also encouraged to consult the books suggested for further reading. To assist self study, solutions to all the exercises are given in Appendix B but students are urged strongly to attempt the problems unaided before consulting the solutions.

It is not necessary to follow the material in the book in a strict linear order. To provide some route maps through the chapters, it should be noted that the material in Chapter 1 is orthogonal to much of the remaining book in that it deals with the more classical topics of utility and the mean-variance approach to portfolio choice, rather than being concerned with derivative pricing, which is the focus in the remainder of the book. If the existence of an individual utility function is taken as given, then this chapter is not required for an understanding of the subsequent material but the chapter is included to give a more rounded view of finance generally. A full understanding of the binomial model, presented in Chapter 2, is central for getting to grips with the pricing of derivatives by self-financing hedging portfolios. It should be possible for the reader to proceed directly from this chapter to the Black–Scholes model in Chapter 5 without studying the general discrete-time model in Chapter 3, having acquired sufficient background on Brownian motion from Chapter 4. For example, if the reader wanted to get quickly to the Black–Scholes formula at a first reading, it would be possible to omit consideration of hitting-time distributions for Brownian motion in Section 4.2, and with the ideas of Sections 4.1, 4.3 and 4.4, proceed to study Sections 5.1 and 5.2; then Section 4.5 on stochastic calculus could be consulted to give the basis for reading Section 5.3 on hedging in the Black–Scholes context. One might then return to look at hitting-time distributions before dealing with path-dependent options in Section 5.4.

I must thank all the former students whose helpful observations contributed to the development of this material, but I am particularly grateful to Bryn Thompson-Clarke and Wenjie Xiang, who gave insightful comments on an early draft of the book, and also to Rob Calver of Chapman & Hall for his encouragement during the project. I am also indebted to an anonymous reviewer who provided very useful and perceptive suggestions. Finally, I want to thank all those in Trinity College who have contributed to making such a congenial, stimulating and beautiful working environment, which I have been privileged to enjoy for thirty-five years.

Douglas Kennedy
Trinity College
Cambridge, CB2 1TQ

Chapter 1

PORTFOLIO CHOICE

1.1 Introduction

The contents of this chapter are somewhat different in approach to much of the remainder of the book. In subsequent chapters we deal principally with the problem of pricing derivative securities; that is, secondary contracts for which the payoff is dependent on the price of an underlying asset, such as a stock. In complete markets, as in the cases of the binomial model of Chapter 2 or the Black–Scholes model of Chapter 5, the price of any derivative contract, or contingent claim, is derived objectively through 'hedging'; the payoff of the contract may be duplicated exactly by trading in the underlying asset and a bank account so that there is no inherent risk to the seller of the contract. The price of the contract is just the initial cost of setting up the trading strategy that duplicates the payoff; it is objective and risk free in that two investors may have different views on how the price of the underlying asset may evolve in the future but they will agree on the price of the contract. Such ideas form the basis of much of modern financial theory.

By contrast, the two topics in this chapter deal with the attitudes of individual investors in relation to investment decisions where these decisions are subjective in nature. The first is the notion of an investor's individual utility function. In a deterministic model it is reasonable to expect that an investor will seek to choose an investment portfolio of assets in order to maximize the final wealth that he achieves. When the model is stochastic the investor's final wealth will typically be a random variable, W, and it would no longer make sense for him to make investment decisions seeking to maximize a random quantity. Instead, he might wish to maximize the expected value of his final wealth, $\mathbb{E}(W)$, so that he achieves the largest wealth on average, or more generally it is often postulated that he seeks to maximize $\mathbb{E}\,v(W)$ for some appropriate function $v(\cdot)$; this function is referred to as the investor's utility function. It is individual to the investor and we show in Section 1.2 that any investor who orders his preferences of random outcomes in a suitably consistent way possesses an essentially unique utility function and that properties of this function may characterize his attitude towards risk.

The second topic of the chapter in Section 1.3 is concerned with mean-variance analysis where, among portfolios giving a fixed mean return, an investor chooses the portfolio with smallest variance of the return. The model is subjective both in its choice of optimality criterion but also in its dependence on the investors beliefs about the means of the returns of the various available assets as well as the covariances between those returns. The capital-asset pricing model in Section 1.3.5 considers the implications for the whole market of the actions of individual investors.

The material of this chapter represents a significant step in the development of mathematical models in finance; its importance was recognized by the award of the Nobel Prize in Economics in 1990 to Harry Markowitz, for his contributions to the theory of portfolio choice, and to William Sharpe, for his work on the capital-asset pricing model.

1.2 Utility

1.2.1 Preferences and utility

We begin by discussing the classical justification for assuming that an investor who may order his preferences for investments in a consistent manner has an essentially unique utility function; furthermore, the properties of this utility function characterize his attitude to risk. We outline the axiomatic approach to showing the existence of such a utility function when it is assumed that the investor's preferences satisfy certain axioms.

Let Γ be a sample space representing the set of possible outcomes of some gambles with random payoffs. Let $\mathcal{P}$ be a set of probabilities on Γ, so that an element $\mathbb{A} \in \mathcal{P}$ is a real-valued function defined on subsets (events) of Γ satisfying the following three conditions:

1. $0 \leqslant \mathbb{A}(G) \leqslant 1$, for all $G \subseteq \Gamma$;
2. $\mathbb{A}(\Gamma) = 1$; and
3. for a finite or infinite collection of disjoint events $\{G_i\}_i$, that is $G_i \cap G_j = \emptyset$ for $i \neq j$, we have
$$\mathbb{A}\left(\bigcup_i G_i\right) = \sum_i \mathbb{A}\left(G_i\right).$$

We refer to an element $\mathbb{A} \in \mathcal{P}$ as a **gamble** where $\mathbb{A}$ may be thought of as the probability distribution of the outcome of the gamble. We assume that the set $\mathcal{P}$ is closed under convex combinations so that for any $\mathbb{A}, \mathbb{B} \in \mathcal{P}$ and $0 \leqslant p \leqslant 1$ we assume that $p\mathbb{A} + (1-p)\mathbb{B} \in \mathcal{P}$. The gamble $p\mathbb{A} + (1-p)\mathbb{B}$ takes the value $p\mathbb{A}(G) + (1-p)\mathbb{B}(G)$ for events $G \subseteq \Gamma$ and it is of course a probability on Γ; this gamble would correspond to the situation where the investor tosses a coin with probability p of 'heads' and $1-p$ of 'tails' and chooses gamble $\mathbb{A}$ or gamble $\mathbb{B}$ according to whether the outcome is heads or tails. It is an immediate consequence of this assumption that for any gambles $\mathbb{A}_1, \ldots, \mathbb{A}_k \in \mathcal{P}$ and for real numbers $p_i \geqslant 0$, $1 \leqslant i \leqslant k$ with $\sum_{i=1}^k p_i = 1$, by induction on k we see that
$$p_1\mathbb{A}_1 + \cdots + p_k\mathbb{A}_k \in \mathcal{P}.$$

We will assume that an investor (or gambler) has a preference relation $\succ$ on $\mathcal{P}$; this corresponds to some given subset $\mathcal{S} \subseteq \mathcal{P} \times \mathcal{P}$ with $\mathbb{A} \succ \mathbb{B}$ if and only if $(\mathbb{A}, \mathbb{B}) \in \mathcal{S}$ for gambles $\mathbb{A}, \mathbb{B} \in \mathcal{P}$. Read $\mathbb{A} \succ \mathbb{B}$ as $\mathbb{A}$ **is preferred to** $\mathbb{B}$. This defines a relation $\sim$ on $\mathcal{P}$ by setting $\mathbb{A} \sim \mathbb{B}$ when $\mathbb{A} \not\succ \mathbb{B}$ and $\mathbb{B} \not\succ \mathbb{A}$ for $\mathbb{A}, \mathbb{B} \in \mathcal{P}$, that is when $(\mathbb{A}, \mathbb{B}) \notin \mathcal{S}$ and $(\mathbb{B}, \mathbb{A}) \notin \mathcal{S}$. We will refer to $\sim$ as an indifference relation and say that the investor is **indifferent between** $\mathbb{A}$ and $\mathbb{B}$ when $\mathbb{A} \sim \mathbb{B}$. We will assume here that the relations $\succ$ and $\sim$ satisfy some plausible axioms which would imply rational consistency on the part of the investor in ordering his preferences.

Axioms

1. For any $\mathbb{A}, \mathbb{B} \in \mathcal{P}$ exactly one of the following holds:

   (i) $\mathbb{A} \succ \mathbb{B}$; (ii) $\mathbb{B} \succ \mathbb{A}$; or (iii) $\mathbb{A} \sim \mathbb{B}$.

2. The relation $\sim$ is an equivalence relation on $\mathcal{P}$; that is,

   (i) $\mathbb{A} \sim \mathbb{A}$ for all $\mathbb{A} \in \mathcal{P}$;
   (ii) for any $\mathbb{A}, \mathbb{B} \in \mathcal{P}$, if $\mathbb{A} \sim \mathbb{B}$ then $\mathbb{B} \sim \mathbb{A}$; and
   (iii) for any $\mathbb{A}, \mathbb{B}, \mathbb{C} \in \mathcal{P}$, if $\mathbb{A} \sim \mathbb{B}$ and $\mathbb{B} \sim \mathbb{C}$ then $\mathbb{A} \sim \mathbb{C}$.

3. For any $\mathbb{A}, \mathbb{B}, \mathbb{C} \in \mathcal{P}$, if $\mathbb{A} \succ \mathbb{B}$ and $\mathbb{B} \succ \mathbb{C}$ then $\mathbb{A} \succ \mathbb{C}$.

4. For any $\mathbb{A}, \mathbb{B}, \mathbb{C} \in \mathcal{P}$,

   (i) if $\mathbb{A} \succ \mathbb{B}$ and $\mathbb{B} \sim \mathbb{C}$ then $\mathbb{A} \succ \mathbb{C}$; and
   (ii) if $\mathbb{A} \sim \mathbb{B}$ and $\mathbb{B} \succ \mathbb{C}$ then $\mathbb{A} \succ \mathbb{C}$.

5. For any $\mathbb{A}, \mathbb{C} \in \mathcal{P}$ and $p \in [0, 1]$, if $\mathbb{A} \sim \mathbb{C}$ and $\mathbb{B} \in \mathcal{P}$ then $p\mathbb{A} + (1-p)\mathbb{B} \sim p\mathbb{C} + (1-p)\mathbb{B}$.

6. For any $\mathbb{A}, \mathbb{C} \in \mathcal{P}$ and $p \in (0, 1]$, if $\mathbb{A} \succ \mathbb{C}$ and $\mathbb{B} \in \mathcal{P}$ then $p\mathbb{A} + (1-p)\mathbb{B} \succ p\mathbb{C} + (1-p)\mathbb{B}$.

7. For any $\mathbb{A}, \mathbb{B}, \mathbb{C} \in \mathcal{P}$, if $\mathbb{A} \succ \mathbb{C} \succ \mathbb{B}$ then there exists $p \in [0, 1]$ with $p\mathbb{A} + (1-p)\mathbb{B} \sim \mathbb{C}$.

We observe first that the p in Axiom 7 is unique.

Lemma 1.1 *Suppose that $\mathbb{A}, \mathbb{B}, \mathbb{C} \in \mathcal{P}$ with $\mathbb{A} \succ \mathbb{C} \succ \mathbb{B}$ and $p\mathbb{A} + (1-p)\mathbb{B} \sim \mathbb{C}$ then $0 < p < 1$ and p is unique.*

Proof. Trivially $p \neq 0$ or 1. Suppose that p is not unique so that there exists q with $q\mathbb{A} + (1-q)\mathbb{B} \sim \mathbb{C}$. Without loss of generality assume that $q < p$ so that we have $0 < p - q < 1 - q$. But

$$\mathbb{B} = \left(\frac{p-q}{1-q}\right)\mathbb{B} + \left(\frac{1-p}{1-q}\right)\mathbb{B} \quad \text{and} \quad \mathbb{A} \succ \mathbb{B},$$

then by Axiom 6

$$\left(\frac{p-q}{1-q}\right)\mathbb{A} + \left(\frac{1-p}{1-q}\right)\mathbb{B} \succ \mathbb{B}.$$

However

$$p\mathbb{A} + (1-p)\mathbb{B} = q\mathbb{A} + (1-q)\left[\left(\frac{p-q}{1-q}\right)\mathbb{A} + \left(\frac{1-p}{1-q}\right)\mathbb{B}\right],$$

and by Axiom 6 again, this implies that

$$p\mathbb{A} + (1-p)\mathbb{B} \succ q\mathbb{A} + (1-q)\mathbb{B}$$

which gives a contradiction. ∎

We may now establish the existence and linearity of a function which quantifies the preferences when those preferences are formulated consistently in that they obey the Axioms.

Theorem 1.1 *There exists a real-valued function $f : \mathcal{P} \to \mathbb{R}$ with*

$$f(\mathbb{A}) > f(\mathbb{B}) \text{ if and only if } \mathbb{A} \succ \mathbb{B}, \tag{1.1}$$

and

$$f(p\mathbb{A} + (1-p)\mathbb{B}) = pf(\mathbb{A}) + (1-p)f(\mathbb{B}) \tag{1.2}$$

for any $\mathbb{A}, \mathbb{B} \in \mathcal{P}$ and $0 \leqslant p \leqslant 1$. Furthermore, f is unique up to affine transformations; that is, if g is any other such function satisfying (1.1) *and* (1.2) *then there exist real numbers $\alpha > 0$ and β with $g(\mathbb{A}) = \alpha f(\mathbb{A}) + \beta$, for all $\mathbb{A} \in \mathcal{P}$.*

Proof. If we have $\mathbb{A} \sim \mathbb{B}$ for all $\mathbb{A}$, $\mathbb{B} \in \mathcal{P}$ then take $f(\mathbb{A}) \equiv 0$ and the conclusions are immediate. So suppose that there exists a pair $\mathbb{C}, \mathbb{D} \in \mathcal{P}$ with $\mathbb{C} \succ \mathbb{D}$. By the axioms, for any $\mathbb{A} \in \mathcal{P}$ there are five possibilities: (a) $\mathbb{A} \succ \mathbb{C}$, (b) $\mathbb{A} \sim \mathbb{C}$, (c) $\mathbb{C} \succ \mathbb{A} \succ \mathbb{D}$, (d) $\mathbb{A} \sim \mathbb{D}$ and (e) $\mathbb{D} \succ \mathbb{A}$. First define $f(\mathbb{C}) = 1$ and $f(\mathbb{D}) = 0$. We define $f(\mathbb{A})$ for $\mathbb{A}$ satisfying each case in turn. For case (a) there exists a unique $p \in (0, 1)$ with $p\mathbb{A} + (1-p)\mathbb{D} \sim \mathbb{C}$; define $f(\mathbb{A}) = 1/p$. For case (b) set $f(\mathbb{A}) = 1$. For case (c) there exists a unique $q \in (0, 1)$ with $q\mathbb{C} + (1-q)\mathbb{D} \sim \mathbb{A}$; define $f(\mathbb{A}) = q$. For case (d) set $f(\mathbb{A}) = 0$. Finally, for case (e) there exists a unique $r \in (0, 1)$ with $r\mathbb{C} + (1-r)\mathbb{A} \sim \mathbb{D}$ and define $f(\mathbb{A}) = -r/(1-r)$.

To check that f satisfies (1.1) and (1.2) for all $\mathbb{A}$ and $\mathbb{B}$ requires checking fifteen different cases for $\mathbb{A}$ and $\mathbb{B}$; these correspond to the five instances where both $\mathbb{A}$ and $\mathbb{B}$ satisfy one of the five cases (a)-(e), together with the $\binom{5}{2} = 10$ instances when $\mathbb{A}$ and $\mathbb{B}$ are in different cases of (a)-(e).

We give the details in just one situation when both $\mathbb{A}$ and $\mathbb{B}$ are in case (c) so that $\mathbb{C} \succ \mathbb{A} \succ \mathbb{D}$ and $\mathbb{C} \succ \mathbb{B} \succ \mathbb{D}$. We have $f(\mathbb{A}) = q_1$ and $f(\mathbb{B}) = q_2$, say, where

$$\mathbb{A} \sim q_1\mathbb{C} + (1-q_1)\mathbb{D} \quad \text{and} \quad \mathbb{B} \sim q_2\mathbb{C} + (1-q_2)\mathbb{D}.$$

When $q_1 = q_2$ then $\mathbb{A} \sim \mathbb{B}$ and (1.1) holds. When $q_1 > q_2$ then, as in the proof of Lemma 1.1,

$$q_1\mathbb{C} + (1 - q_1)\mathbb{D} \succ q_2\mathbb{C} + (1 - q_2)\mathbb{D}$$

and thus $\mathbb{A} \succ \mathbb{B}$ giving (1.1); similarly, when $q_1 < q_2$ it follows that $\mathbb{B} \succ \mathbb{A}$. To see that (1.2) holds, let $p \in (0, 1)$ and then by Axiom 5

$$p\mathbb{A} + (1 - p)\mathbb{B} \sim [p\,(q_1\mathbb{C} + (1 - q_1)\mathbb{D}) + (1 - p)\,(q_2\mathbb{C} + (1 - q_2)\mathbb{D})]$$

which may be rearranged to show that

$$p\mathbb{A} + (1 - p)\mathbb{B} \sim [(pq_1 + (1 - p)q_2)\,\mathbb{C} + (p(1 - q_1) + (1 - p)(1 - q_2))\,\mathbb{D}]\,.$$

It follows from the definition of f that

$$f(p\mathbb{A} + (1 - p)\mathbb{B}) = pq_1 + (1 - p)q_2 = pf(\mathbb{A}) + (1 - p)f(\mathbb{B}),$$

which establishes (1.2) in this case.

To verify that f is unique up to affine transformations, suppose that g is any other function satisfying (1.1) and (1.2). Because $\mathbb{C} \succ \mathbb{D}$ we must have $g(\mathbb{C}) > g(\mathbb{D})$, then define $\beta = g(\mathbb{D})$ and $\alpha = g(\mathbb{C}) - g(\mathbb{D}) > 0$. Now suppose that $\mathbb{A}$ is in case (c) so that $\mathbb{C} \succ \mathbb{A} \succ \mathbb{D}$. If $f(\mathbb{A}) = q$ then $\mathbb{A} \sim q\mathbb{C} + (1 - q)\mathbb{D}$ and it follows that

$$\begin{aligned} g(\mathbb{A}) &= g(q\mathbb{C} + (1 - q)\mathbb{D}) = qg(\mathbb{C}) + (1 - q)g(\mathbb{D}) \\ &= q(\alpha + \beta) + (1 - q)\beta = q\alpha + \beta = \alpha f(\mathbb{A}) + \beta. \end{aligned}$$

The other cases follow in a similar fashion. □

This result establishes that for an investor with a consistent set of preferences there exists a function f, unique up to affine transformations, which quantifies the ordering of his preferences in the sense of (1.1). Note that it is an immediate consequence of (1.2) that for gambles $\mathbb{A}_1, \ldots, \mathbb{A}_k \in \mathcal{P}$ and $p_i \geqslant 0$, $1 \leqslant i \leqslant k$, with $\sum_{i=1}^k p_i = 1$ the function f satisfies

$$f\left(\sum_{i=1}^{k} p_i \mathbb{A}_i\right) = \sum_{i=1}^{k} p_i f(\mathbb{A}_i); \tag{1.3}$$

this is established by induction on k.

Suppose that $\Gamma = \{\gamma_1, \ldots, \gamma_n\}$ has only a finite number of outcomes. Let $\mathbb{A}_i$ be the probability that assigns 1 to the outcome γ_i, $i = 1, \ldots, n$, and 0 to the other outcomes and assume that $\mathbb{A}_i \in \mathcal{P}$ for each i. Let $\mathbb{A} = (p_1, \ldots, p_n)$ be the probability distribution assigning the probability p_i to γ_i where $p_i \geqslant 0$ and $\sum_{i=1}^n p_i = 1$. From (1.3), it follows that

$$f(\mathbb{A}) = \sum_{i=1}^{n} p_i f(\mathbb{A}_i)$$

so that $f(\mathbb{A})$ is the expected value of the random variable which takes the value $f(\mathbb{A}_i)$ when the outcome is γ_i.

To put these ideas into the context in which they are typically encountered in finance, consider an investor who is faced with a range of investments each of which yields a payoff which is a real-valued random variable defined on some underlying probability sample space Ω, which is equipped with a probability (measure) $\mathbb{P}$. Let $\mathcal{X}$ be the set of real-valued random variables defined on Ω and, for each random variable $X \in \mathcal{X}$, let $\mathbb{P}^X$ denote the probability distribution on $\mathbb{R}$ induced by X. Here we will take the sample space Γ in the above description to be $\Gamma = \mathbb{R}$.

Suppose that the investor has a preference system ($\succ$ and $\sim$) that orders the gambles (or investments) $\mathcal{P} = \{\mathbb{P}^X : X \in \mathcal{X}\}$ in a consistent way according to the Axioms, then we know that there exists a function f so that $\mathbb{P}^X \succ \mathbb{P}^Y$ (or we may write $X \succ Y$, equivalently) if and only if $f(\mathbb{P}^X) > f(\mathbb{P}^Y)$. Let us consider the case where each random variable takes on a finite number of values so that the range of X is $\mathcal{R}(X) = \{x_1, \ldots, x_m\}$, say; then for $x \in \mathbb{R}$ we have

$$\mathbb{P}^X(\{x\}) = \begin{cases} \mathbb{P}(X = x) & \text{for} \quad x \in \mathcal{R}(X), \\ 0 & \text{for} \quad x \notin \mathcal{R}(X). \end{cases}$$

For any $x \in \mathbb{R}$ denote by $\mathbb{P}^x$ the probability distribution which assigns 1 to the point x and 0 to all other points of $\mathbb{R}$ and define a function $v : \mathbb{R} \to \mathbb{R}$ by setting $v(x) = f(\mathbb{P}^x)$, for $x \in \mathbb{R}$. With this notation, the relation (1.3) is the statement that

$$f(\mathbb{P}^X) = \sum_{i=1}^{m} f(\mathbb{P}^{x_i})\mathbb{P}(X = x_i) = \sum_{i=1}^{m} v(x_i)\mathbb{P}(X = x_i) = \mathbb{E}\, v(X).$$

The conclusion (1.1) then becomes

$$\mathbb{E}\, v(X) > \mathbb{E}\, v(Y) \quad \text{if and only if} \quad X \succ Y. \tag{1.4}$$

The function $v(\cdot)$ is known as the investor's **utility** function; it is unique up to the affine transformation implied by Theorem 1.1; that is, it is unique up to transformations of the form $\overline{v}(x) = av(x) + b$ for constants $a > 0$ and b, and it is determined by his individual preference system. The relation (1.4) implies that when the investor is faced with a number of investments with random payoff he will choose the one with largest expected utility; in subsequent sections we will refer to an individual acting in this way as a **utility-maximizing** investor. We will see in the next section that properties of the utility function indicate details of the attitude of the investor towards risk.

The discussion that leads to (1.4) was restricted to the situation where the random variables take only finitely many values. The result may be extended to arbitrary random variables but it requires consideration of closure properties of the set of gambles and consistency of the preference Axioms under countable convex combinations of gambles.

1.2.2 Utility and risk aversion

We will assume here that the outcome of an investment is described by a random variable X (defined on some sample space Ω with probability $\mathbb{P}$) and that the preferences of an investor may be described as in the previous section by a utility function $v : \mathbb{R} \to \mathbb{R}$ with the investor preferring investments with higher expected utility. Denote by $\mathbb{E}_{\mathbb{P}}$ the expectation taken with the probability $\mathbb{P}$. We say that the investor is **risk averse** when

$$\mathbb{E}_{\mathbb{P}} v(X) \leqslant v\left(\mathbb{E}_{\mathbb{P}} X\right), \tag{1.5}$$

for all random variables X and all probabilities $\mathbb{P}$. The investor is risk averse if and only if his utility function is concave. To see this, for two fixed values $x, y \in \mathbb{R}$ and $\lambda \in [0, 1]$ suppose that the probability $\mathbb{P}$ is such that $\mathbb{P}(X = x) = \lambda$ and $\mathbb{P}(X = y) = 1 - \lambda$ then (1.5) implies that

$$\lambda v(x) + (1 - \lambda) v(y) \leqslant v(\lambda x + (1 - \lambda) y)$$

for all $x, y \in \mathbb{R}$ and $0 \leqslant \lambda \leqslant 1$ which is the statement that v is concave; conversely, when v is concave then (1.5) is just Jensen's inequality. The investor being risk averse implies that he prefers a certain (that is, deterministic) outcome of μ, say, to an investment X with mean $\mathbb{E}_{\mathbb{P}} X = \mu$.

The investor is **risk neutral** when $\mathbb{E}_{\mathbb{P}} v(X) = v(\mathbb{E}_{\mathbb{P}} X)$ for all $\mathbb{P}$ and X; risk neutrality is equivalent to the utility function v being affine and it means that the investor is indifferent between a random outcome with mean μ and a certain outcome of μ.

The investor is **risk preferring** when $\mathbb{E}_{\mathbb{P}}\, v(X) \geqslant v(\mathbb{E}_{\mathbb{P}} X)$ for all $\mathbb{P}$ and X and it corresponds to the utility function v being convex.

To induce a risk-averse investor to undertake an investment with payoff X and probability $\mathbb{P}$ then a **compensatory risk premium**, α, would have to be offered where α would satisfy

$$\mathbb{E}\, v(\alpha + X) = v(\mu) \quad \text{with} \quad \mu = \mathbb{E} X.$$

We have now suppressed the dependence on the underlying probability $\mathbb{P}$ in the notation. Here the quantity α represents the (deterministic) amount that would have to be added to the payoff of a risky investment X with mean μ to make the investor indifferent between the enhanced risky investment and the certain amount μ.

A related notion is that of an **insurance risk premium**, β, defined by

$$\mathbb{E}\, v(X) = v(\mu - \beta). \tag{1.6}$$

The quantity β is the amount that the risk-averse investor would be willing to pay to avoid the 'fair' investment X with mean μ. Note that when X and Y are two investments with the same mean $\mathbb{E} X = \mathbb{E} Y = \mu$ and $v(\cdot)$ is a strictly increasing function then $X \succ Y$ if and only if $\beta_X < \beta_Y$, where β_X and β_Y are the respective insurance risk premiums; this follows because

$$v(\mu - \beta_X) = \mathbb{E}\, v(X) > \mathbb{E}\, v(Y) = v(\mu - \beta_Y)$$

if and only if $\beta_X < \beta_Y$.

When we expand on the left-hand side of (1.6) using Taylor's Theorem we have

$$\begin{aligned}\mathbb{E}\,v(X) &= \mathbb{E}\left[v(\mu) + (X-\mu)v'(\mu) + \frac{(X-\mu)^2}{2}v''(\mu) + \cdots\right]\\ &= v(\mu) + \frac{\mathbb{V}ar X}{2}v''(\mu) + \cdots\end{aligned}$$

since $\mathbb{E}X = \mu$. Perform a similar expansion on the right-hand side of (1.6) to see that

$$v(\mu - \beta) = v(\mu) - \beta v'(\mu) + \cdots .$$

and when we equate these two expressions, ignoring β^2 and higher-order terms in β as well as the terms $\mathbb{E}\,|X-\mu|^k$ for $k \geqslant 3$, we obtain the approximation

$$\beta \approx \frac{1}{2}\left[\frac{-v''(\mu)}{v'(\mu)}\right]\mathbb{V}ar X.$$

The quantity $-v''(\mu)/v'(\mu)$ is known as the **Arrow-Pratt absolute risk aversion**; it is a measure of how averse the investor is to any investment with mean μ. A related measure of risk aversion is the quantity $-\mathbb{E}\,v''(X)/\mathbb{E}\,v'(X)$, known as the **global absolute risk aversion** which is measuring the investor's aversion to the particular investment X.

The most important source of examples of utility functions is the class of **hyperbolic absolute risk aversion** functions (HARA functions) which have the form

$$v(x) = \frac{1-\gamma}{\gamma}\left(\frac{ax}{1-\gamma} + b\right)^{\gamma}, \tag{1.7}$$

for constants a, b and γ; the range of definition is for values of x for which the term $ax/(1-\gamma) + b \geqslant 0$, so usually we have $b \geqslant 0$. Note that the Arrow–Pratt absolute risk aversion for the function in (1.7) is

$$-\frac{v''(x)}{v'(x)} = \left(\frac{x}{1-\gamma} + \frac{b}{a}\right)^{-1}.$$

The following utility functions that will be used in subsequent chapters may be viewed as special cases or limiting cases of possibly affine transformations of (1.7); they are often chosen for their mathematical tractability.

(a) Quadratic: $v(x) = x - \frac{1}{2}\theta x^2$; take $\gamma = 2$, $a = \sqrt{\theta}$, $ab = 1$.

(b) Exponential: $v(x) = -e^{-ax}$; let $\gamma \to -\infty$. Note that this function has constant absolute risk aversion, a.

(c) Power: $v(x) = x^{\gamma}$ with $\gamma > 0$. Note that this is strictly concave only when $\gamma < 1$. The case $\gamma = 1$ gives the **risk-neutral** utility.

(d) Logarithmic: $v(x) = \ln x$. This follows from (1.7), by using l'Hôpital's rule to see that as $\gamma \to 0$, $(x^{\gamma} - 1)/\gamma \to \ln x$.

1.3 Mean-variance analysis

1.3.1 Introduction

Consider a model evolving for one time period from time 0 to time 1. We first assume that there are s assets, $i = 1, \ldots, s$ for which the prices at time 0 are given by the vector $\boldsymbol{S}_0 = (S_{1,0}, \ldots, S_{s,0})^\top \not\equiv 0$ which is just a constant vector in $\mathbb{R}^s$. The prices at time 1 are given by the random vector $\boldsymbol{S}_1 = (S_{1,1}, \ldots, S_{s,1})^\top$ with the value of $\boldsymbol{S}_1$ not observed until time 1. An investor constructs a portfolio at time 0 by choosing a vector $\boldsymbol{x} = (x_1, \ldots, x_s)^\top$ where x_i is the proportion of his time-0 wealth that he invests in asset i, $1 \leqslant i \leqslant s$, so that $\sum_{i=1}^s x_i = 1$. We allow the possibility that $x_i < 0$ which is known as being **short** in asset i, or having a short position in asset i; this involves borrowing an amount $|x_i|$ of asset i at time 0 which must be repaid at time 1. By contrast, holding a positive amount of an asset is referred to as being **long** in the asset, or having a long position in the asset.

Let $\boldsymbol{R} = (R_1, \ldots, R_s)^\top$ be the random vector representing the rates of return on the assets, so that $R_i = S_{i,1}/S_{i,0}$. Without loss of generality we may assume that each $S_{i,0} \neq 0$, for otherwise we could replace asset i by asset i' where one unit of i' is formed by taking one unit of asset i together with one unit of asset j, for some j, at the price $S_{i,0} + S_{j,0} \neq 0$. When w represents the initial wealth of an investor who forms the portfolio determined by $\boldsymbol{x}$ at time 0, then his wealth at time 1 is the random variable

$$W = \Big(\sum_{i=1}^{s} x_i R_i\Big) w = \big(\boldsymbol{x}^\top \boldsymbol{R}\big)\, w.$$

The rate of return on his portfolio $\boldsymbol{x}$ is given by $W/w = \boldsymbol{x}^\top \boldsymbol{R}$.

1.3.2 All risky assets

We assume that $\boldsymbol{R}$ is a random vector with mean vector $\boldsymbol{r} = \mathbb{E}\,\boldsymbol{R}$, where $\boldsymbol{r} = (r_1, \ldots, r_s)^\top$ with $\mathbb{E}\,R_i = r_i$, and covariance matrix

$$\boldsymbol{V} = \mathbb{C}ov\,(\boldsymbol{R}) = \mathbb{E}\,\left[(\boldsymbol{R} - \boldsymbol{r})(\boldsymbol{R} - \boldsymbol{r})^\top\right].$$

Necessarily $\boldsymbol{V}$ is a symmetric non-negative definite $s \times s$ matrix. Recall that a matrix $\boldsymbol{V}$ is non-negative definite (or positive semi-definite) when $\boldsymbol{z}^\top \boldsymbol{V}\boldsymbol{z} \geqslant 0$ for all vectors $\boldsymbol{z} \in \mathbb{R}^s$ and it is positive definite when $\boldsymbol{z}^\top \boldsymbol{V}\boldsymbol{z} > 0$ for all $\boldsymbol{z} \neq 0$. In this case, since $\boldsymbol{V}$ is a covariance matrix it is symmetric so that its eigenvalues are real, and they are strictly positive when the matrix is positive definite.

The case when an eigenvalue of $\boldsymbol{V}$ is zero corresponds to the existence of a riskless asset, that is one for which the variance of the return is zero, which may be formed by taking a linear combination of the original s assets. In this section we assume that $\boldsymbol{V}$ is positive definite so that there is no riskless asset. Note that since $\boldsymbol{V}$ is positive definite then $\boldsymbol{V}^{-1}$ is also positive definite.

The approach that we will adopt is to assume that for some fixed mean rate of return $\mu = \mathbb{E}(\boldsymbol{x}^\top \boldsymbol{R}) = \boldsymbol{x}^\top \boldsymbol{r}$ the investor seeks to minimize the variance of the return over portfolios $\boldsymbol{x}$. The variance is given by

$$\sigma^2 = \mathbb{V}ar(\boldsymbol{x}^\top \boldsymbol{R}) = \mathbb{E}\left[\boldsymbol{x}^\top (\boldsymbol{R}-\boldsymbol{r})(\boldsymbol{R}-\boldsymbol{r})^\top \boldsymbol{x}\right] = \boldsymbol{x}^\top \boldsymbol{V}\boldsymbol{x}.$$

The problem for the investor then reduces to solving the quadratic programming problem:

$$\text{minimize} \quad \frac{1}{2}\boldsymbol{x}^\top \boldsymbol{V}\boldsymbol{x} \qquad \text{subject to} \quad \begin{cases} \boldsymbol{x}^\top \boldsymbol{e} = 1 \\ \boldsymbol{x}^\top \boldsymbol{r} = \mu, \end{cases} \tag{1.8}$$

where $\boldsymbol{e} = (1, 1, \ldots, 1)^\top$. Here the 1/2 is added in front of the objective function to simplify subsequent algebra slightly. Form the Lagrangian $\mathcal{L}$ for the problem, with Lagrange multipliers λ and ν,

$$\mathcal{L} = \frac{1}{2}\boldsymbol{x}^\top \boldsymbol{V}\boldsymbol{x} + \lambda(1 - \boldsymbol{x}^\top \boldsymbol{e}) + \nu(\mu - \boldsymbol{x}^\top \boldsymbol{r}).$$

Denote the gradient of $\mathcal{L}$ with respect to the vector $\boldsymbol{x}$ by

$$\frac{\partial \mathcal{L}}{\partial \boldsymbol{x}} = \left(\frac{\partial \mathcal{L}}{\partial x_1}, \ldots, \frac{\partial \mathcal{L}}{\partial x_s}\right)^\top,$$

and then to minimize $\mathcal{L}$ set this equal to zero to obtain

$$\frac{\partial \mathcal{L}}{\partial \boldsymbol{x}} = \boldsymbol{V}\boldsymbol{x} - \lambda \boldsymbol{e} - \nu \boldsymbol{r} = 0;$$

here we have used the fact that the matrix $\boldsymbol{V}$ is symmetric. This gives the minimizing value

$$\boldsymbol{x} = \lambda\left(\boldsymbol{V}^{-1}\boldsymbol{e}\right) + \nu\left(\boldsymbol{V}^{-1}\boldsymbol{r}\right). \tag{1.9}$$

That the expression in (1.9) does indeed give a minimum follows because $\mathcal{L}$ is convex in $\boldsymbol{x}$ since $\boldsymbol{V}$, which is the Hessian matrix (of second derivatives) of $\mathcal{L}$, is positive definite. Now substitute back into the constraints in (1.8) to obtain two simultaneous equations for λ and ν,

$$\lambda\alpha + \nu\beta = 1 \quad \text{and} \quad \lambda\beta + \nu\gamma = \mu \tag{1.10}$$

where

$$\alpha = \boldsymbol{e}^\top \boldsymbol{V}^{-1}\boldsymbol{e}, \quad \beta = \boldsymbol{e}^\top \boldsymbol{V}^{-1}\boldsymbol{r} \quad \text{and} \quad \gamma = \boldsymbol{r}^\top \boldsymbol{V}^{-1}\boldsymbol{r}. \tag{1.11}$$

The solution of (1.10) is given by

$$\lambda = \frac{\gamma - \beta\mu}{\delta}, \quad \nu = \frac{\alpha\mu - \beta}{\delta} \quad \text{where} \quad \delta = \alpha\gamma - \beta^2, \tag{1.12}$$

when $\delta \neq 0$. Note that since $\boldsymbol{V}^{-1}$ is positive definite, $\delta \geqslant 0$ by the Cauchy–Schwarz inequality and $\delta > 0$ provided $\boldsymbol{r} \neq c\boldsymbol{e}$ for some $c \in \mathbb{R}$. To see this from first principles when $\boldsymbol{r} \neq c\boldsymbol{e}$ for all $c \in \mathbb{R}$, note that $(\boldsymbol{r} - c\boldsymbol{e})^\top V^{-1}(\boldsymbol{r} - c\boldsymbol{e}) > 0$; expanding out gives $\alpha c^2 - 2\beta c + \gamma > 0$ for all $c \in \mathbb{R}$ so the discriminant of this quadratic must be negative, that is $\beta^2 < \alpha\gamma$. When $\boldsymbol{r} = c\boldsymbol{e}$ for any $c \in \mathbb{R}$, then all the assets would have the same expected rate of return, a possibility which we will exclude since if this is the case the investor would just put all of his resources into the portfolio with the smallest overall variance (see below).

We may obtain an expression for the minimum variance when the expected rate of return is μ as

$$\begin{aligned}\sigma^2 &= \boldsymbol{x}^\top V\boldsymbol{x} = \boldsymbol{x}^\top V\left[\lambda\left(V^{-1}\boldsymbol{e}\right) + \nu\left(V^{-1}\boldsymbol{r}\right)\right]\\ &= \lambda\left(\boldsymbol{x}^\top\boldsymbol{e}\right) + \nu\left(\boldsymbol{x}^\top\boldsymbol{r}\right) = \lambda + \nu\mu\\ &= \left(\alpha\mu^2 - 2\beta\mu + \gamma\right)/\delta.\end{aligned}$$

Plotting this equation in the (σ, μ)-plane, as in Figure 1.1, gives a hyperbola known as the **mean-variance efficient frontier**. The asymptotes of the hyperbola are the lines $\mu = \beta/\alpha \pm \sigma\sqrt{\delta/\alpha}$.

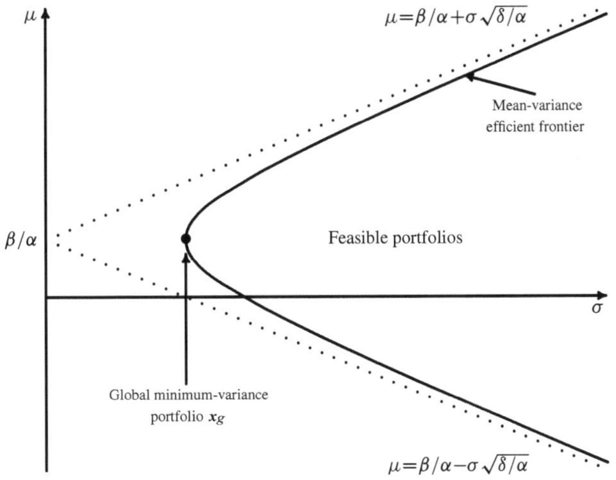

Figure 1.1: The case of all risky assets

The points (σ, μ) on the hyperbola correspond to portfolios with minimum variance σ^2 for expected return μ. Values of (σ, μ) that may be attained by some portfolio lie on, or inside, the hyperbola.

The **global minimum-variance portfolio** $\boldsymbol{x}_g$ is the portfolio we obtain when we set $d\sigma/d\mu = 0$. This has expected return $\mu_g = \beta/\alpha$ with the corresponding $\lambda = 1/\alpha$ and $\nu = 0$, so that $\boldsymbol{x}_g = (1/\alpha)V^{-1}\boldsymbol{e}$. We identify also a second portfolio, known as the **diversified portfolio**, given by $\boldsymbol{x}_d = (1/\beta)V^{-1}\boldsymbol{r}$ with expected return $\mu_d = \boldsymbol{x}_d^\top\boldsymbol{r} = \gamma/\beta$. Then, from (1.9), it follows that any minimum-variance portfolio may be expressed as

$$x = (\lambda\alpha)x_g + (\nu\beta)x_d, \tag{1.13}$$

where

$$\lambda\alpha + \nu\beta = \left(\frac{\gamma - \beta\mu}{\delta}\right)\alpha + \left(\frac{\alpha\mu - \beta}{\delta}\right)\beta = \frac{\alpha\gamma - \beta^2}{\delta} = 1,$$

so that the relation (1.13) defines a portfolio (or affine) combination of the two portfolios x_g and x_d. This conclusion is an example of what is known as a **mutual-fund** theorem; it shows that any minimum-variance portfolio is equivalent to investing in just the global minimum-variance portfolio and the diversified portfolio; the investor puts the proportion $\lambda\alpha$ of his capital in x_g and the proportion $\nu\beta$ in x_d.

Example 1.1 *Uncorrelated returns.* Consider the case where V is diagonal so that the returns on the assets $1, \ldots, s$ are uncorrelated,

$$V = \begin{pmatrix} \sigma_1^2 & 0 & \cdots & 0 \\ 0 & \sigma_2^2 & \cdots & 0 \\ \vdots & \vdots & \ddots & \vdots \\ 0 & 0 & \cdots & \sigma_s^2 \end{pmatrix} \quad \text{and} \quad V^{-1} = \begin{pmatrix} 1/\sigma_1^2 & 0 & \cdots & 0 \\ 0 & 1/\sigma_2^2 & \cdots & 0 \\ \vdots & \vdots & \ddots & \vdots \\ 0 & 0 & \cdots & 1/\sigma_s^2 \end{pmatrix}.$$

It is immediate that

$$\alpha = e^\top V^{-1} e = \sum_{j=1}^{s} (1/\sigma_j^2) \quad \text{and} \quad \beta = e^\top V^{-1} r = \sum_{j=1}^{s} (r_j/\sigma_j^2),$$

which implies that

$$(x_g)_i = \frac{1}{\alpha}\left(V^{-1}e\right)_i = \frac{1/\sigma_i^2}{\sum_{j=1}^{s} 1/\sigma_j^2} \quad \text{and} \quad (x_d)_i = \frac{1}{\beta}\left(V^{-1}r\right)_i = \frac{r_i/\sigma_i^2}{\sum_{j=1}^{s} r_j/\sigma_j^2}.$$

For this case, in constructing the global minimum-variance portfolio, the returns are ignored and the proportion invested in asset i is proportional to the reciprocal of the variance while for the diversified portfolio it is proportional to the expected return times the reciprocal of the variance. The larger the expected return for an asset in relation to the variance of the return the higher the proportion invested in that asset in the diversified portfolio. □

The diversified portfolio may be characterized by the fact that it is the portfolio that maximizes the quantity $s(x) = x^\top r / \sqrt{x^\top V x}$, which is the expected return for the portfolio x per unit of 'risk', where here risk is the standard deviation of the return of the portfolio. This quantity is a special case of the Sharpe ratio which is discussed in the next section in the context of the model where there is a riskless asset. To confirm this characterization, consider the problem of maximizing $s(x)$, or equivalently maximizing $\ln(s(x))$, in x subject to the constraint $x^\top e = 1$. By first considering those portfolios $\boldsymbol{x}$ with fixed mean return $\mu = \boldsymbol{x}^\top \boldsymbol{r}$, it is clear that the optimal portfolio must be a minimum-variance portfolio and, using the form of the mean-variance frontier derived above, the problem reduces to maximizing the function $f(\mu) = \ln\left(\mu/\sqrt{(\alpha\mu^2 - 2\beta\mu + \gamma)/\delta}\right)$ in $\mu > 0$. Calculate the derivative of f as

$$f'(\mu) = \frac{\gamma - \beta\mu}{\mu\left[\alpha(\mu - \beta/\alpha)^2 + \delta/\alpha\right]}$$

which shows that there is just one turning point of the function which occurs at the value $\mu = \gamma/\beta = \mu_d$, which it may be seen is a maximum of the function f. This shows that the portfolio maximizing $s(\boldsymbol{x})$ is $\boldsymbol{x} = \boldsymbol{x}_d$, the diversified portfolio.

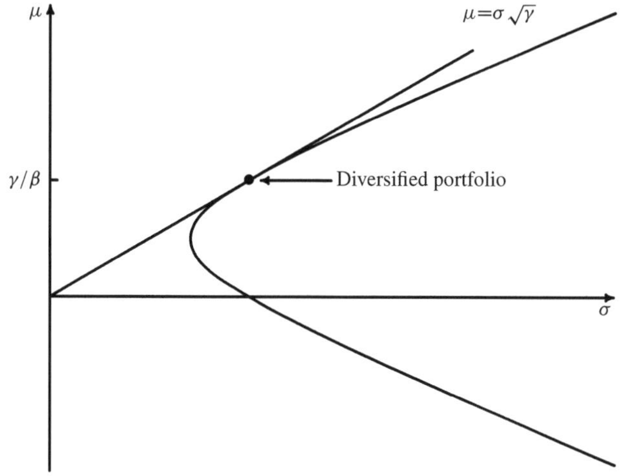

Figure 1.2: The diversified portfolio

This calculation is equivalent to finding the largest value θ such that a line of the form $\mu = \sigma\theta$ meets the hyperbola $\sigma^2 = (\alpha\mu^2 - 2\beta\mu + \gamma)/\delta$; this occurs at the value $\theta = \sqrt{\gamma}$. It may then be seen that the point on the mean-variance frontier corresponding to the diversified portfolio is where the line $\mu = \sigma\sqrt{\gamma}$ is tangent to the hyperbola; this occurs at the point $(\sqrt{\gamma}/\beta, \gamma/\beta)$ and it is illustrated in Figure 1.2.

Furthermore, note that for any minimum-variance portfolio $\boldsymbol{x}$ the covariance between the return of the global minimum-variance portfolio and that of $\boldsymbol{x}$ is constant since, from (1.9),

$$\begin{aligned}\mathbb{C}ov\left(\boldsymbol{x}_g^\top \boldsymbol{R}, \boldsymbol{x}^\top \boldsymbol{R}\right) &= \boldsymbol{x}_g^\top \boldsymbol{V}\boldsymbol{x} = \boldsymbol{x}_g^\top \boldsymbol{V}\left[\lambda\left(\boldsymbol{V}^{-1}\boldsymbol{e}\right) + \nu\left(\boldsymbol{V}^{-1}\boldsymbol{r}\right)\right] \\ &= \lambda\left(\boldsymbol{x}_g^\top \boldsymbol{e}\right) + \nu\left(\boldsymbol{x}_g^\top \boldsymbol{r}\right) = \frac{\lambda}{\alpha}\left(\boldsymbol{e}^\top \boldsymbol{V}^{-1}\boldsymbol{e}\right) + \frac{\nu}{\alpha}\left(\boldsymbol{e}^\top \boldsymbol{V}^{-1}\boldsymbol{r}\right) \\ &= \frac{\lambda\alpha + \nu\beta}{\alpha} = \frac{1}{\alpha}.\end{aligned}$$

1.3.3 A riskless asset

We now assume that in addition to the s risky assets there is a riskless asset available to the investor, asset 0 say, and that it has return r_0. There is no loss of generality in assuming that there is just one such asset since, if there is more than one, the investor will always choose the riskless asset with the largest return. The risky assets have random returns $\boldsymbol{R}$ with mean vector $\mathbb{E}\,\boldsymbol{R} = \boldsymbol{r}$ and non-singular $\boldsymbol{V}$ as before. A portfolio is now described by the pair x_0 and $\boldsymbol{x}$ where x_0 represents the proportion of wealth invested in the riskless asset and $\boldsymbol{x} = (x_1, \ldots, x_s)^\top$ gives the proportions invested in the risky assets. The problem of determining the portfolio of minimum variance for given expected return μ may now be formulated as

$$\text{minimize} \quad \frac{1}{2}\boldsymbol{x}^\top \boldsymbol{V}\boldsymbol{x} \qquad \text{subject to} \quad \begin{cases} x_0 + \boldsymbol{x}^\top \boldsymbol{e} & = 1 \\ x_0 r_0 + \boldsymbol{x}^\top \boldsymbol{r} & = \mu. \end{cases} \tag{1.14}$$

Form the Lagrangian for the problem (1.14) as

$$\mathcal{L} = \frac{1}{2}\boldsymbol{x}^\top \boldsymbol{V}\boldsymbol{x} + \overline{\lambda}(1 - x_0 - \boldsymbol{x}^\top \boldsymbol{e}) + \overline{\nu}(\mu - x_0 r_0 - \boldsymbol{x}^\top \boldsymbol{r}),$$

where $\overline{\lambda}$ and $\overline{\nu}$ are the Lagrange multipliers for the two constraints in (1.14). To minimize $\mathcal{L}$ first consider the terms in x_0. To have a finite minimum of $\mathcal{L}$ we must have $x_0(\overline{\lambda} + \overline{\nu} r_0) = 0$ at the optimum implying that $\overline{\nu} = -\overline{\lambda}/r_0$. Then, as before, setting the gradient of $\mathcal{L}$ with respect to $\boldsymbol{x}$ equal to zero,

$$\frac{\partial \mathcal{L}}{\partial \boldsymbol{x}} = \boldsymbol{V}\boldsymbol{x} - \overline{\lambda}\boldsymbol{e} - \overline{\nu}\boldsymbol{r} = 0$$

gives the minimizing value

$$\boldsymbol{x} = \overline{\lambda}\left(\boldsymbol{V}^{-1}\boldsymbol{e}\right) + \overline{\nu}\left(\boldsymbol{V}^{-1}\boldsymbol{r}\right). \tag{1.15}$$

Substituting back into the constraints in (1.14) yields two simultaneous equations for x_0 and $\overline{\lambda}$,

$$x_0 + \overline{\lambda}\left(\alpha - \frac{\beta}{r_0}\right) = 1 \quad \text{and} \quad x_0 r_0 + \overline{\lambda}\left(\beta - \frac{\gamma}{r_0}\right) = \mu,$$

where α, β and γ are given by (1.11). Solving these, we obtain

$$\overline{\lambda} = \frac{1}{\epsilon^2}\left(r_0 - \mu\right) r_0 \quad \text{and} \quad \overline{\nu} = -\frac{1}{\epsilon^2}\left(r_0 - \mu\right),$$

where

$$\epsilon^2 = \alpha r_0^2 - 2\beta r_0 + \gamma = \alpha (r_0 - \beta/\alpha)^2 + \delta/\alpha.$$

Note that $\epsilon^2 > 0$ provided either $r_0 \neq \beta/\alpha$ or $\delta > 0$ and recall that a sufficient condition for the latter is that not all risky assets have the same expected return. The proportion of wealth invested in the riskless asset is

$$x_0 = (\alpha\mu r_0 - \beta r_0 + \gamma - \beta\mu)/\epsilon^2. \tag{1.16}$$

When we substitute for $\boldsymbol{x}$ from (1.15) and use the fact that x_0 and $\boldsymbol{x}$ satisfy the constraints in (1.14) it follows that the minimum variance is given by

$$\begin{aligned}\sigma^2 &= \boldsymbol{x}^\top \boldsymbol{V}\boldsymbol{x} = \boldsymbol{x}^\top \boldsymbol{V}\left(\overline{\lambda}\boldsymbol{V}^{-1}\boldsymbol{e} + \overline{\nu}\boldsymbol{V}^{-1}\boldsymbol{r}\right)\\ &= \overline{\lambda}\left(\boldsymbol{x}^\top\boldsymbol{e}\right) + \overline{\nu}\left(\boldsymbol{x}^\top\boldsymbol{r}\right) = \overline{\lambda}(1 - x_0) + \overline{\nu}(\mu - x_0 r_0) = \overline{\lambda} + \overline{\nu}\mu\\ &= (\mu - r_0)^2/\epsilon^2,\end{aligned}$$

since $\overline{\lambda} + r_0\overline{\nu} = 0$. This shows that the mean-variance efficient frontier when there is a riskless asset is given by the two straight lines $\mu = r_0 \pm \epsilon\sigma$ in the (σ, μ)-plane.

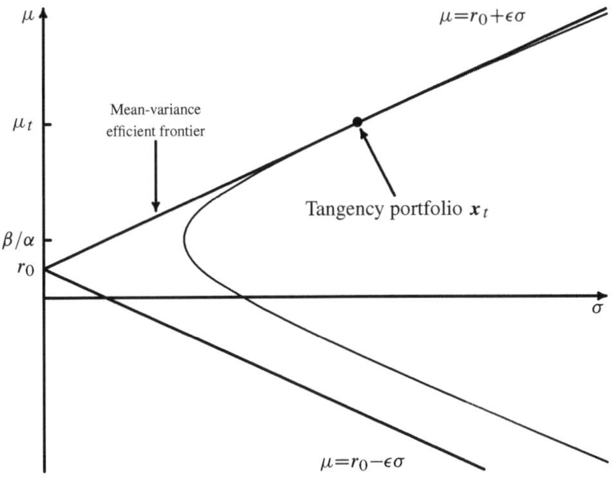

Figure 1.3: The case of a riskless asset

Notice that when $r_0 < \beta/\alpha$ the line $\mu = r_0 + \epsilon\sigma$ touches the hyperbola of the previous section at the point $\sigma = \epsilon/(\beta - \alpha r_0)$, $\mu = (\gamma - \beta r_0)/(\beta - \alpha r_0)$; to see this, check that the point lies on both the line and the hyperbola and the slope of the hyperbola at the point (σ, μ) is

$$\frac{d\mu}{d\sigma} = \frac{\sigma\delta}{\alpha\mu - \beta},$$

which equals ϵ at the given values of σ and μ. In Figure 1.3 the situation has been illustrated for the case when $r_0 < \beta/\alpha$ which corresponds to the intuitively plausible assumption that the riskless return is less than the expected return of the global minimum-variance portfolio of risky assets. When $r_0 > \beta/\alpha$ then the lower line touches the hyperbola. Let us assume in the following that $r_0 < \beta/\alpha$. From (1.15) we find that the optimal mix of the risky assets for expected return μ is

$$\boldsymbol{x} = \overline{\nu}\, V^{-1}\left(\boldsymbol{r} - r_0\boldsymbol{e}\right) = (1 - x_0)\boldsymbol{x}_t$$

where

$$\boldsymbol{x}_t = \left(\frac{1}{\beta - \alpha r_0}\right) V^{-1}\left(\boldsymbol{r} - r_0\boldsymbol{e}\right) = \left(\frac{\beta}{\beta - \alpha r_0}\right)\boldsymbol{x}_d - \left(\frac{\alpha r_0}{\beta - \alpha r_0}\right)\boldsymbol{x}_g \tag{1.17}$$

is known as the **tangency** portfolio of risky assets; check that this does represent a portfolio of risky assets since

$$\boldsymbol{e}^\top\boldsymbol{x}_t = \left(\frac{\beta}{\beta - \alpha r_0}\right)\boldsymbol{e}^\top\boldsymbol{x}_d - \left(\frac{\alpha r_0}{\beta - \alpha r_0}\right)\boldsymbol{e}^\top\boldsymbol{x}_g = \frac{\beta}{\beta - \alpha r_0} - \frac{\alpha r_0}{\beta - \alpha r_0} = 1.$$

The tangency portfolio $\boldsymbol{x}_t$ corresponds to the point where the minimum-variance efficient frontier touches the hyperbola, since its expected return is

$$\begin{aligned}\mu_t &= \left(\frac{\beta}{\beta - \alpha r_0}\right)\left(\boldsymbol{r}^\top\boldsymbol{x}_d\right) - \left(\frac{\alpha r_0}{\beta - \alpha r_0}\right)\left(\boldsymbol{r}^\top\boldsymbol{x}_g\right)\\ &= \left(\frac{\beta}{\beta - \alpha r_0}\right)\mu_d - \left(\frac{\alpha r_0}{\beta - \alpha r_0}\right)\mu_g = \frac{\gamma - \beta r_0}{\beta - \alpha r_0},\end{aligned} \tag{1.18}$$

where $\mu_d = \gamma/\beta$ and $\mu_g = \beta/\alpha$ are the expected returns of the diversified and global minimum-variance portfolios respectively. Observe that when $r_0 < \beta/\alpha$ we have $\mu_t > \mu_d$ and $\mu_t > \beta/\alpha$; when $r_0 > \beta/\alpha$ it is the case that $\mu_t < \mu_d$ and $\mu_t < \beta/\alpha$. This shows that the diversified portfolio corresponds to a point on the hyperbola lying between the points corresponding to the global minimum-variance and the tangency portfolios. Note that the tangency portfolio reduces to the diversified portfolio in the case when $r_0 = 0$.

What this analysis demonstrates is that an investor wishing to obtain a given expected return with minimum variance divides his wealth into a proportion x_0 given by (1.16), which is invested in the riskless asset, and the remaining proportion $1 - x_0$, which is invested in the risky assets in the proportions of the tangency portfolio. Note that when $\mu = \mu_t$ the investment in the riskless asset $x_0 = 0$ from (1.16) and (1.18).

The tangency portfolio may be characterized in a similar fashion to that given in the previous section for the diversified portfolio. For a portfolio, $\boldsymbol{x}$, of risky assets the **Sharpe ratio** is

$$s(\boldsymbol{x}) = \frac{\boldsymbol{x}^\top\boldsymbol{r} - r_0}{\sqrt{\boldsymbol{x}^\top V\boldsymbol{x}}}$$

and it determines the excess expected return of the portfolio over the riskless rate of return per unit of risk as measured by the standard deviation of the return. The tangency portfolio maximizes the Sharpe ratio among all portfolios of risky assets. The proof follows the lines of the previous case; by first considering for fixed μ those portfolios $\boldsymbol{x}$ for which the expected return is $\boldsymbol{x}^\top \boldsymbol{r} = \mu$, it is again clear that the optimal portfolio must be a minimum-variance portfolio of risky assets so the problem reduces to maximizing the function $f(\mu) = \ln\big((\mu - r_0)/\sqrt{(\alpha\mu^2 - 2\beta\mu + \gamma)/\delta}\big)$ in $\mu > r_0$. Calculate the derivative of f as

$$f'(\mu) = \frac{(\gamma - \beta r_0) - (\beta - \alpha r_0)\mu}{(\mu - r_0)(\alpha\mu^2 - 2\beta\mu + \gamma)},$$

then it may be seen that this function has only one turning point, at

$$\mu = \frac{\gamma - \beta r_0}{\beta - \alpha r_0} = \mu_t,$$

which gives a maximum of f. It follows that the tangency portfolio maximizes the Sharpe ratio.

Notice that the covariance between the return on risky asset i and the return of the tangency portfolio is

$$\mathbb{C}ov\left(R_i, \boldsymbol{x}_t^\top \boldsymbol{R}\right) = (V\boldsymbol{x}_t)_i = \frac{1}{\beta - \alpha r_0}(r_i - r_0),$$

and the variance of the return on the tangency portfolio is

$$\mathbb{V}ar\left(\boldsymbol{x}_t^\top \boldsymbol{R}\right) = \boldsymbol{x}_t^\top V \boldsymbol{x}_t = \frac{\mu_t - r_0}{\beta - \alpha r_0}.$$

Define the vector $\boldsymbol{\beta}_t = (\beta_{1,t}, \ldots, \beta_{s,t})^\top$ by setting

$$\beta_{i,t} = \frac{\mathbb{C}ov\left(R_i, \boldsymbol{x}_t^\top \boldsymbol{R}\right)}{\mathbb{V}ar\left(\boldsymbol{x}_t^\top \boldsymbol{R}\right)} \quad \text{or} \quad \boldsymbol{\beta}_t = \frac{1}{\mu_t - r_0}(\boldsymbol{r} - r_0\boldsymbol{e}).$$

Somewhat unimaginatively, this vector is known as the **tangency beta**. The expected return on the risky assets may now be expressed as

$$\boldsymbol{r} = r_0\boldsymbol{e} + (\mu_t - r_0)\,\boldsymbol{\beta}_t. \tag{1.19}$$

The equation (1.19) is the **mean-variance pricing equation** and it states that for asset i, the excess return over the riskless rate, $r_i - r_0$, is proportional to $\beta_{i,t}$ which may be expressed in terms of the correlation between the return on asset i and the tangency portfolio since

$$\beta_{i,t} = \mathbb{C}orr\left(R_i, \boldsymbol{x}_t^\top \boldsymbol{R}\right)\sqrt{\frac{\mathbb{V}ar(R_i)}{\mathbb{V}ar\left(\boldsymbol{x}_t^\top \boldsymbol{R}\right)}}.$$

1.3.4 Mean-variance analysis and expected utility

In certain cases the mean-variance approach as developed in this chapter may be justified in terms of expected-utility maximization. Suppose that the investor's expected utility of final wealth may be written as a function $f(\sigma, \mu)$ just of the expected return, μ, of the portfolio used, and the standard deviation of the return, σ. That is, when W is his final wealth and w his initial wealth, we have

$$f(\sigma, \mu) = \mathbb{E}v(W) \quad \text{where} \quad W = \left(x_0 r_0 + \boldsymbol{x}^\top \boldsymbol{R}\right) w,$$

and $\mu = x_0 r_0 + \boldsymbol{x}^\top \boldsymbol{r}$, $\sigma^2 = \boldsymbol{x}^\top \boldsymbol{V}\boldsymbol{x}$. As before the pair x_0 and $\boldsymbol{x}$ specify his portfolio and give the proportions of initial wealth invested in the riskless asset and the risky assets respectively. Assume that

$$\frac{\partial f}{\partial \sigma} < 0 \quad \text{and} \quad \frac{\partial f}{\partial \mu} > 0,$$

so that the expected utility is decreasing in the standard deviation of the return and it is increasing in the expected return. After using the fact that $x_0 + \boldsymbol{x}^\top \boldsymbol{e} = 1$, it follows that the investor seeks to maximize the function

$$f\left(\sqrt{\boldsymbol{x}^\top \boldsymbol{V}\boldsymbol{x}}, r_0 + \boldsymbol{x}^\top (\boldsymbol{r} - r_0 \boldsymbol{e})\right) \quad \text{in} \quad \boldsymbol{x} = (x_1, \ldots, x_s)^\top.$$

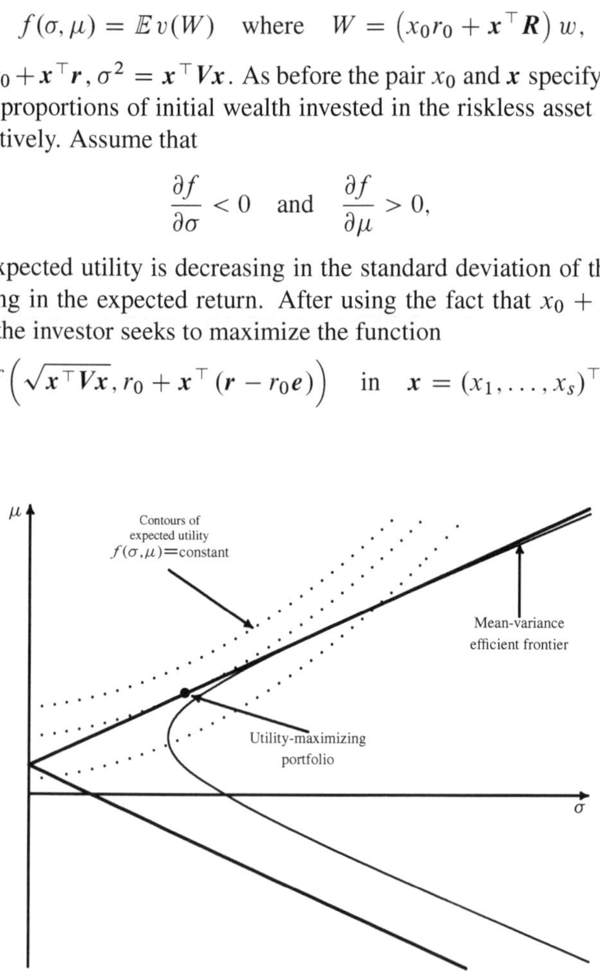

Figure 1.4: Determining the utility-maximizing portfolio

Set the gradient of f with respect to $\boldsymbol{x}$ equal to zero to obtain

$$\frac{\partial f}{\partial \boldsymbol{x}} = \frac{1}{\sigma}\left(\frac{\partial f}{\partial \sigma}\right) \boldsymbol{V}\boldsymbol{x} + \left(\frac{\partial f}{\partial \mu}\right)(\boldsymbol{r} - r_0 \boldsymbol{e}) = 0,$$

which demonstrates that the optimal investment in the risky assets is proportional to the tangency portfolio since the maximizing $\boldsymbol{x}$ satisfies

$$\boldsymbol{x} = -\left(\sigma \frac{\partial f}{\partial \mu} \Big/ \frac{\partial f}{\partial \sigma}\right) V^{-1} (\boldsymbol{r} - r_0 \boldsymbol{e}) \propto \boldsymbol{x}_t,$$

and hence the optimal portfolio corresponds to a point on the mean-variance efficient frontier. When the function f is concave in σ and μ then sets of the form $\{(\sigma, \mu) : f(\sigma, \mu) \geqslant c\}$ are convex which implies that this point representing the optimal portfolio is where an indifference contour of the function f touches the mean-variance efficient frontier, as illustrated in Figure 1.4; an **indifference** contour is a curve corresponding to portfolios yielding the same utility. Two cases where $\mathbb{E}\, v(W) = f(\sigma, \mu)$, that is the expected utility of final wealth is a function only of the mean and standard deviation of the return, are the following.

Example 1.2 *Quadratic utility.* Suppose that the utility function v is quadratic so that $v(x) = ax + bx^2$, where a and b are constants and $b \leqslant 0$ would be required for concavity. Then

$$\mathbb{E}\, v(W) = \mathbb{E}\, v\left(\left(x_0 r_0 + \boldsymbol{x}^\top \boldsymbol{R}\right) w\right) = aw\mu + bw^2(\mu^2 + \sigma^2) = f(\sigma, \mu),$$

is a function of σ and μ only. □

Example 1.3 *Normally distributed returns.* When the returns on the risky assets, $\boldsymbol{R}$, have a multivariate normal distribution, $N(\boldsymbol{r}, V)$, then it follows that the linear combination $\boldsymbol{x}^\top \boldsymbol{R}$ has the $N(\boldsymbol{x}^\top \boldsymbol{r}, \boldsymbol{x}^\top V \boldsymbol{x})$-distribution. That gives

$$\mathbb{E}\, v(W) = \mathbb{E}\, v\left((\mu + \sigma Y)\, w\right)$$

where Y has the standard $N(0, 1)$-distribution, which implies immediately that the right-hand side is a function of σ and μ only. □

1.3.5 Equilibrium: the capital-asset pricing model

Suppose now that we consider a market made up of a number of investors, indexed by $j \in \mathcal{J}$. Let $x_{0,j}$ and $\boldsymbol{x}_j = \left(x_{1,j}, \ldots, x_{s,j}\right)^\top$ be the proportions of his wealth that investor j puts in asset 0 and in assets $1, \ldots, s$, respectively. When each investor has a utility of the form considered in the previous section, then the optimal holding of risky assets for investor j is $\boldsymbol{x}_j \propto \boldsymbol{x}_t$ so that $\boldsymbol{x}_j = (1 - x_{0,j})\boldsymbol{x}_t$, for each $j \in \mathcal{J}$, where $\boldsymbol{x}_t$ is the tangency portfolio defined in (1.17). When w_j denotes the initial wealth of investor j then the total value of the demand for risky asset i is

$$\sum_{j \in \mathcal{J}} w_j x_{i,j} = \left(\sum_{j \in \mathcal{J}} (1 - x_{0,j}) w_j\right) (\boldsymbol{x}_t)_i\,.$$

The **market portfolio** of risky assets, $\boldsymbol{x}_m$, is defined by

$$(\boldsymbol{x}_m)_i = \frac{\text{The total value of the supply of risky asset } i}{\text{The total value of the supply of all risky assets}},$$

so that $\boldsymbol{x}_m^\top \boldsymbol{e} = 1$. In economics, the term 'equilibrium' is usually used to denote the situation in a model where overall supply equals overall demand, so that the two are balanced; thus in equilibrium in this context to have supply equal to demand would require that

$$\begin{aligned}(\boldsymbol{x}_m)_i &= \frac{\left(\sum_{j\in\mathcal{J}} w_j(1-x_{0,j})\right)(\boldsymbol{x}_t)_i}{\sum_{j\in\mathcal{J}}\sum_{k=1}^{s} w_j x_{k,j}} \\ &= \frac{\left(\sum_{j\in\mathcal{J}} w_j(1-x_{0,j})\right)(\boldsymbol{x}_t)_i}{\left(\sum_{j\in\mathcal{J}} w_j\left(1-x_{0,j}\right)\right)\sum_{k=1}^{s}(\boldsymbol{x}_t)_k} = (\boldsymbol{x}_t)_i\,,\end{aligned}$$

since $\boldsymbol{x}_t^\top \boldsymbol{e} = 1$. That is, in equilibrium the market portfolio coincides with the tangency portfolio. This allows the pricing equation (1.19) to be rewritten as

$$\boldsymbol{r} = r_0\boldsymbol{e} + (\mu_m - r_0)\,\boldsymbol{\beta}_m, \tag{1.20}$$

where μ_m is the expected return on the market portfolio and

$$\boldsymbol{\beta}_m = (\beta_{1,m},\ldots,\beta_{s,m})^\top \quad \text{with} \quad \beta_{i,m} = \mathbb{C}ov\,(R_i, \boldsymbol{x}_m^\top \boldsymbol{R})/\mathbb{V}ar(\boldsymbol{x}_m^\top \boldsymbol{R})$$

being the **market beta** for the market portfolio. The equation (1.20) is usually referred to as the **capital-asset-pricing equation**.

It should be noted that while mean-variance analysis provides a useful framework for thinking about the issues of portfolio choice, its usefulness in applications depends on the availability of good estimates of mean returns of assets and of the covariance between those returns, which may not be easy to obtain. Similar problems arise with the capital-asset pricing model; for example, when it is viewed in a dynamic setting, changes over time in estimates of parameters in the model from market data may lead to instability in estimates of the market beta.

1.4 Exercises

Exercise 1.1 Suppose that an investment X has either (i) the uniform distribution $U[0, 2\mu]$ or (ii) the exponential distribution with mean $\mathbb{E}X = \mu$, and the investor has a utility function which is either (a) logarithmic, $v(x) = \ln x$ or (b) power form,

$v(x) = x^{\theta}$. Show that both the compensatory risk premium and the insurance risk premium are proportional to μ in all four possible cases.

For the case of logarithmic utility and the uniform distribution determine the risk premiums as explicitly as possible.

Exercise 1.2 An investor has utility function $v(x) = \sqrt{x}$ and is considering three investments with random outcomes X, Y and Z. Here, X has the uniform distribution $U[0, a]$, Y has the gamma distribution $\Gamma(\gamma, \lambda)$ with probability density function $e^{-\lambda y}\lambda^{\gamma} y^{\gamma-1}/\Gamma(\gamma)$, for $y > 0$, where $\gamma > 0, \lambda > 0$ and Z is log-normal, that is, $\ln Z$ has a $N(\nu, \sigma^2)$-distribution. The parameters of the distributions are such that $\mathbb{E}X = \mathbb{E}Y = \mathbb{E}Z = \mu$, say, and $\mathbb{V}ar(X) = \mathbb{V}ar(Y) = \mathbb{V}ar(Z)$.

Recall that the gamma function $\Gamma(\gamma) = \int_0^{\infty} u^{\gamma-1}e^{-u}du$ satisfies $\Gamma(\gamma + 1) = \gamma\Gamma(\gamma)$ and $\Gamma(\frac{1}{2}) = \sqrt{\pi}$.

Determine the investor's preference ordering of X, Y and Z for all values of μ.

Exercise 1.3 Suppose that an investor has the utility function $v(x) = 1 - e^{-ax}$ where $a > 0$ and the outcome of an investment is a random variable X with mean μ, finite variance and finite moment-generating function $\psi(a) = \mathbb{E}\left(e^{-aX}\right)$, for $a > 0$.

Show that the compensatory risk premium and the insurance risk premium have the same value, α say, and express α in terms of μ and the moment-generating function ψ.

In this case both the Arrow-Pratt and global risk aversions are a. Confirm directly that as $a \downarrow 0$, $\alpha = a\mathbb{V}ar(X)/2 + o(a)$. Under what circumstances is it true that $\alpha = a\mathbb{V}ar(X)/2$ for all $a > 0$?

Prove that $\psi''\psi - (\psi')^2 \geq 0$, and hence that α is an increasing function of a. This shows that the more risk-averse that the investor is, the higher the value of the premium that is required.

Exercise 1.4 Consider a one-period investment model in which there are only two assets, both of which are risky. The returns on these assets have means 3, 4 respectively and variances 2, 3 respectively with the covariance between the returns being 2. From first principles, calculate the mean-variance efficient frontier and the minimum-variance portfolio in terms of the mean return. Calculate the mean return of the global minimum-variance portfolio and of the diversified portfolio. Check your answers using the results in Section 1.3.

Now suppose that in addition to the two risky assets there is a riskless asset with return 3/2. Find the minimum-variance portfolio in terms of the mean return and hence calculate the mean return of the tangency portfolio.

Exercise 1.5 Suppose that v is a concave function, X is a random variable with the $N\left(\mu, \sigma^2\right)$-distribution and set $f(\sigma, \mu) = \mathbb{E}\, v(X)$. Here, you may assume that the function v is twice differentiable and that the expectation $\mathbb{E}v(X)$ is finite. Show that

$$\frac{\partial f}{\partial \mu} > 0 \quad \text{when } v \text{ is strictly increasing, and} \quad \frac{\partial f}{\partial \sigma} \leqslant 0.$$

Hence show in the context of mean-variance analysis that, when all returns are jointly normally distributed, an investor maximizing the expected utility of his final wealth will choose a mean-variance-efficient optimal portfolio.

Show that f is concave in μ and σ; that is, check that the matrix of second derivatives is negative semi-definite. Deduce that this optimal portfolio corresponds to a point in the (σ, μ)-plane where an indifference contour is tangent to the efficient frontier.

Exercise 1.6 In the framework of Section 1.3.3, suppose that an investor has a concave utility function v. The investor wishes to maximize $\mathbb{E}\, v(W)$ where W is his final wealth given by

$$W = w\left(x_0 r_0 + \sum_{j=1}^{s} x_j R_j\right);$$

w is his initial wealth and x_j is the proportion he invests in asset j.

Show that when $\overline{W}$ is his optimal final wealth then

$$\mathbb{E}\left[v'\left(\overline{W}\right)\left(R_j - r_0\right)\right] = 0 \quad \text{for each} \quad j = 1, \ldots, s.$$

Deduce, using the properties of the normal distribution given in Appendix A.3, that when $\boldsymbol{R}$ has a multivariate normal distribution then for each $j = 1, \ldots, s$

$$r_j - r_0 = \alpha\, \mathbb{C}ov\left(\overline{W}, R_j\right),$$

where $r_j = \mathbb{E}\, R_j$ and $\alpha = -\mathbb{E}\left[v''\left(\overline{W}\right)\right] / \mathbb{E}\left[v'\left(\overline{W}\right)\right]$ is his global risk aversion.

Now suppose that the market is determined by investors $i = 1, \ldots, n$, where investor i has concave utility v_i, initial wealth w_i, optimal final wealth $\overline{W}_i$ and corresponding α_i. With the same normality assumption, deduce that

$$\mathbb{E}\, M - r_0 = \overline{w}\,\overline{\alpha}\, \mathbb{V}ar\,(M) \tag{1.21}$$

where M is the market rate of return given by $M = \sum_{i=1}^{n} \overline{W}_i / \sum_{i=1}^{n} w_i$, $\overline{w} = \sum_{i=1}^{n} w_i / n$ is the average initial wealth of investors and $\overline{\alpha}$ is the harmonic mean of the α_i.

Exercise 1.7 Consider an investor with the utility function $v(x) = 1 - e^{-ax}$, where $a > 0$, who is faced with a riskless asset with return r_0 and s risky assets with returns $\boldsymbol{R}$ which have the multivariate normal $N(\boldsymbol{r}, V)$-distribution. Show directly that when he acts to maximize the expected utility of his final wealth he will hold the risky assets in the same proportion as the tangency portfolio. Furthermore, show that when $\beta > \alpha r_0$, with α and β as in Section 1.3.2, the more risk averse that he is the smaller the amount of his wealth that he invests in the risky assets.

Exercise 1.8 Consider an investor in the context of Section 1.3.2 when there are s risky assets with returns $\boldsymbol{R} = (R_1, R_2, \ldots, R_s)^\top$ where $R_1, R_2, \ldots, R_s$ are independent random variables with R_i having a gamma distribution (see Exercise 1.2) with $\mathbb{E} R_i = r_i$ and $\mathbb{V}ar(R_i) = \sigma_i^2$. Suppose that he has the exponential utility function $v(x) = 1 - e^{-ax}$, where $a > 0$, and that he acts to maximize the expected utility of his final wealth, when his initial wealth is $w > 0$.

Show that he divides his wealth between the diversified portfolio and a second portfolio, which should be identified, and determine the amounts that he invests in each.

When he also may invest in a riskless asset with return r_0 as in Section 1.3.3, show that he will again divide his investment between these two portfolios and the riskless asset and determine the amounts that he puts into each. Give a necessary and sufficient condition, expressed in terms of the parameters r_i, $0 \leqslant i \leqslant s$, and σ_j^2, $1 \leqslant j \leqslant s$, that he is long in the risky assets.

Chapter 2

THE BINOMIAL MODEL

2.1 One-period model

2.1.1 Introduction

We begin our study of the ideas underlying the elimination or minimization of risk and the pricing of assets in financial markets by considering the binomial model. While at first sight it might seem too simple mathematically for any practical application, we will see that it is rich enough to motivate many of the ideas that we will develop in subsequent chapters; also it is an important tool for approximating results in the more realistic models which come later.

The first case to consider is the one-period model operating from time 0 to time 1. There are just two assets: the first may be thought of as a bank account (or bond) which is riskless in that 1 unit of wealth at time 0 held in the bank account becomes $1 + \rho$ with certainty at time 1, where $\rho \geqslant 0$ is a constant and may be interpreted as the interest rate on the bank account; the second asset is a stock for which the price at time 0 is S_0, where $S_0 > 0$ is constant, and its price at time 1 is a random variable S_1 taking just two possible values uS_0 and dS_0 where $u > 0$ and $d > 0$ are given constants. The random variable S_1 is defined on the underlying probability sample space $\Omega = \{\omega_1, \omega_2\}$ with $S_1(\omega_1) = uS_0$ and $S_1(\omega_2) = dS_0$. We will assume that u and d are distinct values and take $u > d$.

We denote the underlying probabilities for the two outcomes ω_1 and ω_2 by

$$p_1 = \mathbb{P}(\{\omega_1\}) = \mathbb{P}(S_1 = uS_0) \quad \text{and} \quad p_2 = \mathbb{P}(\{\omega_2\}) = \mathbb{P}(S_1 = dS_0),$$

where $p_i > 0$, $i = 1, 2$ and, of course, $p_1 + p_2 = 1$.

A useful way to represent the change in the stock price between time 0 and time 1 is by a binomial tree (or binary tree)

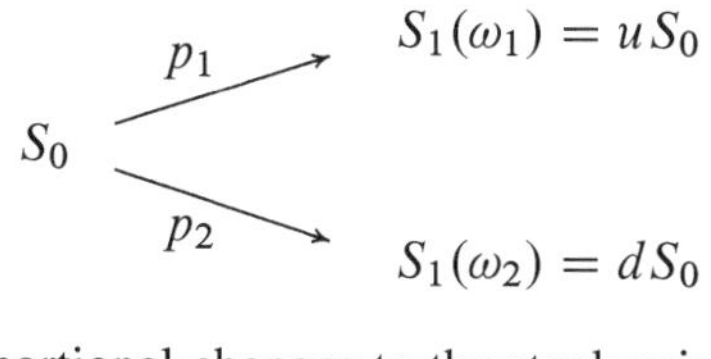

with u and d being proportional changes to the stock price; u is often regarded as being an 'up' jump of the stock and d as a 'down' jump although we do not assume necessarily that $u > 1$ and $d < 1$. We may think of the points ω_1 and ω_2 as the two possible governing states of nature in this model; when ω_1 is the 'true' or prevailing state of nature then the stock price performs the up jump to uS_0 while when ω_2 is the prevailing state of nature the stock price has a down jump to dS_0 at time 1.

One further notation that we will require is to set $\alpha = 1/(1+\rho)$; we will refer to α as the **discount factor** for the model. Alternatively, α is the price at time 0 of a bond paying 1 with certainty at time 1; such a bond with deterministic payoff is often referred to as a **riskless bond**. Multiplying prices at time 1 by α gives their 'values' at time 0, since to meet a charge of c at time 1 requires an amount αc in the bank account at time 0.

2.1.2 Hedging

We will consider a **contingent claim** C which is just a random variable defined on the underlying probability space Ω. It takes two values: $C(\omega_1)$, in the case when ω_1 is the true state of nature (in which case the stock price becomes uS_0 at time 1) and $C(\omega_2)$, when ω_2 is the governing state of nature (and then the stock price takes the value dS_0 at time 1). A contingent claim may be thought of as the payoff at time 1 of a contract where the payoff is random (and in this model it may be regarded as a function of the stock price at time 1). The holder of the claim, who has bought it at time 0 with no knowledge of which of ω_1 or ω_2 is the prevailing state of nature, receives the random payoff C at time 1. A central question of finance is how much should an individual pay at time 0 to hold the claim C, or equivalently how much should be charged by the seller of the claim who will have the liability to pay out the amount C at time 1?

Consider the problem from the point of view of the individual who has sold the claim and wishes to 'hedge' against his liability to pay C at time 1. Suppose that he forms a portfolio at time 0 which consists of an amount y in the bank account, $-\infty < y < \infty$, and x units of stock, $-\infty < x < \infty$. A negative value of y corresponds to borrowing from the bank while a negative value of x corresponds to being **short** in the stock, or holding a short position in the stock; that is, $|x|$ units of stock are borrowed at time 0 (and they must be paid back at time 1). A portfolio where $x > 0$ is said to be **long** in the stock, or holds a long position in the stock. At time 1 he will have $(1+\rho)y$ in the bank and the value xS_1 in the stock. If he chooses x and y in such a way that at time 1 his portfolio will yield exactly the required amount to pay the claim C (irrespective of which of ω_1 or ω_2 is the prevailing state of nature) then we obtain the two equations

$$\begin{aligned} xuS_0 + (1+\rho)y &= C(\omega_1) \\ xdS_0 + (1+\rho)y &= C(\omega_2). \end{aligned} \tag{2.1}$$

Because $u \neq d$, these two equations are linearly independent and may always be solved to obtain

$$x = \frac{C(\omega_1) - C(\omega_2)}{S_0(u-d)} \quad \text{and} \quad y = \frac{1}{1+\rho}\left[\frac{uC(\omega_2) - dC(\omega_1)}{u-d}\right]. \tag{2.2}$$

The cost of setting up the portfolio (x, y) at time 0, with the amounts given in (2.2), is

$$\begin{aligned} xS_0 + y &= \frac{C(\omega_1) - C(\omega_2)}{u - d} + \frac{1}{1+\rho}\left[\frac{uC(\omega_2) - dC(\omega_1)}{u-d}\right] \\ &= \frac{1}{1+\rho}\left[\left(\frac{1+\rho-d}{u-d}\right)C(\omega_1) + \left(\frac{u-(1+\rho)}{u-d}\right)C(\omega_2)\right] \\ &= \alpha\left[q_1 C(\omega_1) + q_2 C(\omega_2)\right], \end{aligned} \tag{2.3}$$

when we let

$$q_1 = \frac{1+\rho-d}{u-d} \quad \text{and} \quad q_2 = \frac{u-(1+\rho)}{u-d}. \tag{2.4}$$

Notice that q_1 and q_2 satisfy

$$q_1 + q_2 = 1. \tag{2.5}$$

The portfolio (x, y) specified in (2.2) is said to **replicate** (or **hedge**) the claim C; it eliminates the exposure to risk on the part of the seller of the claim in that, no matter which state of nature prevails, at time 1 the portfolio provides exactly the amount required to pay the claim. The initial cost of setting up this portfolio, $\alpha\left[q_1 C(\omega_1) + q_2 C(\omega_2)\right]$, given in (2.3) determines the minimum price at which the seller would be prepared to sell the claim; any amount more than that sum would yield a riskless profit to the seller. On the other hand, the amount in (2.3) is the maximum amount that the buyer would be prepared to pay to hold the claim; if the claim was priced at an amount below that in (2.3), he would sell the claim and form the portfolio and take the riskless profit. We may conclude that in this one-period binomial model the sum specified in (2.3) is the 'fair' price for the claim.

Notice that when $q_i \geqslant 0$, $i = 1, 2$ then, since they add to 1 from (2.5), these quantities may be interpreted as probabilities. The exact circumstances under which they will be both strictly positive have important financial (and mathematical) implications and will be characterized in the next section. When they are, the amount $\alpha\left[q_1 C(\omega_1) + q_2 C(\omega_2\right]$ may be interpreted as the expectation of the discounted payoff, αC, of the claim under these probabilities. In this case we normally write $\mathbb{Q} = \{q_1, q_2\}$ to represent this new probability distribution and $\mathbb{E}_{\mathbb{Q}}$ for expectations under these probabilities so that the price of the claim is

$$\alpha \mathbb{E}_{\mathbb{Q}} C = \alpha\left[q_1 C(\omega_1) + q_2 C(\omega_2)\right]. \tag{2.6}$$

We may represent the process of calculating the price of the claim C graphically as

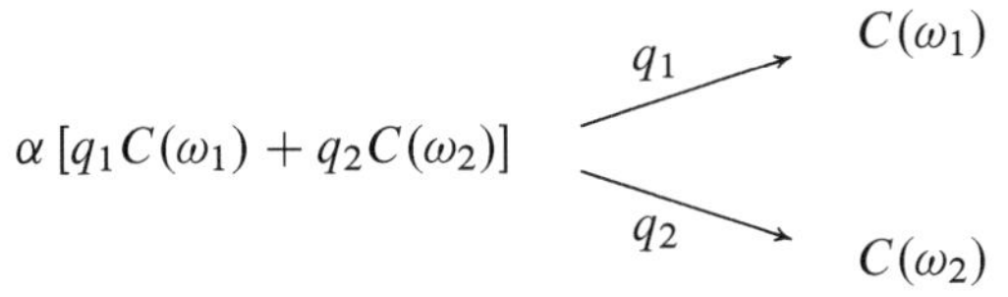

Notice that the original probabilities p_1 and p_2 play no role in determining the price of the claim; indeed, the buyer and seller of the claim may have different ideas about

what the values of the probabilities $\{p_i\}$ might be but they will agree on the price specified in (2.3).

Furthermore, observe that q_1 and q_2 are specified uniquely by the underlying parameters in the model, u, d and ρ, and do not depend on the specific claim C. There is a further property that they satisfy (and from which they will derive their name) which is that

$$\begin{aligned}\alpha \mathbb{E}_{\mathbb{Q}} S_1 &= \alpha \left[q_1 S_1(\omega_1) + q_2 S_1(\omega_2)\right] \\ &= \alpha \left[\left(\frac{1+\rho-d}{u-d}\right) u S_0 + \left(\frac{u-(1+\rho)}{u-d}\right) d S_0\right] = S_0.\end{aligned} \tag{2.7}$$

The relation (2.7) means that the prices of the stock at times 0 and 1 discounted to time-0 values, namely $(S_0, \alpha S_1)$, form a martingale under the probabilities $\mathbb{Q}$ (see the Appendix); for this reason we refer to them as the **martingale** probabilities for the model. When q_1 and q_2 are not both non-negative (so that they do not form a probability distribution on Ω) then we may refer to them as a martingale **measure** for the model.

It is also important to note that in this model all claims C may be hedged; there are two simultaneous equations in (2.1) which may always be solved to give the values for x and y in (2.2); the model, or the market it describes, is said to be **complete**.

2.1.3 Arbitrage

One of the principal requirements in formulating a realistic model of a financial market is to build in to the specification that there are no opportunities for investors to make riskless profits. If such an opportunity exists, then the model is inherently unstable (and unrealistic) since all investors will take advantage of it and invest unlimited amounts in the opportunity with corresponding unlimited rewards. The mathematical characterization of the situation in which such opportunities are precluded has important implications for the pricing of claims.

In the binomial model, we say that a portfolio (x, y), holding x units of stock and y in the bank account at time 0, is an **arbitrage** if $xS_0 + y = 0$, and

$$xuS_0 + (1+\rho)y \geqslant 0, \quad xdS_0 + (1+\rho)y \geqslant 0, \tag{2.8}$$

with at least one of the inequalities in (2.8) being strict.

An arbitrage here is an initial investment which is worth 0 at time 0 but which becomes non-negative with certainty at time 1 and has positive probability of being strictly positive at time 1 (since both $p_1 > 0$ and $p_2 > 0$). There is a very simple characterization of the situation where no arbitrage exists.

Theorem 2.1 *There is no arbitrage in the binomial model if and only if*

$$u > 1 + \rho > d. \tag{2.9}$$

Proof. Firstly, suppose that (2.9) does not hold; when $1 + \rho \geqslant u > d$ we obtain an arbitrage by taking $x = -1/S_0$, $y = 1$, which corresponds to shorting the stock and putting the proceeds in the bank. The initial value of this portfolio is 0, while at time 1 the value is either $1 + \rho - u \geqslant 0$ or $1 + \rho - d > 0$. On the other hand when (2.9) does not hold and $u > d \geqslant 1 + \rho$, then we obtain an arbitrage by setting $x = 1/S_0$, $y = -1$, that is, we borrow from the bank and invest in stock; the value at time 1 is either $u - 1 - \rho > 0$ or $d - 1 - \rho \geqslant 0$.

Alternatively, if (x, y) is an arbitrage, then $y = -xS_0 \neq 0$, and the values in (2.8) are

$$y\,(1 + \rho - u) \geqslant 0 \quad \text{and} \quad y\,(1 + \rho - d) \geqslant 0$$

so that either $1 + \rho \geqslant u > d$ or $1 + \rho \leqslant d < u$ as required. □

We may now see from (2.4) and (2.9) that there is no arbitrage in the model if and only if $q_1 > 0$ and $q_2 > 0$. In this case we refer to q_1 and q_2 as the **equivalent martingale probabilities** for the model. (We will discuss the notion of 'equivalent' probabilities in more detail in the next chapter; here it just means that $\mathbb{Q} = \{q_1, q_2\}$ attaches positive probabilities to ω_1 and ω_2, respectively when the underlying probability distribution $\mathbb{P} = \{p_1, p_2\}$ does.) We will assume from now on that the condition (2.9) holds so that there is no arbitrage.

2.1.4 Utility maximization

Consider an investor in the binomial model with initial wealth $w_0 > 0$, and utility function v, which will be assumed to be a differentiable concave increasing function. Suppose that his terminal wealth at time 1 is W, then he seeks to choose W so as to solve the optimization problem

$$\text{maximize} \quad \mathbb{E}\,v(W) \quad \text{subject to} \quad \mathbb{E}_{\mathbb{Q}}\,(\alpha W) = w_0, \tag{2.10}$$

since $\mathbb{E}_{\mathbb{Q}}\,(\alpha W)$ will be the time-0 value of the 'claim' W, where $\mathbb{Q}$ is the martingale probability. Set $w_i = W(\omega_i)$, $i = 1, 2$, then this may be written as the optimization problem in w_1 and w_2 given as

$$\begin{aligned} &\text{maximize} \quad p_1 v(w_1) + p_2 v(w_2) \\ &\text{subject to} \quad \alpha\,(q_1 w_1 + q_2 w_2) = w_0, \end{aligned} \tag{2.11}$$

where q_1 and q_2 are the martingale probabilities specified in (2.4). Let λ be a Lagrange multiplier for the constraint in the problem (2.11) and form the Lagrangian

$$\mathcal{L} = p_1 v(w_1) + p_2 v(w_2) + \lambda\,(w_0 - \alpha q_1 w_1 - \alpha q_2 w_2)\,;$$

then set

$$\frac{\partial \mathcal{L}}{\partial w_i} = p_i v'(w_i) - \lambda \alpha q_i = 0,$$

to give

$$v'(w_i) = \lambda \alpha (q_i / p_i), \quad \text{for} \quad i = 1, 2. \tag{2.12}$$

Solve (2.12) for $w_i = w_i(\lambda)$ as a function of λ and then choose the value $\lambda = \overline{\lambda}$ so that $w_i(\overline{\lambda})$ satisfies the constraint in (2.11). This procedure is shown in detail in Example 2.1.

Notice that, in general, the original probabilities p_1 and p_2 will play a role here in the choice of W, the optimal investment, as will be illustrated in the example. As we have seen in this section, given a contingent claim, the fair price of the claim is made without reference to the original probabilities and is determined using the martingale probabilities q_1 and q_2. However, which claim or final wealth the investor will choose will depend upon the original (or 'subjective' probabilities) p_1 and p_2 and the investor's utility function $v(\cdot)$.

Example 2.1 *Logarithmic utility.* Consider the case when $v(x) = \ln x$, then equation (2.12) becomes

$$v'(w_i) = 1/w_i = \lambda\alpha(q_i/p_i)$$

to give $w_i = p_i/(\lambda\alpha q_i)$; now substitute into the constraint in (2.11) to obtain

$$\alpha\left[q_1\left(\frac{p_1}{\lambda\alpha q_1}\right) + q_2\left(\frac{p_2}{\lambda\alpha q_2}\right)\right] = \frac{1}{\lambda}(p_1 + p_2) = \frac{1}{\lambda} = w_0.$$

This gives $\overline{\lambda} = 1/w_0$ and hence the optimal

$$w_i = W(\omega_i) = \frac{w_0 p_i}{\alpha q_i} \quad \text{for} \quad i = 1, 2.$$

Substituting the expressions for q_1 and q_2 from (2.4) shows that

$$w_1 = \frac{w_0 p_1}{\alpha}\left(\frac{u-d}{1/\alpha - d}\right) \quad \text{and} \quad w_2 = \frac{w_0 p_2}{\alpha}\left(\frac{u-d}{u - 1/\alpha}\right).$$

The corresponding replicating portfolio may be obtained from (2.2); for example, the holding in stock is

$$\begin{aligned} x &= \frac{1}{S_0}\left(\frac{w_1 - w_2}{u-d}\right) = \frac{1}{S_0}\left(\frac{w_0 p_1}{1-\alpha d} - \frac{w_0 p_2}{\alpha u - 1}\right) \\ &= \frac{w_0}{S_0}\left[\frac{\alpha(p_1 u + p_2 d) - 1}{(1-\alpha d)(\alpha u - 1)}\right] = \frac{w_0}{S_0^2}\left[\frac{\mathbb{E}(\alpha S_1) - S_0}{(1-\alpha d)(\alpha u - 1)}\right], \end{aligned}$$

since $\mathbb{E}(\alpha S_1) = \alpha S_0(p_1 u + p_2 d)$. As we are assuming that there is no arbitrage we have $u > 1/\alpha > d$ so that x, the amount held in stock, satisfies $x > 0$ or $x < 0$ according as the expected discounted stock price at time 1, $\mathbb{E}(\alpha S_1)$, satisfies $\mathbb{E}(\alpha S_1) > S_0$ or $\mathbb{E}(\alpha S_1) < S_0$, while $x = 0$ in the case that $\mathbb{E}(\alpha S_1) = S_0$; the last case corresponds to the situation when the original probabilities $\{p_i\}$ coincide with the martingale probabilities $\{q_i\}$.

These observations agree with the intuition that the investor will have a positive holding in stock when he expects the price to go up relative to the bank account, that is the case $\mathbb{E}(\alpha S_1) > S_0$, while he will be short in the stock when he expects its price to go down in relation to the bank account, the case $\mathbb{E}(\alpha S_1) < S_0$. □

2.2 Multi-period model

2.2.1 Introduction

Consider the simplest multi-period form of the model presented in Section 2.1 which operates over times $0, 1, \ldots, n$, where n is a fixed positive integer and is the **terminal time** of the model. Again, there are just two assets: a stock for which the price evolves randomly from period to period and a bank account paying a constant rate of interest, $\rho \geqslant 0$, per period.

Suppose that the price of the stock at time r is S_r where we can write

$$S_r = S_0 \prod_{i=1}^{r} Z_i, \quad \text{for} \quad r = 1, 2, \ldots, n,$$

with $Z_1, Z_2, \ldots, Z_n$ being independent, identically distributed random variables with

$$\mathbb{P}(Z_i = u) = p_1 = 1 - \mathbb{P}(Z_i = d) = 1 - p_2,$$

for $1 \leqslant i \leqslant n$. The price of the stock at time 0, S_0 is taken to be a constant. The random variable Z_i represents the proportional change to the stock price between time $i-1$ and time i and it takes just the two values u and d with probabilities $p_1 > 0$ and $p_2 > 0$, respectively, where we will assume that $u > d > 0$. For most of the remainder of this chapter, we will deal with what might be termed the standard multi-period binomial model, in which the values u and d taken on by the random variable (Z_i) are the same for each i, so that the random variables are identically distributed; in Section 2.2.7 we will mention the case where this assumption is relaxed. It is customary to represent the evolution of the values of the stock price as being on the nodes of a binary tree, which is illustrated in Figure 2.1 for the values corresponding to times $0, 1, 2, 3$.

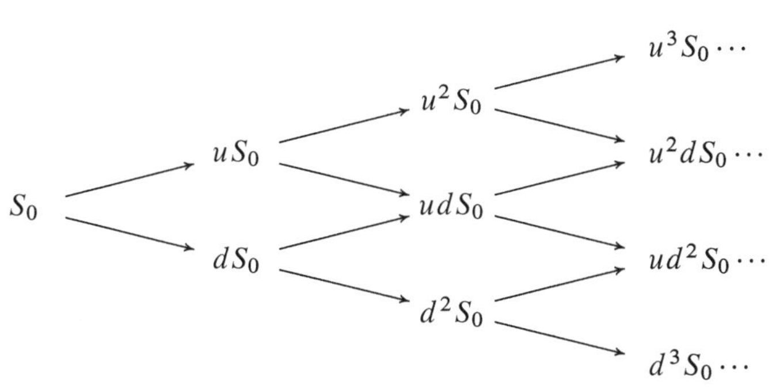

Figure 2.1: The binary tree for the stock price

We see that S_r takes on $r+1$ possible values

$$s_{0,r} \leqslant s_{1,r} \leqslant \cdots \leqslant s_{r,r} \quad \text{where} \quad s_{i,r} = u^i d^{r-i} S_0,$$

and if we set $p = p_1 = 1 - p_2$, the random variable S_r has a binomial distribution over these values with

$$\mathbb{P}(S_r = s_{i,r}) = \binom{r}{i} p^i (1-p)^{r-i}, \quad i = 0, 1, \ldots, r. \tag{2.13}$$

It is useful to think of the node in the binary tree corresponding to the stock price $S_r = u^i d^{r-i} S_0$ as (i, r) so that the binary tree may be represented as in Figure 2.2, where there will be a total of $1 + 2 + \cdots + (n+1) = \frac{1}{2}(n+1)(n+2)$ nodes in the full tree.

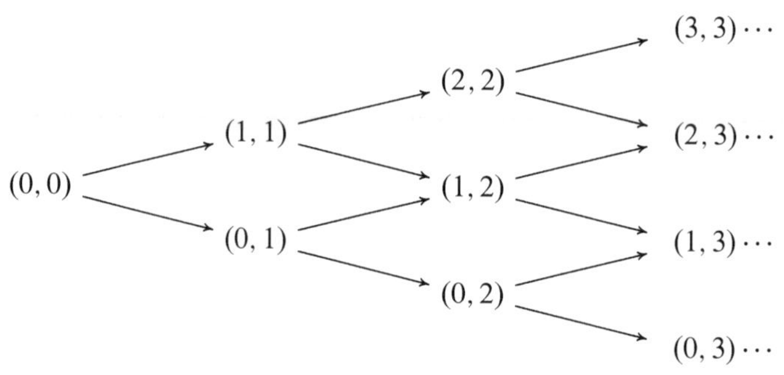

Figure 2.2: The binary tree with the labels for the nodes

One unit invested at time 0 in the bank account yields $(1+\rho)^r$ at time r, or equivalently, when we set $\alpha = 1/(1+\rho)$ to be the discount factor per period, α^r invested at time 0 in the bank account yields 1 unit at time r. The quantity α^r is the r-period discount factor that converts values at time r to time-0 values. We may view α^{n-r} as the price at time r of a riskless bond which pays 1 unit with certainty at time n, so that the holding in the bank account may be regarded as trading in this riskless bond.

Throughout, assume that the condition $u > 1 + \rho > d$ in (2.9) holds so that there will be no arbitrage opportunities in the model; that this condition is necessary and sufficient to preclude opportunities for riskless profits is immediate from the discussion of the one-period model.

Consider the problem of hedging and pricing a contingent claim in this model and we will deal with the case of a contingent claim paying off an amount C at time n, where C is a random variable which is some function of S_n. We will see later that the treatment we give extends easily to the case where C may depend on the whole sequence of values of the stock price, $S_1, \ldots, S_n$, but for the moment consider the case of a **terminal-value claim** where $C = f(S_n)$ is just a function of the stock price at the terminal time n, for some f.

An important special case is that of a **European call option**; this contract entitles, but does not require, the holder to purchase one unit of stock at a pre-determined fixed price c, the **strike price**, at the **expiry time** of the option, here taken to be n. The holder hopes that the market price of the stock at the expiry time S_n will exceed the strike price c in which case he will exercise the option, buying the stock for c and selling it in the market for S_n, realizing a gain of $S_n - c$; when $S_n < c$, then the holder will not exercise the option and he will receive 0. Thus the payoff of this contract is $C = (S_n - c)_+$, where $x_+ = \max(x, 0)$ denotes the positive part of x; this corresponds to $C = f(S_n)$ for the function $f(x) = (x - c)_+$.

Related to this is the **European put option** which entitles, but does not require, the holder to sell one unit of stock at the strike price c at the expiry time, n. In this case the holder hopes that the market price will end below the strike price, that is $S_n < c$, in which event the holder will buy one unit at the market price and sell at the strike price realizing $c - S_n$; when $c \leqslant S_n$ then the holder will not exercise the option and he will receive 0. Thus for the put option $C = f(S_n)$ where $f(x) = (c - x)_+$.

2.2.2 Dynamic hedging

Building on the arguments given in the one-period model, we will demonstrate that the claim $C = f(S_n)$ may be hedged by trading dynamically in this market in the following way. A fixed amount of money is used to form a portfolio of stock and money in the bank account at time 0; the amounts of the holdings in the portfolio in stock and the bank are then adjusted dynamically as the stock price goes up or down randomly through the nodes of the tree in such a way that no money is either injected or withdrawn from the portfolio at times $1, 2, \dots, n-1$ and at time n the portfolio is worth exactly the same as the claim C.

A trading strategy (or portfolio) is said to be **self-financing** when no money is injected or withdrawn between setting up the strategy and the terminal time. A self-financing trading strategy **replicates** the claim when its value matches that of the claim at the payoff time of the claim; in this case, that is at the terminal time n. We will demonstrate that for any claim C a replicating trading strategy exists and show how it is calculated.

Denote the number of units of stock held between time r and $r+1$ for the trading strategy by X_r and the holding in the bank account from r to $r+1$ by Y_r; here, X_r and Y_r are random variables and we will see that both are functions of the stock price S_r at time r. We will determine the appropriate values of these random variables by using dynamic programming, which is the technique of backward induction, to calculate the values (X_r, Y_r) successively for $r = n-1, n-2, \dots, 0$. We calculate first the values that the pair (X_{n-1}, Y_{n-1}) should take at all possible nodes of the form $(i, n-1)$, $0 \leqslant i \leqslant n-1$, at time $n-1$, that is for all possible values of the stock price S_{n-1} at time $n-1$, then we move on to calculate the values (X_{n-2}, Y_{n-2}), and so on, as we step backwards through the binary tree.

Since the value of the portfolio held at time n is to match the value of the claim C then we must have

$$X_{n-1} S_n + (1+\rho) Y_{n-1} = C. \tag{2.14}$$

The condition that the trading strategy is self-financing is

$$X_{r-1} S_r + (1+\rho) Y_{r-1} = X_r S_r + Y_r, \quad \text{for} \quad 1 \leqslant r \leqslant n-1, \tag{2.15}$$

since the left-hand side of (2.15) represents the total value at time r of the holdings (X_{r-1}, Y_{r-1}) brought into the time period r and the right-hand side represents the value at time r of the holdings (X_r, Y_r) that will be carried forward to time $r+1$. For $r \leqslant n-1$, let

$$C_r = X_{r-1} S_r + (1+\rho) Y_{r-1} = X_r S_r + Y_r \tag{2.16}$$

be the common value on both sides of (2.15) which must be the value of the trading strategy at time r, and set $C_n = X_{n-1} S_n + (1+\rho) Y_{n-1} \equiv C$.

Let $c_{i,n} = f\left(u^i d^{n-i} S_0\right) = f(s_{i,n})$ denote the value taken on by the random variable C when $S_n = u^i d^{n-i} S_0 = s_{i,n}$, for $0 \leqslant i \leqslant n$. Let $x_{i,r}$ denote the value that X_r should take on at the node (i, r) for $0 \leqslant i \leqslant r$ and $0 \leqslant r \leqslant n-1$, corresponding to the stock price $S_r = u^i d^{r-i} S_0$ at time r; similarly, let $y_{i,r}$ be the value that Y_r should take on at the same node. Furthermore, for $r \leqslant n-1$ we will denote by $c_{i,r}$ the value of this portfolio at time r at the node (i, r); thus

$$c_{i,r} = x_{i,r} s_{i,r} + y_{i,r}, \tag{2.17}$$

because the stock price $S_r = u^i d^{r-i} S_0 = s_{i,r}$ at the node (i, r). The value of the random variable C_r at the node (i, r) is then $c_{i,r}$.

Consider first the part of the tree corresponding to the calculation required at the node $(i, n-1)$ as illustrated in Figure 2.3. To satisfy the equation (2.14) at the nodes $(i+1, n)$ and (i, n) requires exactly the same calculation as for the one-period model that was set out in (2.1) with $x = x_{i,n-1}$ and $y = y_{i,n-1}$.

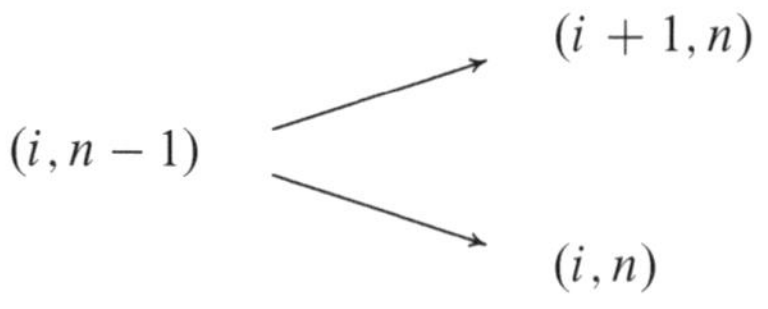

Figure 2.3: The node $(i, n-1)$

Recalling the expressions in (2.2), we have that

$$x_{i,n-1} = \frac{c_{i+1,n} - c_{i,n}}{s_{i,n-1}(u-d)} \quad \text{and} \quad y_{i,n-1} = \frac{1}{1+\rho}\left[\frac{u c_{i,n} - d c_{i+1,n}}{u-d}\right];$$

we also have from the calculation in (2.3) that

$$c_{i,n-1} = x_{i,n-1} s_{i,n-1} + y_{i,n-1} = \alpha\left(q_1 c_{i+1,n} + q_2 c_{i,n}\right),$$

where q_1 and q_2 are the martingale probabilities given in (2.4). Now, having calculated all of the $\{c_{i,n-1}\}$ (as well as all of the $\{x_{i,n-1}\}$ and $\{y_{i,n-1}\}$) in terms of the $\{c_{i,n}\}$ the argument may be repeated to calculate successively all the $\{c_{i,r}\}$ in terms of $\{c_{i,r+1}\}$ for $r = n-2, n-3, \ldots, 0$, by using the same calculations at each stage

$$x_{i,r} = \frac{c_{i+1,r+1} - c_{i,r+1}}{s_{i,r}(u-d)} \quad \text{and} \quad y_{i,r} = \frac{1}{1+\rho}\left[\frac{uc_{i,r+1} - dc_{i+1,r+1}}{u-d}\right]; \tag{2.18}$$

furthermore,

$$c_{i,r} = x_{i,r}s_{i,r} + y_{i,r} = \alpha\left(q_1 c_{i+1,r+1} + q_2 c_{i,r+1}\right). \tag{2.19}$$

The calculation of $c_{i,r}$ in terms of the values $c_{i+1,r+1}$ and $c_{i,r+1}$ is illustrated in Figure 2.4. When we have determined $c_{0,0}$, the initial value of the trading strategy,

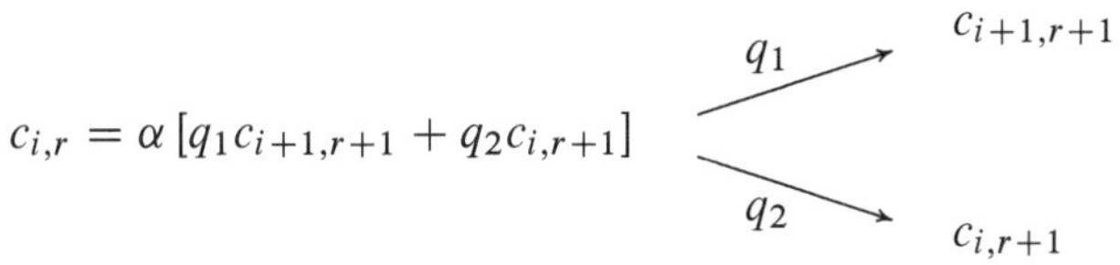

Figure 2.4: The calculation at the node (i, r)

we see that this must be the (unique) price of the claim at time 0. To see that is the case, if the price was less than this value an investor would buy the claim and sell the trading strategy and thus make a riskless profit, or alternatively, if the price is greater than this value he would sell the claim and hedge it using the trading strategy, again having a riskless profit; in either case he would have an arbitrage which is ruled out by our assumption that $u > 1 + \rho > d$.

To explore the probabilistic interpretation of these calculations we will denote by $\mathcal{Q}$ the probability distribution under which the random variables $Z_1, Z_2, \ldots, Z_n$ are independent and identically distributed with

$$\mathcal{Q}(Z_i = u) = q_1 = 1 - \mathcal{Q}(Z_i = d) = 1 - q_2, \quad \text{for} \quad 1 \leqslant r \leqslant n.$$

Set $q = q_1 = 1 - q_2$, then under the distribution $\mathcal{Q}$ the random variable S_r has a binomial distribution over the values $\{s_{i,r}\}$ with

$$\mathcal{Q}(S_r = s_{i,r}) = \binom{r}{i} q^i (1-q)^{r-i}, \quad i = 0, 1, \ldots, r, \tag{2.20}$$

so that when making calculations with the probability distribution $\mathcal{Q}$ the probability p in (2.13) is replaced by q. When taking expectations under the probability distribution $\mathcal{Q}$ we write the expectation as $E_{\mathcal{Q}}$.

For $r \geqslant 1$, denote by $\mathcal{F}_r$ the information obtained by observing the random variables $S_1, \ldots, S_r$ or, equivalently, observing $Z_1, \ldots, Z_r$; formally, $\mathcal{F}_r$ is known as the σ-field generated by the random variables $S_1, \ldots, S_r$ and is often written as $\mathcal{F}_r = \sigma(S_1, \ldots, S_r)$, see the Appendix, but for the present circumstances it may just be thought of as a shorthand for writing down conditional expectations given the generating random variables. Thus for some random variable W, the conditional expectation $\mathbb{E}[W \mid \mathcal{F}_r]$ is just the same as $\mathbb{E}[W \mid S_1, \ldots, S_r]$ if we are taking expectations with the probability $\mathbb{P}$; if we are taking expectations using the probability $\mathbb{Q}$ then $\mathbb{E}_{\mathbb{Q}}[W \mid \mathcal{F}_r]$ is the same as $\mathbb{E}_{\mathbb{Q}}[W \mid S_1, \ldots, S_r]$. We will take $\mathcal{F}_0$ to be the trivial case where no random variables are observed so conditional expectations given $\mathcal{F}_0$ are the same as ordinary (unconditional) expectations. Note that under either $\mathbb{P}$ or $\mathbb{Q}$ the random variables $Z_{r+1}, \ldots, Z_n$ are independent of $\mathcal{F}_r$ since $\mathcal{F}_r$ depends only on $Z_1, \ldots, Z_r$. Also, $\mathcal{F}_r$ would be the information available to an investor at time r who is trading with no knowledge of future prices, as he must assemble his portfolio to carry forward to time $r+1$ just on the basis of knowing the prices up to time r (and not anticipating or foreseeing the prices after time r).

Recall (2.7), then we see that

$$s_{i,r} = \alpha\left(q s_{i+1,r+1} + (1-q) s_{i,r+1}\right) \quad \text{for} \quad 0 \leqslant i \leqslant r,\ 0 \leqslant r < n. \tag{2.21}$$

Rewrite this statement as

$$S_r = \mathbb{E}_{\mathbb{Q}}(\alpha S_{r+1} \mid \mathcal{F}_r) \quad \text{for} \quad 0 \leqslant r < n, \tag{2.22}$$

and, after multiplying both sides of (2.22) by the constant α^r, we obtain

$$\alpha^r S_r = \mathbb{E}_{\mathbb{Q}}\left(\alpha^{r+1} S_{r+1} \mid \mathcal{F}_r\right) \quad \text{for} \quad 0 \leqslant r < n, \tag{2.23}$$

which shows that the process $\{\alpha^r S_r, \mathcal{F}_r, 0 \leqslant r \leqslant n\}$ is a martingale under the probability $\mathbb{Q}$. For this reason $\mathbb{Q}$ is known as the **martingale probability** for the model.

In the same way that (2.22) follows from (2.21), from (2.19) we see that

$$C_r = \mathbb{E}_{\mathbb{Q}}(\alpha C_{r+1} \mid \mathcal{F}_r) \quad \text{for} \quad 0 \leqslant r < n; \tag{2.24}$$

again, multiplying through (2.24) by α^r gives

$$\alpha^r C_r = \mathbb{E}_{\mathbb{Q}}\left(\alpha^{r+1} C_{r+1} \mid \mathcal{F}_r\right) \quad \text{for} \quad 0 \leqslant r < n, \tag{2.25}$$

which shows that the process $\{\alpha^r C_r, \mathcal{F}_r, 0 \leqslant r \leqslant n\}$ is a martingale under the probability $\mathbb{Q}$. From the martingale property (2.25), we may conclude that, for $r < n$, $\alpha^r C_r = \mathbb{E}_{\mathbb{Q}}(\alpha^n C_n \mid \mathcal{F}_r)$ which gives

$$C_r = \mathbb{E}_{\mathbb{Q}}(\alpha^{n-r} C_n \mid \mathcal{F}_r) = \mathbb{E}_{\mathbb{Q}}(\alpha^{n-r} C \mid \mathcal{F}_r), \tag{2.26}$$

because $C_n \equiv C$.

The random variable C_r is the value at time r of the trading strategy replicating the claim C and so it must be the price at time r of C; equation (2.26) shows that we calculate that price by discounting the payoff C by the discount factor α^{n-r} from time n to time r and then taking the conditional expectation, given the information $\mathcal{F}_r$ at time r, using the martingale probability $\mathbb{Q}$. In particular, the price at time 0 is

$$c_{0,0} = C_0 = \mathbb{E}_{\mathbb{Q}}\left(\alpha^n C \mid \mathcal{F}_0\right) = \mathbb{E}_{\mathbb{Q}}\left(\alpha^n C\right). \tag{2.27}$$

Using the representation $C = f(S_n)$ and (2.20), we may calculate the expression given in (2.27), as

$$\begin{aligned} C_0 &= \mathbb{E}_{\mathbb{Q}}\left(\alpha^n f(S_n)\right) = \alpha^n \sum_{i=0}^{n} f(s_{i,n})\mathbb{Q}(S_n = s_{i,n}) \\ &= \alpha^n \sum_{i=0}^{n} f(s_{i,n}) \binom{n}{i} q^i (1-q)^{n-i} \\ &= \alpha^n \sum_{i=0}^{n} f\left(u^i d^{n-i} S_0\right) \binom{n}{i} q^i (1-q)^{n-i}. \end{aligned} \tag{2.28}$$

When we observe that for $r < n$, $S_n = S_r \prod_{r+1}^{n} Z_i$ and recall that $Z_{r+1}, \dots, Z_n$ are independent of $\mathcal{F}_r$ under the probability $\mathbb{Q}$, we may write down the expression corresponding to (2.28) for the case of C_r as

$$\begin{aligned} C_r &= \mathbb{E}_{\mathbb{Q}}\left(\alpha^{n-r} f(S_n) \mid \mathcal{F}_r\right) \\ &= \alpha^{n-r} \sum_{i=0}^{n-r} f\left(u^i d^{n-r-i} S_r\right) \binom{n-r}{i} q^i (1-q)^{n-r-i}. \end{aligned} \tag{2.29}$$

Note that for $0 \leqslant r \leqslant n$, C_r is a function of the stock price S_r, so that $C_r = f_r(S_r)$, where the function $f_r(\cdot)$ is determined from the expression in (2.29); of course, $f_n = f$.

A final observation on the replicating strategy for the claim is to note that the holding in the bank account, Y_r, satisfies

$$Y_r = \mathbb{E}_{\mathbb{Q}}(\alpha Y_{r+1} \mid \mathcal{F}_r), \quad \text{for} \quad 0 \leqslant r < n-1. \tag{2.30}$$

This follows by observing that $Y_r = y_{i,r}$ when $S_r = s_{i,r}$, and using (2.18) and (2.19), we obtain

$$\begin{aligned} y_{i,r} &= \frac{\alpha}{u-d}\left[u c_{i,r+1} - d c_{i+1,r+1}\right] \\ &= \frac{\alpha^2}{u-d}\left[u\left(q c_{i+1,r+2} + (1-q) c_{i,r+2}\right) - d\left(q c_{i+2,r+2} + (1-q) c_{i+1,r+2}\right)\right] \\ &= \alpha\left[q y_{i+1,r+1} + (1-q) y_{i,r+1}\right], \end{aligned}$$

which is (2.30). In the same way that (2.25) follows from (2.24), we may deduce that $\{\alpha^r Y_r, \mathcal{F}_r, 0 \leqslant r \leqslant n-1\}$ is a martingale under the probability $\mathbb{Q}$.

We may also see that the value of the holding in stock at time r in the replicating portfolio, $X_r S_r$, satisfies

$$X_r S_r = \mathbb{E}_{\mathbb{Q}}(\alpha X_{r+1} S_{r+1} \mid \mathcal{F}_r), \quad \text{for} \quad 0 \leqslant r < n-1. \tag{2.31}$$

This follows from (2.16), (2.24) and (2.30), by noting that

$$\begin{aligned} X_r S_r = C_r - Y_r &= \mathbb{E}_{\mathbb{Q}}(\alpha C_{r+1} \mid \mathcal{F}_r) - \mathbb{E}_{\mathbb{Q}}(\alpha Y_{r+1} \mid \mathcal{F}_r) \\ &= \mathbb{E}_{\mathbb{Q}}(\alpha (C_{r+1} - Y_{r+1}) \mid \mathcal{F}_r) = \mathbb{E}_{\mathbb{Q}}(\alpha X_{r+1} S_{r+1} \mid \mathcal{F}_r). \end{aligned}$$

We conclude that $\{\alpha^r X_r S_r, \mathcal{F}_r, 0 \leqslant r \leqslant n-1\}$ is also a martingale under $\mathbb{Q}$.

Note that the conclusions that $\alpha^r C_r$, $\alpha^r X_r$ and $\alpha^r Y_r S_r$ are martingales under the probability $\mathbb{Q}$ follow from the self-financing property; the replicating property only serves to determine the terminal value C_n of the portfolio (and consequently X_{n-1} and Y_{n-1}).

As an application of (2.31), suppose that the function $f(x)$ is non-decreasing in x, then it follows that the holding in stock in the replicating portfolio, X_r, is always non-negative at each stage $r = 0, 1, \ldots, n-1$; that is, the portfolio is long in the stock. To see this, since $S_r > 0$, if the random variable $X_{r+1} \geqslant 0$, then (2.31) implies that $X_r \geqslant 0$; the conclusion follows from (2.18) by backward induction on r, when we observe that

$$X_{n-1} = \frac{f(uS_{n-1}) - f(dS_{n-1})}{S_{n-1}(u-d)} \geqslant 0.$$

In the case of the European call option, $f(x) = (x-c)_+$ is non-decreasing so the replicating portfolio is long in stock. In the same way, when $f(x)$ is non-increasing in x, as would be the case for the European put option $f(x) = (c-x)_+$, then we may conclude that $X_r \leqslant 0$ and the replicating portfolio is short in stock. Exercise 2.4 has a related result for the holding in stock in the replicating portfolio.

We conclude this section by considering a numerical example which illustrates the calculations that have been introduced.

Example 2.2 Consider the case of a two-period model, $n = 2$, in which the stock price moves on the binomial tree illustrated in Figure 2.5. Note that the proportional up jumps, $u = 3$, and the proportional down jumps, $d = \frac{1}{2}$, are the same in each period, so that this is a standard binomial model. We will not specify the underlying probability p, $0 < p < 1$, of an up jump as we have seen that it does not enter into the calculations to be made. Let us assume that the interest rate in each period is $\rho = \frac{1}{2}$, so that the discount factor is $\alpha = 1/(1+\rho) = \frac{2}{3}$. Suppose that we wish to price a European put option at strike price 11, with expiry time 2, and also find the replicating portfolio. The payoff at time 2 of this option is $C = (11 - S_2)_+ = 0$, 5 or 10 according as the stock price $S_2 = 36$, 6 or 1; these values are indicated in Figure 2.6.

First compute the martingale probability of an up jump

$$q_1 = q = \frac{1 + \rho - d}{u - d} = \tfrac{2}{5}$$

from (2.4); then, using the expressions in (2.19) and (2.18), we calculate (i) the value of the option, (ii) the number of units of stock and (iii) the amount in the bank account

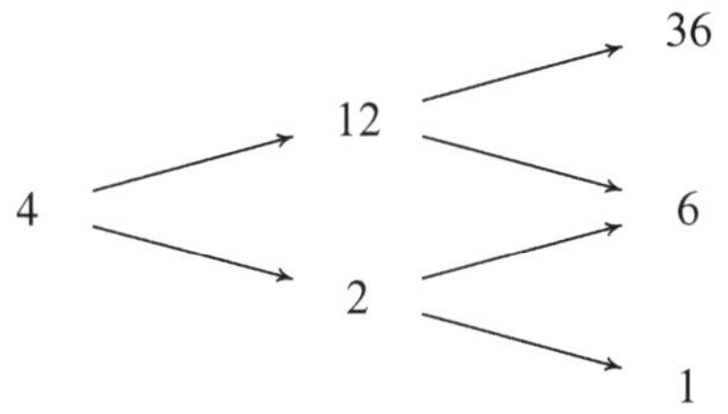

Figure 2.5: The stock price at times $r = 0, 1, 2$

held in the replicating portfolio successively at each of the nodes $(1, 1)$, $(0, 1)$ and $(0, 0)$ (corresponding to the stock prices $S_1 = 12$, $S_1 = 2$ and $S_0 = 4$, respectively). These quantities are displayed as the corresponding triple in Figure 2.6 at each of the three nodes and the martingale probabilities are shown for the first period on the edges of the tree; the up and down probabilities will of course be the same for the second period in this example. Thus, for example at node $(0, 1)$ when the stock price is $S_1 = 2$, because

$$\begin{aligned}\alpha \mathbb{E}_{\mathbb{Q}}(C \mid S_1 = 2) &= \alpha \left[5\mathbb{Q}(C = 5) + 10\mathbb{Q}(C = 10)\right] \\ &= \tfrac{2}{3}\left[5 \times \tfrac{2}{5} + 10 \times \tfrac{3}{5}\right] = \tfrac{16}{3},\end{aligned}$$

the value of the put is $\frac{16}{3}$; the replicating portfolio holds $\frac{22}{3}$ in the bank account from the calculation

$$y_{0,1} = \alpha\left(\frac{uc_{0,2} - dc_{1,2}}{u - d}\right) = \tfrac{2}{3}\left(\frac{3\times 10 - \frac{1}{2}\times 5}{3 - \frac{1}{2}}\right) = \tfrac{22}{3},$$

while the holding of -1 unit of stock worth -2 follows because

$$x_{0,1}s_{0,1} = \frac{c_{1,2} - c_{0,2}}{u - d} = \frac{5 - 10}{3 - \frac{1}{2}} = -2,$$

which gives $x_{0,1} = -1$ since $s_{0,1} = 2$.

We may also note the martingale property (under the martingale probability $\mathbb{Q}$) satisfied by the holding in the bank account; recall that we observed that $(\alpha^r Y_r)$ will be

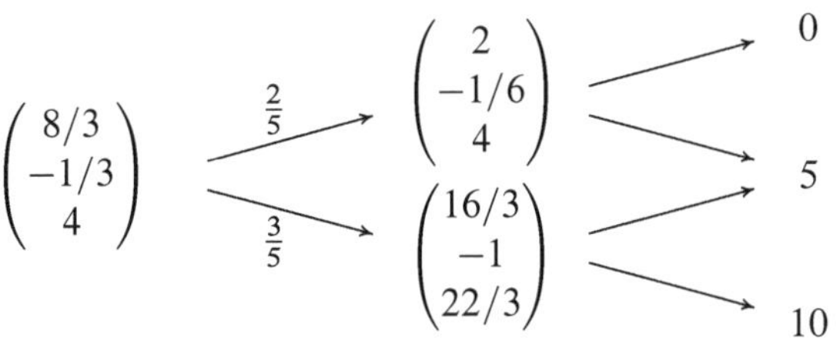

Figure 2.6: The prices of the European put and the replicating portfolio

a martingale, where Y_r is the holding in the bank account. For example, here $Y_0 = 4$, $Y_1 = 4$ when $S_1 = 12$ and $Y_1 = \frac{22}{3}$ when $S_1 = 2$, so that $Y_0 = \alpha \mathbb{E}_{\mathbb{Q}}(Y_1 \mid \mathcal{F}_0)$ holds since

$$\begin{aligned}\alpha \mathbb{E}_{\mathbb{Q}}\left(Y_1 \mid \mathcal{F}_0\right) &= \alpha\left[4\mathbb{Q}(Y_1 = 4) + \tfrac{22}{3}\mathbb{Q}(Y_1 = \tfrac{22}{3})\right] \\ &= \tfrac{2}{3}\left[4 \times \tfrac{2}{5} + \tfrac{22}{3} \times \tfrac{3}{5}\right] = 4 = Y_0.\end{aligned}$$

In a similar way, it may be verified that the discounted value of the holding in stock in the replicating portfolio, $(\alpha^r X_r S_r)$ is also a martingale under the probability $\mathbb{Q}$. □

2.2.3 Change of probability

As the previous section shows, the martingale probability $\mathbb{Q}$ is central for the calculation of prices and for the determination of the hedging portfolio in the binomial model; we will investigate further how calculations with $\mathbb{Q}$ are made. Underlying the discussion, but not referred to explicitly as is usually the case in probability, is the sample space Ω on which the random variables $S_1, \ldots, S_n$ are defined. Here

$$\Omega = \{(\omega_1, \ldots, \omega_n) : \omega_r = 1 \text{ or } 0, \quad 1 \leqslant r \leqslant n\}, \tag{2.32}$$

where $\omega_r = 1$ corresponds to the rth jump being an up jump, u, and $\omega_r = 0$ to the rth jump being a down jump, d.

We then calculate the probability of the point $(\omega_1, \ldots, \omega_n)$ in Ω for the underlying probability distribution $\mathbb{P}$ as

$$\mathbb{P}(\omega_1, \ldots, \omega_n) = p^{\omega_1 + \cdots + \omega_n}(1-p)^{n-(\omega_1+\cdots+\omega_n)}$$

while for the martingale probability $\mathbb{Q}$ we have

$$\mathbb{Q}(\omega_1, \ldots, \omega_n) = q^{\omega_1 + \cdots + \omega_n}(1-q)^{n-(\omega_1+\cdots+\omega_n)}.$$

Write $\boldsymbol{\omega} = (\omega_1, \ldots, \omega_n)$, and consider the random variable L, defined on the sample space Ω, given by

$$L(\boldsymbol{\omega}) = \left(\frac{q}{p}\right)^{\omega_1+\cdots+\omega_n}\left(\frac{1-q}{1-p}\right)^{n-(\omega_1+\cdots+\omega_n)} = \frac{\mathbb{Q}(\boldsymbol{\omega})}{\mathbb{P}(\boldsymbol{\omega})}; \tag{2.33}$$

L is usually known as the **Radon–Nikodym** derivative of the probability $\mathbb{Q}$ with respect to the probability $\mathbb{P}$ and written as $L = \dfrac{d\mathbb{Q}}{d\mathbb{P}}$. Note that L is the likelihood ratio of the random variables $S_1, \ldots, S_n$ (or equivalently of the random variables $Z_1, \ldots, Z_n$, or of the point $\boldsymbol{\omega} \in \Omega$) under the two probabilities $\mathbb{P}$ and $\mathbb{Q}$. Then for any event $A \subseteq \Omega$ we may express

$$\mathbb{Q}(A) = \sum_{\boldsymbol{\omega} \in A} \mathbb{Q}(\boldsymbol{\omega}) = \sum_{\boldsymbol{\omega} \in A} L(\boldsymbol{\omega})\, \mathbb{P}(\boldsymbol{\omega}) = \mathbb{E}\left[L I_A\right]$$

where I_A is the indicator of the event A, which is the random variable that takes the value 1 when the event A occurs and 0 otherwise; the unsubscripted $\mathbb{E}$ denotes the expectation taken with respect to the underlying probability $\mathbb{P}$. For any random variable X, from (2.33) we have that

$$\begin{aligned}\mathbb{E}_{\mathbb{Q}}(X) &= \sum_{\omega\in\Omega} X(\omega)\,\mathbb{Q}(\omega)\\ &= \sum_{\omega\in\Omega} L(\omega)\,X(\omega)\,\mathbb{P}(\omega) = \mathbb{E}\,(LX) = \mathbb{E}\left(\frac{d\mathbb{Q}}{d\mathbb{P}}\,X\right).\end{aligned}$$

We may express L in terms of the stock price S_n at the terminal time n; suppose that

$$S_n = u^k d^{n-k} S_0 = (u/d)^k\, d^n S_0,$$

so there have been k up jumps and $n-k$ down jumps, it follows that

$$k = \ln\left(\frac{S_n}{d^n S_0}\right) \Big/ \ln\left(\frac{u}{d}\right). \tag{2.34}$$

From (2.33) and (2.34) we see that

$$\begin{aligned}\frac{d\mathbb{Q}}{d\mathbb{P}} = L &= \left(\frac{1-q}{1-p}\right)^n \left[\frac{q(1-p)}{p(1-q)}\right]^k\\ &= \left(\frac{1-q}{1-p}\right)^n \left[\frac{S_n}{S_0 d^n}\right]^{\left(\frac{\ln(q/p)-\ln((1-q)/(1-p))}{\ln u-\ln d}\right)}\end{aligned} \tag{2.35}$$

which shows, in particular, that the Radon–Nikodym derivative for this model depends only on S_n and not on the intermediate values of the stock price, $S_1, \ldots, S_{n-1}$. We will write

$$\frac{d\mathbb{Q}}{d\mathbb{P}} = L = \left(\frac{1-q}{1-p}\right)^n \left[\frac{S_n}{S_0 d^n}\right]^\theta, \tag{2.36}$$

where

$$\theta = \frac{\ln(q/p) - \ln\left((1-q)/(1-p)\right)}{\ln u - \ln d}. \tag{2.37}$$

Notice that

$$\left(\frac{u}{d}\right)^\theta = \frac{q(1-p)}{p(1-q)}. \tag{2.38}$$

Now suppose that we set

$$L_r = \left(\frac{1-q}{1-p}\right)^r \left[\frac{S_r}{S_0 d^r}\right]^\theta, \quad \text{for} \quad 0 \leqslant r \leqslant n,$$

so that $L_n = L$. We observe that $\{L_r, \mathcal{F}_r, 0 \leqslant r \leqslant n\}$ is a martingale (under the probability $\mathbb{P}$), since conditional on S_r, $S_{r+1} = uS_r$ or $S_{r+1} = dS_r$ with probabilities p and $1-p$ respectively, we have

$$\begin{aligned}\mathbb{E}\left(L_{r+1}\mid\mathcal{F}_r\right)&=p\left(\frac{1-q}{1-p}\right)^{r+1}\left(\frac{uS_r}{S_0d^{r+1}}\right)^{\theta}+(1-p)\left(\frac{1-q}{1-p}\right)^{r+1}\left(\frac{dS_r}{S_0d^{r+1}}\right)^{\theta}\\&=\frac{1-q}{1-p}\left[p\left(\frac{u}{d}\right)^{\theta}+1-p\right]L_r=L_r,\end{aligned}$$

after using the relation (2.38). From the martingale property we have that

$$L_r=\mathbb{E}\left(L\mid\mathcal{F}_r\right)=\mathbb{E}\left(\frac{d\mathbb{Q}}{d\mathbb{P}}\,\middle|\,\mathcal{F}_r\right)$$

and we may also note that L_r is just the Radon–Nikodym derivative when the model is restricted to the time periods $0,1,\dots,r$, so it is the likelihood ratio for the random variables $S_1,\dots,S_r$ under the probabilities $\mathbb{Q}$ and $\mathbb{P}$.

2.2.4 Utility maximization

One application of the explicit form for the Radon–Nikodym derivative given in the previous section occurs when we consider the problem of an investor with initial capital w_0, at time 0, who wishes to trade in the binomial model so as to maximize the expected utility of his final wealth, W, at time n when his utility function is $v(\cdot)$. The investor achieves his final wealth by trading in the stock and the bank account at the times $0,1,\dots,n-1$; he sets up an initial holding in the bank account and in stock at time 0 and then re-adjusts the amounts in his portfolio at times $1,\dots,n-1$, without injecting or withdrawing funds. This problem is equivalent to determining that contingent claim, with payoff C at time n, which may be purchased at time 0 at a cost of w_0 so as to maximize $\mathbb{E}\,v(C)$, and then the final wealth $W\equiv C$; this is because the initial price of any claim is determined through considering how it is replicated exactly through trading at the intermediate times. As we have seen in (2.27), this initial price is $\mathbb{E}_{\mathbb{Q}}\left(\alpha^nC\mid\mathcal{F}_0\right)=\mathbb{E}_{\mathbb{Q}}\left(\alpha^nC\right)$, since S_0 is a constant, where $\mathbb{Q}$ is the equivalent martingale probability.

The problem may be formulated mathematically as that of choosing a random variable C, which might in general depend on S_j, $0\leqslant j\leqslant n$, so as to

$$\text{maximize}\quad\mathbb{E}\,v(C)\qquad\text{subject to}\quad\mathbb{E}_{\mathbb{Q}}\left(\alpha^nC\right)=w_0.\tag{2.39}$$

As in the one-period case, this is a constrained optimization problem which again may be solved by considering the Lagrangian,

$$\begin{aligned}\mathcal{L}&=\mathbb{E}\,v(C)+\lambda\left[w_0-\mathbb{E}_{\mathbb{Q}}\left(\alpha^nC\right)\right]\\&=\mathbb{E}\,v(C)+\lambda\left[w_0-\alpha^n\mathbb{E}\left(\frac{d\mathbb{Q}}{d\mathbb{P}}C\right)\right]\qquad(2.40)\\&=\mathbb{E}\left[v(C)-\lambda\alpha^n\frac{d\mathbb{Q}}{d\mathbb{P}}C\right]+\lambda w_0,\qquad(2.41)\end{aligned}$$

for suitable scalar Lagrange multiplier λ. Maximizing inside the expectation yields

$$v'(C) = \lambda\alpha^n \left(\frac{d\mathbb{Q}}{d\mathbb{P}}\right); \tag{2.42}$$

assuming that v is concave, so that v' is non-increasing, then for fixed λ, a solution C $(= C(\lambda))$ of the equation (2.42) will yield a maximum of the Lagrangian for that value of λ and it will be a maximal solution of (2.39) if $\lambda = \overline{\lambda}$ is chosen so that $C(\overline{\lambda})$ satisfies the constraint $\mathbb{E}_{\mathbb{Q}}\left(\alpha^n C(\overline{\lambda})\right) = w_0$. That this procedure yields the maximum follows from the standard Lagrangian argument, since for any C satisfying $\mathbb{E}_{\mathbb{Q}}(\alpha^n C) = w_0$, we have

$$\begin{aligned} \mathbb{E}\,v(C) &= \mathbb{E}\,v(C) + \overline{\lambda}\left[w_0 - \mathbb{E}_{\mathbb{Q}}(\alpha^n C)\right] \\ &= \mathbb{E}\left[v(C) - \overline{\lambda}\alpha^n \frac{d\mathbb{Q}}{d\mathbb{P}} C\right] + \overline{\lambda} w_0 \\ &\leqslant \mathbb{E}\left[v(C(\overline{\lambda})) - \overline{\lambda}\alpha^n \frac{d\mathbb{Q}}{d\mathbb{P}} C(\overline{\lambda})\right] + \overline{\lambda} w_0 = \mathbb{E}\,v\left(C(\overline{\lambda})\right); \end{aligned}$$

here we have used the rearrangement that relates (2.40) to (2.41) twice and also the fact that $C(\overline{\lambda})$ satisfies the constraint in (2.39). Recall the expression for the Radon–Nikodym derivative given in (2.36)

$$\frac{d\mathbb{Q}}{d\mathbb{P}} = L = \left(\frac{1-q}{1-p}\right)^n \left[\frac{S_n}{S_0 d^n}\right]^\theta,$$

where θ is defined in (2.37). As was noted, the Radon–Nikodym derivative is a function of S_n. Because of this, we see from equation (2.42) that any investor maximizing the utility of his final wealth will always choose to invest his capital in a terminal-value claim; that is a claim for which the payoff is just a function of S_n, the final stock price at time n.

Example 2.3 *Logarithmic utility.* Consider again the case considered in Example 2.1 where $v(x) = \ln x$, then

$$1/C = \lambda\alpha^n \frac{d\mathbb{Q}}{d\mathbb{P}} = \lambda\alpha^n \left(\frac{1-q}{1-p}\right)^n \left[\frac{S_n}{S_0 d^n}\right]^\theta$$

where θ is defined in (2.37); substituting for C into the constraint to determine the value $\lambda = \overline{\lambda}$ yields

$$w_0 = \mathbb{E}_{\mathbb{Q}}(\alpha^n C) = \frac{1}{\lambda}\mathbb{E}\left[\frac{d\mathbb{Q}}{d\mathbb{P}}\left(1/\left(\frac{d\mathbb{Q}}{d\mathbb{P}}\right)\right)\right] = \frac{1}{\lambda}$$

which gives $\overline{\lambda} = 1/w_0$ and and we see that

$$C = \frac{w_0}{\alpha^n}\left(\frac{1-p}{1-q}\right)^n \left[\frac{S_0 d^n}{S_n}\right]^\theta$$

is the optimal final wealth. □

As was observed in the one-period case, the original (or subjective) probability, p, in the model does not enter into the pricing of any contingent claim but, in general, it will enter into the choice of claim that a utility-maximizing investor will buy.

2.2.5 Path-dependent claims

The method outlined in Section 2.2.2 was applied to the case of a terminal-value claim paying $C = f(S_n)$ at time n. The technique of calculating the price of the claim at time r, using (2.29), and the composition of the hedging portfolio were derived using the movement of the stock price on the tree illustrated in Figure 2.2. We calculate the price and the composition of the portfolio for nodes of the form (i, r) backwards in time; for given r, assume that the values are determined at all nodes of the form $(i, r+1)$ and then calculate the values for nodes of the form (i, r). This is done successively for $r = n-1, \dots, 0$. At each time r, the price of the claim, C_r, and the constituent amounts, X_r, Y_r, in the replicating portfolio are functions of the stock price S_r, which is a Markov chain; the future evolution of the stock price after time r depends only on the price at that time and not on the prices S_i at times $i < r$.

We may extend the technique to consider claims where the payoff may depend on the whole path of the stock price between times 0 and n, that is on $S_1, \dots, S_n$ or, equivalently, on the proportional stock price changes $Z_1, \dots, Z_n$. Thus the random variable C, which is defined on the probability space Ω set out in Section 2.2.3, is such that $C(\omega_1, \dots, \omega_n)$ may not necessarily just be a function of $\omega_1 + \cdots + \omega_n$ as was the case previously.

We may now think of the process as moving on a general binary tree of the form illustrated in Figure 2.7, for the times $r = 0, 1, 2$. A node at time r is represented

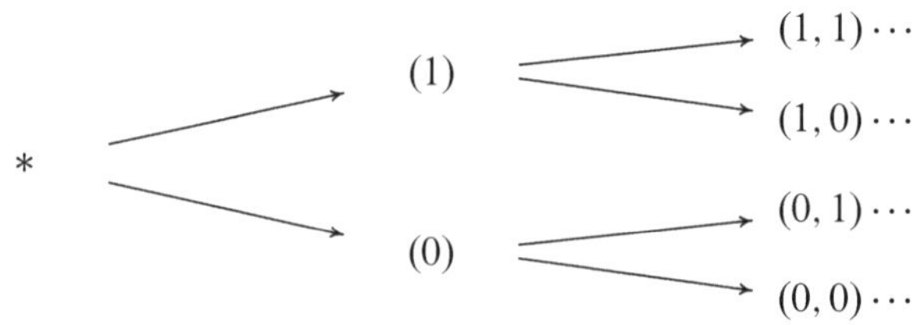

Figure 2.7: The general binary tree

as $(\omega_1, \dots, \omega_r)$ where each $\omega_i = 1$ or 0, corresponding to an up jump, $Z_i = u$, or a down jump, $Z_i = d$, respectively, between time i and $i+1$. The root node at time $r = 0$ is represented as $*$ where the corresponding stock price is the constant, S_0. There are now $1 + 2 + 2^2 + \cdots + 2^n = 2^{n+1} - 1$ nodes in the full tree (by contrast with the $\frac{1}{2}(n+1)(n+2)$ nodes in the previous tree, illustrated in Figure 2.2). The methodology is the same as before in that calculations are made by backwards induction through the tree. The price at time r of the claim C will be a random variable $C_r = C_r(\omega_1, \ldots, \omega_r)$, depending only on the outcomes of the first r moves in the stock price. Similarly, in the hedging portfolio, for the amount of stock held between time r and $r+1$ we have $X_r = X_r(\omega_1, \ldots, \omega_r)$, and $Y_r = Y_r(\omega_1, \ldots, \omega_r)$

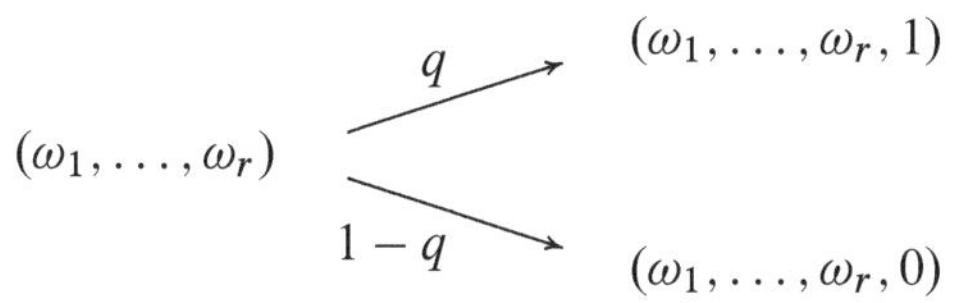

Figure 2.8: The general step

for the holding in the bank account. By considering Figure 2.8, exactly as before we may determine the price of the claim at node $(\omega_1, \ldots, \omega_r)$ in terms of the prices at the two nodes $(\omega_1, \ldots, \omega_r, 1)$ and $(\omega_1, \ldots, \omega_r, 0)$ by calculating

$$C_r(\omega_1, \ldots, \omega_r) = \alpha\left[q C_{r+1}(\omega_1, \ldots, \omega_r, 1) + (1-q) C_{r+1}(\omega_1, \ldots, \omega_r, 0)\right],$$

for $r \geqslant 1$ and $C_0 = \alpha\left[q C_1(1) + (1-q) C_1(0)\right]$; these expressions are of course the statement that

$$C_r = \alpha \mathbb{E}_Q(C_{r+1} \mid \mathcal{F}_r), \quad \text{for} \quad 0 \leqslant r \leqslant n, \tag{2.43}$$

since conditioning on the random variables $S_1, \ldots, S_r$ is equivalent to conditioning on the outcomes $\omega_1, \ldots, \omega_r$. The relation (2.43) leads in the same way to the representation that we had in (2.26),

$$C_r = \alpha^{n-r} \mathbb{E}_Q(C_n \mid \mathcal{F}_r) = \alpha^{n-r} \mathbb{E}_Q(C \mid \mathcal{F}_r), \tag{2.44}$$

and (2.44) may also be written as

$$\begin{aligned} C_r &= C_r(\omega_1, \ldots, \omega_r) \\ &= \sum_{\substack{\omega_{r+1}, \ldots, \omega_n \\ \omega_i = 1 \text{ or } 0}} C(\omega_1, \ldots, \omega_n)\, q^{\omega_{r+1} + \cdots + \omega_n} (1-q)^{n-r-(\omega_{r+1}+\cdots+\omega_n)}. \end{aligned} \tag{2.45}$$

We determine the replicating portfolio by now-familiar calculations, with

$$S_r(\omega_1, \ldots, \omega_r) = (u/d)^{\omega_1 + \cdots + \omega_r} d^r S_0,$$

we have

$$X_r(\omega_1, \ldots, \omega_r) = \frac{C_{r+1}(\omega_1, \ldots, \omega_r, 1) - C_{r+1}(\omega_1, \ldots, \omega_r, 0)}{S_r(\omega_1, \ldots, \omega_r)(u-d)}, \quad \text{and}$$

$$Y_r(\omega_1, \ldots, \omega_r) = \alpha\left[\frac{u C_{r+1}(\omega_1, \ldots, \omega_r, 0) - d C_{r+1}(\omega_1, \ldots, \omega_r, 1)}{u-d}\right]. \tag{2.46}$$

Observe that the replicating portfolio in (2.46) also satisfies the self-financing property given in (2.16).

It should be noted that $\{\alpha^r C_r, \mathcal{F}_r, 0 \leqslant r \leqslant n\}$ will still be a martingale under the martingale probability $\mathbb{Q}$, with $C_r = \mathbb{E}_{\mathbb{Q}}(\alpha^{n-r} C \mid \mathcal{F}_r)$ as seen in (2.44). However, we may see that the discounted bank holding $\{\alpha^r Y_r, \mathcal{F}_r, 0 \leqslant r \leqslant n\}$ and the discounted stock holding, $\{\alpha^r X_r S_r, \mathcal{F}_r, 0 \leqslant r \leqslant n\}$, are not necessarily martingales, as Example 2.4 shows. These latter processes are generally only martingales when the calculations are made on a recombining tree of the sort illustrated in Figure 2.1; a recombining tree is one in which an up jump followed by a down jump gets to the same point as a down jump followed by an up jump.

Example 2.4 Return again to the model in Example 2.2 on page 38 with the stock prices specified by the tree in Figure 2.5, and with the interest rate, $\rho = \frac{1}{2}$ as before; recall that the martingale probability for an up jump was calculated as $q = \frac{2}{5}$ and the discount factor $\alpha = \frac{2}{3}$. Suppose that we wish to price and hedge a claim

$$C(\omega_1, \omega_2) = 30\omega_1 + 15\omega_2,$$

which, at time 2, pays out 30 if the first stock price jump is up together with a payment

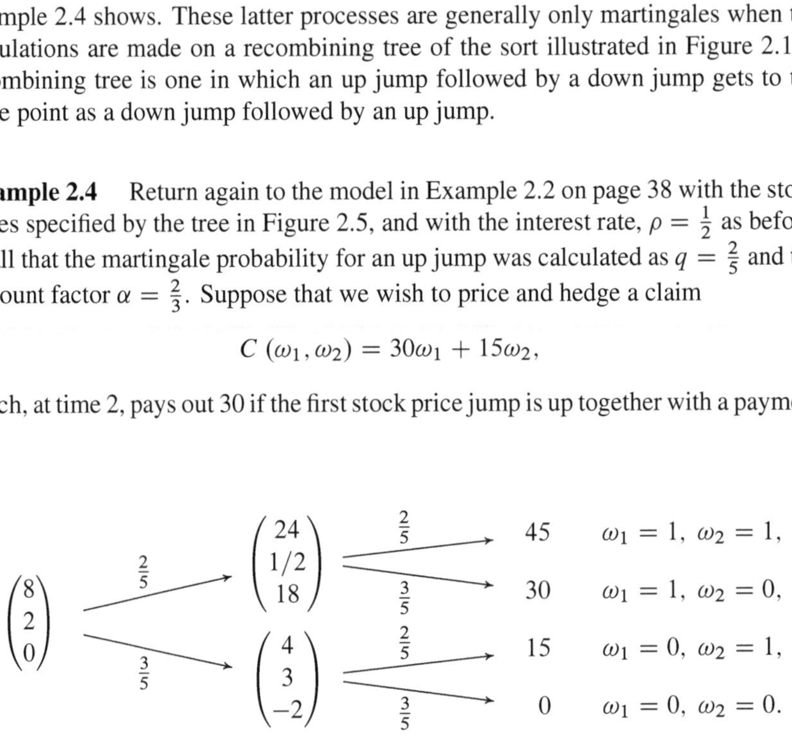

Figure 2.9: The claim $30\omega_1 + 15\omega_2$ of Example 2.4

of 15 if the second jump in the stock price is up. Figure 2.9 illustrates the calculation; the martingale probabilities are indicated on the arcs of the tree, the triple indicated on the nodes at times 0 and 1 are the claim price, the number of units of stock and the holding in the bank account in the hedging portfolio. The calculations are immediate using (2.46). Note that here the discounted holding in the bank account is not a martingale since we have

$$Y_0 = 0 \neq 4 = \tfrac{2}{3}\left[\tfrac{2}{5} \times 18 + \tfrac{3}{5} \times (-2)\right] = \alpha\left[qY_1(1) + (1-q)Y_1(0)\right];$$

it follows that the discounted value of the holding in stock cannot be a martingale in this case also. □

We have been discussing the situation where it is assumed that the payoff of the claim is made solely at time n but only minor changes are required to consider the case where a claim pays dividends $D_0, D_1, \ldots, D_n$ at times $0, 1, \ldots, n$ respectively. A dividend D_r paid at time $r \leqslant n$ may be thought of as being equivalent to an amount $(1+\rho)^{n-r}$ paid at time n, since it may be invested in the riskless bank account from time r to time n at the interest rate ρ per period. Here, D_r is a random variable depending just on $Z_1, \ldots, Z_r$ (or equivalently, $\omega_1, \ldots, \omega_r$), so its value is known after observing the information determining $\mathcal{F}_r$, the history up to and including time $r \geqslant 1$; D_0 would be a constant. Thus the total payout of the dividend-paying claim is equivalent to a payment of

$$C = \sum_{i=0}^{n} (1+\rho)^{n-i} D_i$$

at time n, and so its price at time 0 is

$$\mathbb{E}_{\mathbb{Q}}\left[\alpha^n \sum_{i=0}^{n} (1+\rho)^{n-i} D_i\right] = \mathbb{E}_{\mathbb{Q}}\left[\sum_{i=0}^{n} \alpha^i D_i\right], \tag{2.47}$$

since $\alpha = 1/(1+\rho)$. Using the same considerations, we may see that the value at time r of the payments at times $r, \ldots, n$ is

$$V_r = \mathbb{E}_{\mathbb{Q}}\left[\alpha^{n-r} \sum_{i=r}^{n} (1+\rho)^{n-i} D_i \,\middle|\, \mathcal{F}_r\right] = \mathbb{E}_{\mathbb{Q}}\left[\sum_{i=r}^{n} \alpha^{i-r} D_i \,\middle|\, \mathcal{F}_r\right], \tag{2.48}$$

so it is the conditional expected value of the total present and future discounted dividends given the information available at time r, $\mathcal{F}_r$, using the martingale probability, $\mathbb{Q}$; here, the dividend D_i at time i is discounted to time r by the discount factor α^{i-r}. Note that, since S_0 is a constant, taking the expectation conditional on $\mathcal{F}_0$ is just the unconditioned expected value so that the expression in (2.47) is V_0, when we set $r = 0$ in (2.48).

It should also be noted that $\{\alpha^r V_r, \mathcal{F}_r, 0 \leqslant r \leqslant n\}$ is not, in general, a martingale under the probability $\mathbb{Q}$ but $\{\alpha^r C_r, \mathcal{F}_r, 0 \leqslant r \leqslant n\}$ will be a martingale where

$$C_r = \sum_{i=0}^{r-1} (1+\rho)^{r-i} D_i + V_r, \tag{2.49}$$

is the total value of the claim at time r; it represents the value of the dividends earned before time r re-invested in the bank account together with the value of the present and future dividends. Here $C_0 = V_0$, because we take the empty sum in (2.49) to be 0 when $r = 0$. To see the martingale property observe that, since $D_1, \ldots, D_r$ are known when we condition on $\mathcal{F}_r$, we have from (2.48) that

$$\begin{aligned} C_r &= \sum_{i=0}^{r-1} (1+\rho)^{r-i} D_i + \mathbb{E}_Q\left[\sum_{i=r}^{n} \alpha^{i-r} D_i \,\middle|\, \mathcal{F}_r\right] \\ &= \mathbb{E}_Q\left[\sum_{i=0}^{n} (1+\rho)^{r-i} D_i \,\middle|\, \mathcal{F}_r\right] \\ &= \alpha^{n-r}\mathbb{E}_Q\left[\sum_{i=0}^{n} (1+\rho)^{n-i} D_i \,\middle|\, \mathcal{F}_r\right] = \alpha^{n-r}\mathbb{E}_Q[C \mid \mathcal{F}_r], \end{aligned}$$

giving $\alpha^r C_r = \mathbb{E}_Q[\alpha^n C \mid \mathcal{F}_r]$, which implies the martingale property.

From (2.48) we note also that, $V_n = D_n$ and

$$V_r = D_r + \alpha \mathbb{E}_Q(V_{r+1} \mid \mathcal{F}_r) \quad \text{for} \quad r < n; \tag{2.50}$$

to see this, from (2.48) we have

$$V_{r+1} = \mathbb{E}_Q\left[\sum_{i=r+1}^{n} \alpha^{i-r-1} D_i \,\middle|\, \mathcal{F}_{r+1}\right],$$

so that using the tower property of conditional expectations it follows that

$$\begin{aligned} D_r + \alpha\mathbb{E}_Q(V_{r+1} \mid \mathcal{F}_r) &= D_r + \alpha\mathbb{E}_Q\left(\mathbb{E}_Q\left[\sum_{i=r+1}^{n} \alpha^{i-r-1} D_i \,\middle|\, \mathcal{F}_{r+1}\right] \,\middle|\, \mathcal{F}_r\right) \\ &= D_r + \alpha\mathbb{E}_Q\left[\sum_{i=r+1}^{n} \alpha^{i-r-1} D_i \,\middle|\, \mathcal{F}_r\right] \\ &= \mathbb{E}_Q\left[\sum_{i=r}^{n} \alpha^{i-r} D_i \,\middle|\, \mathcal{F}_r\right] = V_r, \end{aligned}$$

because D_r is fixed when we condition on $\mathcal{F}_r$.

Note that in the case of a dividend-paying claim, the replicating portfolio will no longer satisfy the self-financing condition (2.16) because at time r the dividend D_r is being paid out; in place of (2.16) we must have

$$V_r = X_{r-1}S_r + (1+\rho)Y_{r-1} = D_r + X_r S_r + Y_r, \tag{2.51}$$

for $1 \leqslant r \leqslant n$, with $X_n = Y_n = 0$ and $V_0 = D_0 + X_0 S_0 + Y_0$. To see why (2.51) is true, observe that the pair (X_{r-1}, Y_{r-1}) represents the amount of stock and the amount in the bank carried forward from time $r-1$ to time r; on the other hand, the pair (X_r, Y_r) represents the amounts carried forward from time r to time $r+1$, and D_r is the dividend extracted from the replicating portfolio at time r.

We must modify (2.46) to

$$X_r(\omega_1, \ldots, \omega_r) = \frac{V_{r+1}(\omega_1, \ldots, \omega_r, 1) - V_{r+1}(\omega_1, \ldots, \omega_r, 0)}{S_r(\omega_1, \ldots, \omega_r)(u-d)}, \quad \text{and}$$

$$Y_r(\omega_1, \ldots, \omega_r) = \alpha \left[\frac{u V_{r+1}(\omega_1, \ldots, \omega_r, 0) - d V_{r+1}(\omega_1, \ldots, \omega_r, 1)}{u-d} \right]. \tag{2.52}$$

With the hedging portfolio constructed recursively by computing the amounts V_r, X_r and Y_r using (2.50) and (2.52), for an initial investment of V_0 we may generate a portfolio which will pay with certainty the dividends $D_1, \ldots, D_n$ at times $1, \ldots, n$, respectively, and thus the claim is replicated exactly.

2.2.6 American claims

Up to now we have been considering the situation where the holder, after purchasing a claim, takes no decision until the expiry time when the payoff is determined; such contracts are usually referred to as European claims. By contrast, an American claim entitles the holder to determine a time T, of his choosing, at which the payoff is calculated and paid out and the claim is terminated; the time T may not exceed the pre-determined expiry time of the claim which we will assume is n, as before. Consider the case where the holder receives the payment $f(S_r)$ when he chooses the time $T = r$, where $f(\cdot)$ is a given function.

In the case of an American call option at strike price c, the holder is entitled, but not required, to purchase one unit of stock at the strike price at any time at, or before, the expiry time n. The holder will only exercise the option at a time r where $S_r > c$, so that he can buy the stock for c and sell in the market for S_r and realize the profit $S_r - c$; if he does not exercise before time n and $S_n \leqslant c$ then the option lapses worthless. Thus, in the above context this would correspond to the case where $f(x) = (x - c)_+$. An American put option is defined similarly, except the holder has the right to sell one unit of stock at the strike price at any time at, or before, the expiry time n.

The holder chooses the termination (or exercise) time T without knowledge of the future of the stock price after the termination of the claim, but he may base his decision on observing the stock price movements up to that time; mathematically, this requirement is that the time T should be a **stopping time** relative to the observed histories of the stock price $\mathcal{F}_1, \mathcal{F}_2, \ldots, \mathcal{F}_n$. Formally, T is a stopping time if for each r, $0 \leqslant r \leqslant n$, the event $\{T = r\} \in \mathcal{F}_r$; intuitively, this means that deciding that $T = r$ is based only on observing $S_1, \ldots, S_r$ (having no foreknowledge of $S_{r+1}, \ldots, S_n$) and it implies that the indicator random variable $I_{(T=r)}$ is a function of $S_1, \ldots, S_r$ only.

Note that in the binomial model there are only a finite number of stopping times as here a stopping time T is restricted to lie in the range $0 \leqslant T \leqslant n$ and there are only a finite number, 2^n, of points in the underlying probability space Ω.

In order to study the pricing of an American claim, consider the situation of the holder of such a claim at time 0 who decides to employ the stopping time T; he is in the position of holding a claim that pays dividends

$$D_i^T = f(S_i)\, I_{(T=i)}, \quad \text{at times} \quad i = 0, 1, \ldots, n,$$

since if he chooses to stop at time i, so that $T = i$, then he receives $f(S_i)$ at time i and zero at all other times. From the discussion in the previous section, the value of this claim at time 0 is

$$V_0^T = \mathbb{E}_{\mathbb{Q}}\left[\sum_{i=0}^{n} \alpha^i f(S_i)\, I_{(T=i)}\right] = \mathbb{E}_{\mathbb{Q}}\left[\alpha^T f(S_T)\right].$$

The holder of the claim will choose the stopping time that maximizes his return and we will show below that the value at time 0 of the American claim is

$$\max_T V_0^T = \max_T \mathbb{E}_{\mathbb{Q}}\left[\alpha^T f(S_T)\right]; \tag{2.53}$$

we will also determine the stopping time T^* that achieves the maximum in (2.53). Moreover, if the claim has not already terminated before time r, its value at time r will be

$$\max_{T \geqslant r} V_r^T = \max_{T \geqslant r} \mathbb{E}\left[\alpha^T f(S_T) \mid \mathcal{F}_r\right], \tag{2.54}$$

where $V_r^T = \mathbb{E}\left[\alpha^T f(S_T) \mid \mathcal{F}_r\right]$ for a stopping time $T \geqslant r$. In order to see this define the sequence of random variables $V_1, \ldots, V_n$ recursively by setting $V_n = f(S_n)$ and for $0 \leqslant r < n$ set

$$V_r = \max\left[f(S_r), \alpha \mathbb{E}_{\mathbb{Q}}(V_{r+1} \mid \mathcal{F}_r)\right]. \tag{2.55}$$

The relation (2.55) may be interpreted as the optimality equation for the dynamic programming problem faced by the holder of the claim; at time r he must choose between stopping immediately, in which case he will receive $f(S_r)$ the first term on the right-hand side, and continuing at least until time $r + 1$, in which case the value of the claim to him will be the second term in the maximization, $\alpha \mathbb{E}_{\mathbb{Q}}(V_{r+1} \mid \mathcal{F}_r)$.

Use the relation $\max(a, b) = (a - b)_+ + b$, to see from (2.55) that

$$V_r = D_r + \alpha \mathbb{E}_{\mathbb{Q}}(V_{r+1} \mid \mathcal{F}_r) \quad \text{for} \quad 0 \leqslant r < n, \tag{2.56}$$

where

$$D_r = \left[f(S_r) - \alpha \mathbb{E}_{\mathbb{Q}}(V_{r+1} \mid \mathcal{F}_r)\right]_+, \tag{2.57}$$

and set $D_n = V_n = f(S_n)$. Note that $V_r \geqslant f(S_r)$ and $D_r \geqslant 0$, for each $r < n$.

Consider the dividend-paying claim with dividend sequence $D_1, \ldots, D_n$ where the D_r are specified by (2.57), then by comparing (2.56) with (2.50), we see that V_r is the value at time r of the payments at times $r, \ldots, n$, and consequently the worth of the replicating portfolio. We will see that V_0, the initial value of this dividend-paying claim, is necessarily the price of the American claim. First see that, irrespective of which stopping time T the holder of the American claim uses, the replicating portfolio for the dividend-paying claim will always provide at least the required amount to meet the claim; for suppose that $T = r$, then at time r the replicating portfolio will have paid out non-negative amounts $D_0, D_1, \ldots, D_{r-1}$ and will be worth $V_r \geqslant f(S_r)$; consequently, the initial price of the American claim cannot exceed V_0 the initial value of the replicating portfolio, and

$$V_0 \geqslant \max_T \mathbb{E}_{\mathbb{Q}}\left[\alpha^T f(S_T)\right]. \tag{2.58}$$

To see that we have equality in (2.58), define the stopping time T^* by

$$T^* = \min\{i \geqslant 0 : V_i = f(S_i)\}; \tag{2.59}$$

since $V_n = f(S_n)$, necessarily $T^* \leqslant n$. Note that when $T^* = r$ we see that $D_0 = \cdots = D_{r-1} = 0$ and C_r, defined by (2.49), satisfies $C_r = V_r$; that is,

$$C_{T^*} = V_{T^*} = f(S_{T^*}).$$

Use the martingale property of $\{\alpha^r C_r\}$ under the probability $\mathbb{Q}$ and the Optional Sampling Theorem to argue that

$$V_0 = C_0 = \mathbb{E}_{\mathbb{Q}}\left[\alpha^{T^*} C_{T^*}\right] = \mathbb{E}_{\mathbb{Q}}\left[\alpha^{T^*} f(S_{T^*})\right],$$

showing that (2.58) holds with equality. Furthermore, this argument shows that the price at time 0 of the American claim is $V_0 = V_0^{T^*}$ and that T^* is the optimal choice of stopping time for the holder of the claim.

By considering the situation where the claim has not been terminated before time r, this analysis extends immediately to see that we must have

$$V_r = \max_{T \geqslant r} \mathbb{E}_{\mathbb{Q}}\left[\alpha^{T-r} f(S_T) \mid \mathcal{F}_r\right], \tag{2.60}$$

where the maximum extends over all stopping times t for which $r \leqslant T \leqslant n$. Now V_r will be the value of the claim when the claim has not been exercised before time r. An optimal choice of time to terminate the claim will be given by

$$T_r^* = \min\{i \geqslant r : V_i = f(S_i)\},$$

that is, stop at the first time the current value of the option equals the immediate payoff, and then

$$V_r = \mathbb{E}_{\mathbb{Q}}\left[\alpha^{T_r^*-r} f\left(S_{T_r^*}\right) \mid \mathcal{F}_r\right].$$

We may note from (2.55) that $\{\alpha^r V_r\}$ is a supermartingale under $\mathbb{Q}$ and it dominates the sequence $\{\alpha^r f(S_r)\}$, in that we have

$$\alpha^r V_r \geqslant \mathbb{E}_{\mathbb{Q}}\left(\alpha^{r+1} V_{r+1} \mid \mathcal{F}_r\right) \quad \text{and} \quad \alpha^r V_r \geqslant \alpha^r f(S_r).$$

Furthermore, $\{\alpha^r V_r\}$ is the smallest supermartingale under $\mathbb{Q}$ dominating $\{\alpha^r f(S_r)\}$. That is, if $\{W_r\}$ is a supermartingale under $\mathbb{Q}$ with $W_r \geqslant \alpha^r f(S_r)$ for each r

then $W_r \geqslant \alpha^r V_r$ for each r. The proof proceeds by backward induction on r, for $r = n, n-1, \ldots, 0$. We have $W_n \geqslant \alpha^n f(S_n) = \alpha^n V_n$; now suppose that $W_{r+1} \geqslant \alpha^{r+1} V_{r+1}$, then

$$W_r \geqslant \mathbb{E}_{\mathbb{Q}}\left(W_{r+1} \mid \mathcal{F}_r\right) \geqslant \mathbb{E}_{\mathbb{Q}}\left(\alpha^{r+1} V_{r+1} \mid \mathcal{F}_r\right)$$

so that

$$W_r \geqslant \max\left(\alpha^r f(S_r), \alpha^{r+1} \mathbb{E}_{\mathbb{Q}}\left(V_{r+1} \mid \mathcal{F}_r\right)\right) = \alpha^r V_r,$$

from (2.55), establishing the inductive step.

In the theory of optimal stopping the supermartingale $\{\alpha^r V_r\}$ is known as the **Snell envelope** of the sequence $\{\alpha^r f(S_r)\}$.

Example 2.5 Return to the two-period model of Example 2.2 on page 38 in which prices for the European put option at strike price 11 were calculated. We now consider the American put at the same strike so that the payoff when the option is terminated at time i is $f(S_i) = (11 - S_i)_+$ for $i = 0, 1, 2$. At time 2 the value of the American put will be the same as for the European put and we indicate the cases where the option is exercised by placing (E) on the corresponding nodes in Figure 2.10. We

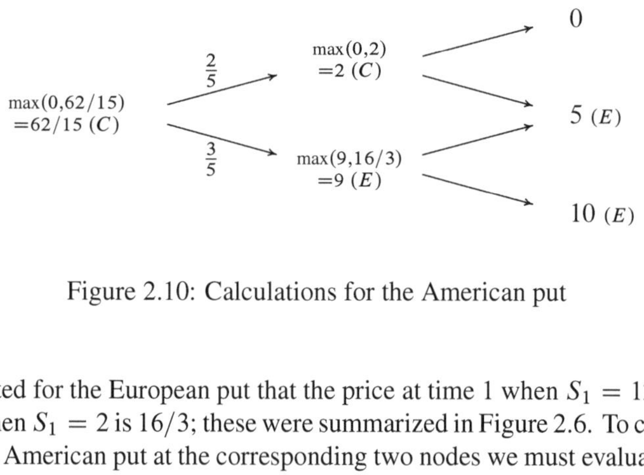

Figure 2.10: Calculations for the American put

had calculated for the European put that the price at time 1 when $S_1 = 12$ is 2 while the price when $S_1 = 2$ is $16/3$; these were summarized in Figure 2.6. To calculate the value of the American put at the corresponding two nodes we must evaluate equation (2.55) in each case; for example when $S_1 = 2$ we have $f(S_1) = 9$, the value of the American put is $\max(9, 16/3) = 9$ and the option would be exercised. When $S_1 = 12$ there is 0 payoff if the option is exercised so it is optimal to continue (denoted (C) at the node) and the value is 2.

The final stage is to determine the value of the option at time 0 if it is held until at least time 1; this is the calculation

$$\alpha \mathbb{E}_{\mathbb{Q}}(V_1) = \tfrac{2}{3}\left[2\mathbb{Q}(S_1 = 12) + 9\mathbb{Q}(S_1 = 2)\right] = \tfrac{2}{3}\left[2 \times \tfrac{2}{5} + 9 \times \tfrac{3}{5}\right] = \tfrac{62}{15},$$

then, since at time 0 if the option were exercised immediately the payoff would be 0, it is optimal to continue and we conclude that the price of the option at time 0 is $\frac{62}{15}$. □

This example shows that for an American put option it may be optimal for the holder to exercise the option prior to the expiry time. However, for the American call option it is always the case that it is optimal to hold the option until the expiry time; consequently, the price of the European call option and the American call option will always be the same. To establish this, we will give conditions on a function $f : [0, \infty) \to \mathbb{R}$ which are sufficient to ensure that, for an American claim which pays off $f(S_r)$ when terminated at time r, it is optimal for the holder to terminate the claim at the expiry time.

Suppose that

$$\begin{aligned} &\text{(i) } f \text{ is a convex function; and} \\ &\text{(ii) for any } \lambda \geqslant 1,\ \lambda f(x) \leqslant f(\lambda x) \quad \text{for all} \quad x \geqslant 0. \end{aligned} \tag{2.61}$$

The conditions (i) and (ii) are sufficient to imply that $\{\alpha^r f(S_r)\}$ is a submartingale under the probability $\mathbb{Q}$ because, using the conditional form of Jensen's inequality and the martingale property of $\{\alpha^r S_r\}$ under $\mathbb{Q}$, we have

$$\begin{aligned} \mathbb{E}_{\mathbb{Q}}\left[\alpha^{r+1} f(S_{r+1}) \mid \mathcal{F}_r\right] &\geqslant \alpha^{r+1} f\left(\mathbb{E}_{\mathbb{Q}}\left[S_{r+1} \mid \mathcal{F}_r\right]\right) \\ &= \alpha^{r+1} f\left(\alpha^{-1} S_r\right) \geqslant \alpha^r f(S_r), \end{aligned} \tag{2.62}$$

with the second inequality coming from (2.61). By the Optional Sampling Theorem for submartingales, for any stopping time T, $r \leqslant T \leqslant n$, it follows that

$$f(S_r) \leqslant \mathbb{E}_{\mathbb{Q}}\left[\alpha^{T-r} f(S_T) \mid \mathcal{F}_r\right] \leqslant \mathbb{E}_{\mathbb{Q}}\left[\alpha^{n-r} f(S_n) \mid \mathcal{F}_r\right];$$

consequently we see that

$$V_r = \max_{T \geqslant r} \mathbb{E}_{\mathbb{Q}}\left[\alpha^{T-r} f(S_T) \mid \mathcal{F}_r\right] = \mathbb{E}_{\mathbb{Q}}\left[\alpha^{n-r} f(S_n) \mid \mathcal{F}_r\right],$$

with $T \equiv n$ achieving the maximum. Thus when the conditions (i) and (ii) hold, it is always optimal to wait until the expiry time to terminate the corresponding American claim.

Note that when f is convex then the condition (ii) in (2.61) is equivalent to the condition that $f(0) \leqslant 0$. To see this, when (ii) holds set $x = 0$, then we see that necessarily $f(0) \leqslant 0$; conversely, when $f(0) \leqslant 0$, for $\lambda \geqslant 1$ write x as a convex combination of λx and 0 as $x = \lambda^{-1}(\lambda x) + (1 - \lambda^{-1})0$ and then convexity implies that

$$f(x) \leqslant \lambda^{-1} f(\lambda x) + \left(1 - \lambda^{-1}\right) f(0) \leqslant \lambda^{-1} f(\lambda x),$$

which gives (ii). The condition (2.61) is discussed further on page 133 and also in Section 5.3.6, in the context of the Black–Scholes model (see also Exercise 2.4).

In the case of an American call option with strike price c, then $f(x) = (x - c)_+$ is a convex function with $f(0) = 0$ so that (2.61) always holds and we may conclude that any American call will be held until expiry.

When we are interested in whether an American claim should be held to expiry for a particular value of the interest rate ρ, we should note that for the argument giving (2.62) to work, it is sufficient that the inequality (2.61) holds for all x for the particular value $\lambda = \alpha^{-1} = 1 + \rho$; indeed, for this value of λ we only need the inequality (2.61) to hold for those values x that S_r, $0 \leqslant r \leqslant n$, may take on with positive $\mathcal{Q}$ probability, that is, $x = S_0 u^i d^{r-i}, 0 \leqslant i \leqslant r \leqslant n$.

For the American put option the corresponding function $f(x) = (c - x)_+$ is convex but (2.61) will not be true for all x when $\lambda > 1$; however, when the interest rate $\rho = 0$, so that $\alpha = 1$, then the inequality in (2.61) holds trivially when $\lambda = 1$. In particular, when $\rho = 0$, we may conclude that the American put will be held until expiry.

2.2.7 The non-standard multi-period model

Hitherto we have been considering the multi-period binomial model where the stock price at time r is represented as $S_r = S_0 \prod_{i=1}^{r} Z_i$, and the random variables (Z_i) are assumed to be independent and identically distributed; much of the discussion of pricing and hedging claims may be extended to the case when both of these assumptions are relaxed. The important feature that must be retained is that the evolution of the model should still take place on a general binary tree as illustrated in Figure 2.7 on page 44 (with the general step shown in Figure 2.8).

We will assume that the underlying probability space is again Ω as given in (2.32) and that the stock price at time r depends only on $(\omega_1, \ldots, \omega_r)$; that is, $S_r = S_r(\omega_1, \ldots, \omega_r)$. The proportional change in the stock price between times r and $r + 1$, for $1 \leqslant r < n$, will be

$$Z_{r+1}(\omega_1, \ldots, \omega_{r+1}) = \frac{S_{r+1}(\omega_1, \ldots, \omega_{r+1})}{S_r(\omega_1, \ldots, \omega_r)}$$

and Z_{r+1} will be assumed to take just two values

$$Z_{r+1}(\omega_1, \ldots, \omega_r, 1) = u_r(\omega_1, \ldots, \omega_r) \quad \text{and}$$
$$Z_{r+1}(\omega_1, \ldots, \omega_r, 0) = d_r(\omega_1, \ldots, \omega_r),$$

corresponding to an up jump, $\omega_{r+1} = 1$, and a down jump, $\omega_{r+1} = 0$, respectively. We take $Z_1(\omega_1) = S_1(\omega_1)/S_0$ where the initial stock price S_0 is a constant with $u_0 = S_1(1)/S_0$ and $d_0 = S_1(0)/S_0$. Observe that u_r and d_r are random variables with values determined at time r, $1 \leqslant r < n$, and assume that $u_r > d_r$; note further that we depart in this instance from the usual convention of employing upper case letters to denote random variables which is followed in most of this book. Here, u_0 and d_0 will be constants with $u_0 > d_0$.

We may allow the interest rate on the bank account between times r and $r + 1$ to be a random variable $\rho_r = \rho_r(\omega_1, \ldots, \omega_r)$. The information available at time r will again be $\mathcal{F}_r = \sigma(Z_1, \ldots, Z_r)$ and it is equivalent to knowing the values of $\omega_1, \ldots, \omega_r$. The interest rate on the bank account for the period r to $r + 1$ is then known at time r when we have observed $\mathcal{F}_r$, so investment in the bank account for that period is riskless; equivalently, if we set $\alpha_r = 1/(1 + \rho_r)$ then α_r is the price of a bond bought at time r yielding 1 unit with certainty at time $r+1$. Assume that on each branch of the binary tree that there is no arbitrage which requires that

$$u_r(\omega_1, \ldots, \omega_r) > 1 + \rho_r(\omega_1, \ldots, \omega_r) > d_r(\omega_1, \ldots, \omega_r); \tag{2.63}$$

the interest rate ρ_0 for the first period will be a constant with $u_0 > 1 + \rho_0 > d_0$.

To specify the underlying probability $\mathbb{P}$ on the sample space Ω, first assume that for $1 \leqslant r < n$, $p_r = p_r(\omega_1, \ldots, \omega_r)$, with $0 < p_r < 1$, denotes the conditional probability of an up jump between r and $r+1$ given the outcomes $\omega_1, \ldots, \omega_r$, with $1 - p_r$ being the conditional probability of a down jump; that is

$$\mathbb{P}(\omega_{r+1} = 1 \mid \mathcal{F}_r) = p_r \quad \text{and} \quad \mathbb{P}(\omega_{r+1} = 0 \mid \mathcal{F}_r) = 1 - p_r;$$

a constant p_0, with $0 < p_0 < 1$, will be the (unconditional) probability of an up jump between times 0 and 1. For $1 \leqslant r < n$, the conditional probability at the rth step is then,

$$\mathbb{P}(\omega_{r+1} \mid \omega_1, \ldots, \omega_r) = p_r^{\omega_{r+1}} (1 - p_r)^{1-\omega_{r+1}}, \tag{2.64}$$

with $\mathbb{P}(\omega_1) = p_0^{\omega_1}(1-p_0)^{1-\omega_1}$. This implies that

$$\mathbb{P}(\omega_1, \ldots, \omega_{r+1}) = \mathbb{P}(\omega_1, \ldots, \omega_r)\, p_r^{\omega_{r+1}} (1 - p_r)^{1-\omega_{r+1}},$$

and by iterating this relation for $r = n-1, \ldots 1$, we see that the full probability for any point $\boldsymbol{\omega} = (\omega_1, \ldots, \omega_n) \in \Omega$ will be

$$\mathbb{P}(\boldsymbol{\omega}) = \prod_{i=1}^{n} \left[p_{i-1}^{\omega_i} (1 - p_{i-1})^{1-\omega_i} \right]. \tag{2.65}$$

We define the martingale probability $\mathbb{Q}$ in the obvious way. For $0 \leqslant r < n$, recalling (2.4) again, let

$$q_r(\omega_1, \ldots, \omega_r) = \frac{1 + \rho_r(\omega_1, \ldots, \omega_r) - d_r(\omega_1, \ldots, \omega_r)}{u_r(\omega_1, \ldots, \omega_r) - d_r(\omega_1, \ldots, \omega_r)} \tag{2.66}$$

be the conditional probability under $\mathbb{Q}$, given $\omega_1, \ldots, \omega_r$, that there will be an up jump between time r and $r+1$. Corresponding to (2.64) we have

$$\mathbb{Q}(\omega_{r+1} \mid \omega_1, \ldots, \omega_r) = q_r^{\omega_{r+1}} (1 - q_r)^{1-\omega_{r+1}}, \tag{2.67}$$

with $\mathbb{Q}(\omega_1) = q_0^{\omega_1}(1-q_0)^{1-\omega_1}$, and in the same way that (2.65) is obtained from (2.64), we see from (2.67) that

$$\mathbb{Q}(\boldsymbol{\omega}) = \prod_{i=1}^{n} \left[q_{i-1}^{\omega_i} (1 - q_{i-1})^{1-\omega_i} \right]. \tag{2.68}$$

Note that the no arbitrage condition (2.63) implies that $0 < q_r < 1$ for each r. From the definition of q_r in (2.66) we have that

$$\begin{aligned} S_r &= S_r(\omega_1, \ldots, \omega_r) = \alpha_r S_r(\omega_1, \ldots, \omega_r)[1 + \rho_r] \\ &= \alpha_r S_r(\omega_1, \ldots, \omega_r)[q_r u_r + (1 - q_r) d_r] \\ &= \alpha_r [q_r S_{r+1}(\omega_1, \ldots, \omega_r, 1) + (1 - q_r) S_{r+1}(\omega_1, \ldots, \omega_r, 0)] \\ &= \mathbb{E}_{\mathbb{Q}}(\alpha_r S_{r+1} \mid \mathcal{F}_r). \end{aligned}$$

Multiply both sides by $\prod_{i=0}^{r-1} \alpha_i$, and observe that we may take this product term inside the conditional expectation because its value is fixed given $\mathcal{F}_r$, or equivalently, when we know $\omega_1, \ldots, \omega_r$; we conclude that

$$\left(\prod_{i=0}^{r-1} \alpha_r\right) S_r = \mathbb{E}_{\mathbb{Q}}\left[\left(\prod_{i=0}^{r} \alpha_i\right) S_{r+1} \,\Big|\, \mathcal{F}_r\right]; \tag{2.69}$$

when $r = 0$ we interpret the empty product on the left-hand side as 1, so that the relation (2.69) holds for $0 \leqslant r < n$ and it shows that $\left\{\left(\prod_{i=0}^{r-1} \alpha_i\right) S_r, \mathcal{F}_r, 0 \leqslant r \leqslant n\right\}$ is a martingale under the probability $\mathbb{Q}$.

Following the discussion in Section 2.2.3, we define the Radon–Nikodym derivative as the random variable

$$L = L(\boldsymbol{\omega}) = \frac{d\mathbb{Q}}{d\mathbb{P}} = \prod_{i=1}^{n}\left[\left(\frac{q_{i-1}}{p_{i-1}}\right)^{\omega_i}\left(\frac{1-q_{i-1}}{1-p_{i-1}}\right)^{1-\omega_i}\right],$$

and as before for any random variable X we will have

$$\mathbb{E}_{\mathbb{Q}}(X) = \mathbb{E}(LX) = \mathbb{E}\left(\frac{d\mathbb{Q}}{d\mathbb{P}} X\right).$$

In general there will not be any straightforward representation of L in terms of the stock price S_n as we had in the standard binomial model, but if we set

$$L_r = L(\omega_1, \ldots, \omega_r) = \prod_{i=1}^{r}\left[\left(\frac{q_{i-1}}{p_{i-1}}\right)^{\omega_i}\left(\frac{1-q_{i-1}}{1-p_{i-1}}\right)^{1-\omega_i}\right],$$

so that $L_n = L$, we may check here that $\{L_r, \mathcal{F}_r, 0 \leqslant r \leqslant n\}$ is a martingale under the probability $\mathbb{P}$, since

$$\begin{aligned}
\mathbb{E}\left[L_{r+1} \mid \mathcal{F}_r\right] &= \mathbb{E}\left[L_r\left(\frac{q_r}{p_r}\right)^{\omega_{r+1}}\left(\frac{1-q_r}{1-p_r}\right)^{1-\omega_{r+1}} \,\Big|\, \mathcal{F}_r\right] \\
&= L_r\left[\frac{q_r}{p_r}\mathbb{P}(\omega_{r+1} = 1 \mid \mathcal{F}_r) + \left(\frac{1-q_r}{1-p_r}\right)\mathbb{P}(\omega_{r+1} = 0 \mid \mathcal{F}_r)\right] \\
&= L_r\left[\frac{q_r}{p_r} p_r + \left(\frac{1-q_r}{1-p_r}\right)(1-p_r)\right] = L_r;
\end{aligned}$$

consequently we have again that

$$L_r = \mathbb{E}(L_n \mid \mathcal{F}_r) = \mathbb{E}(L \mid \mathcal{F}_r) = \mathbb{E}\left(\frac{d\mathbb{Q}}{d\mathbb{P}} \,\Big|\, \mathcal{F}_r\right).$$

We may also note that L_r is just the Radon–Nikodym derivative when the model is restricted to the time periods $0, 1, \ldots, r$ so, as for the standard model, it is the likelihood ratio for the random variables $S_1, \ldots, S_r$ (or equivalently for the outcomes $\omega_1, \ldots, \omega_r$) under the probabilities $\mathbb{Q}$ and $\mathbb{P}$.

Suppose that we wish to price a claim paying $C = C(\omega)$, $\omega \in \Omega$, at time n, then we may construct a replicating portfolio to hedge it exactly by adapting appropriately the procedure described in Section 2.2.5. Suppose that for each r, C_r is the price at time r of this claim then, for $1 \leqslant r \leqslant n$, $C_r = C_r(\omega_1, \ldots, \omega_r)$, C_0 is a constant and $C_n \equiv C$. In the same way that we derived (2.43) by considering the movement on the part of the binary tree which is illustrated in Figure 2.8 on page 45 (and replacing the probability q there by q_r) we may calculate C_r in terms of C_{r+1} as

$$C_r(\omega_1, \ldots, \omega_r) = \alpha_r\left[q_r C_{r+1}(\omega_1, \ldots, \omega_r, 1) + (1 - q_r) C_{r+1}(\omega_1, \ldots, \omega_r, 0)\right];$$

for $r \geqslant 1$ and $C_0 = \alpha_0\left[q_0 C_1(1) + (1 - q_0) C_1(0)\right]$. These expressions may be summarized as

$$C_r = \alpha_r \mathbb{E}_{\mathcal{Q}}\left(C_{r+1} \mid \mathcal{F}_r\right) = \mathbb{E}_{\mathcal{Q}}\left(\alpha_r C_{r+1} \mid \mathcal{F}_r\right), \tag{2.70}$$

because in the present context even though α_r may be a random variable it may taken inside the conditional expectation given $\mathcal{F}_r$ since its value is fixed when we know $\mathcal{F}_r$. By using the relation (2.70) and the tower property of conditional expectations, by backward induction on r we may see that

$$C_r = \mathbb{E}_{\mathcal{Q}}\left[\left(\prod_{i=r}^{n-1} \alpha_i\right) C_n \,\middle|\, \mathcal{F}_r\right] = \mathbb{E}_{\mathcal{Q}}\left[\left(\prod_{i=r}^{n-1} \alpha_i\right) C \,\middle|\, \mathcal{F}_r\right], \tag{2.71}$$

which shows that $\left\{\left(\prod_{i=0}^{r-1} \alpha_i\right) C_r, \mathcal{F}_r, 0 \leqslant r \leqslant n\right\}$ is a martingale under the probability $\mathcal{Q}$; again, interpret the empty product as 1 when $r = 0$. By multiplying both sides of (2.71) by $\prod_{i=0}^{r-1} \alpha_i$ we obtain

$$\left(\prod_{i=0}^{r-1} \alpha_i\right) C_r = \mathbb{E}_{\mathcal{Q}}\left[\left(\prod_{i=0}^{n-1} \alpha_i\right) C \,\middle|\, \mathcal{F}_r\right];$$

as with (2.69), we are able to take the term $\prod_{i=0}^{r-1} \alpha_i$ inside the conditional expectation because its value is known given $\mathcal{F}_r$.

Only minor changes are required to (2.46) to obtain the holdings in the hedging portfolio. At time r, the amount of stock held, $X_r = X_r(\omega_1, \ldots, \omega_r)$, and the amount in the bank account, $Y_r = Y_r(\omega_1, \ldots, \omega_r)$, are given by

$$\begin{aligned} X_r &= \frac{C_{r+1}(\omega_1, \ldots, \omega_r, 1) - C_{r+1}(\omega_1, \ldots, \omega_r, 0)}{S_r(\omega_1, \ldots, \omega_r)(u_r - d_r)}, \quad \text{and} \\ Y_r &= \alpha_r\left[\frac{u_r C_{r+1}(\omega_1, \ldots, \omega_r, 0) - d_r C_{r+1}(\omega_1, \ldots, \omega_r, 1)}{u_r - d_r}\right]; \end{aligned} \tag{2.72}$$

recall that here u_r, d_r and α_r will be functions of $(\omega_1, \ldots, \omega_r)$ in general.

Example 2.6 Consider a non-standard two-period model where the stock price moves on the tree illustrated in Figure 2.11. This example may be thought of as

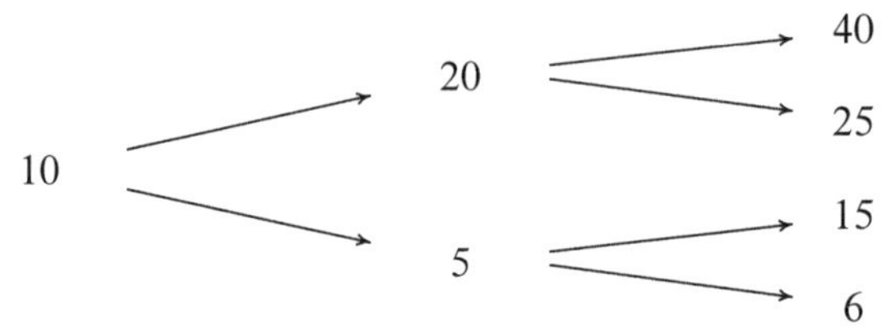

Figure 2.11: The binary tree for the stock price of Example 2.6

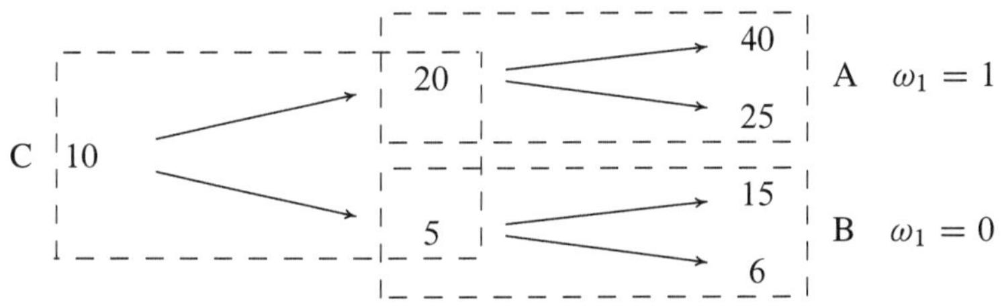

Figure 2.12: The three embedded one-period models of Example 2.6

splitting into three one-period models, A, B, and C shown in Figure 2.12. Here C corresponds to the transition from time 0 to time 1; A corresponds to the next time period when $\omega_1 = 1$ so that there has been an up jump in the first period; and B corresponds to the case $\omega_1 = 0$, when there has been a down jump. We see that in C we have $u_0 = 2$ while $d_0 = \frac{1}{2}$; for the part of the tree represented by A we have $u_1(1) = 2$, $d_1(1) = \frac{5}{4}$ and for the part in B, $u_1(0) = 3$ and $d_1(0) = \frac{6}{5}$. Suppose that the interest rates are given by $\rho_0 = \frac{1}{2}$, $\rho_1(1) = \frac{1}{3}$ and $\rho_1(0) = \frac{4}{5}$ so that $\alpha_0 = \frac{2}{3}$, $\alpha_1(1) = \frac{3}{4}$ and $\alpha_1(0) = \frac{5}{9}$.

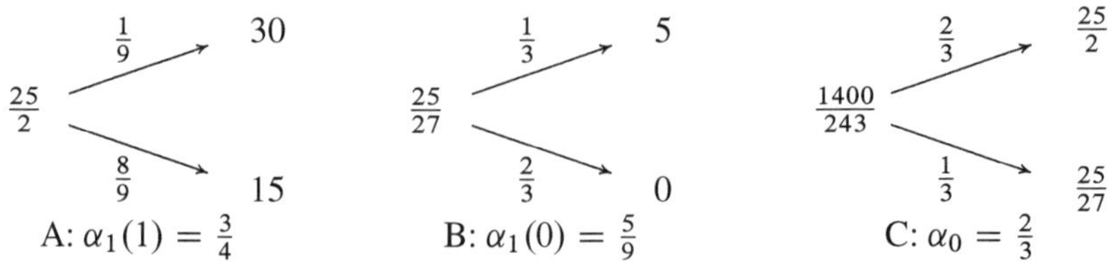

Figure 2.13: The calculations for the European call of Example 2.6

Consider the pricing of a European call option with expiry time 2 and strike price $c = 10$, so that the payoffs corresponding to $S_2 = 40$, 25, 15, and 6 are 30, 15, 5, and 0 respectively. The calculations are illustrated in Figure 2.13. The order of the calculations is to consider first the sub-model A, then B and finally C, where the prices calculated in A and B at time 1 become the entries on the terminal nodes in C. The martingale probabilities are shown on the respective arcs in Figure 2.13.

The composition of the replicating portfolio for each of A, B and C may be made using (2.72); these show that the amounts in the portfolio in the three cases are given by

$$\mathrm{A} : \left(1, -\tfrac{15}{2}\right), \quad \mathrm{B} : \left(\tfrac{5}{9}, -\tfrac{50}{27}\right) \quad \text{and } \mathrm{C} : \left(\tfrac{125}{162}, -\tfrac{475}{243}\right),$$

with the first entry being the amount of stock held and the second the holding in the bank account in each case. □

2.3 Exercises

Exercise 2.1 Consider a utility-maximizing investor in the one-period binomial model who has initial wealth $w_0 > 0$ and utility function $v(x) = \sqrt{x}$. Find his optimal terminal wealth and verify that in the corresponding replicating portfolio the holding in stock is positive, negative or zero according as $\mathbb{E}\,(\alpha S_1) > S_0$, $\mathbb{E}\,(\alpha S_1) < S_0$ or $\mathbb{E}\,(\alpha S_1) = S_0$, where S_0 and S_1 are the stock prices at times 0 and 1, respectively.

Exercise 2.2 Suppose that for a binomial model operating over two periods the stock price moves as shown on the tree:

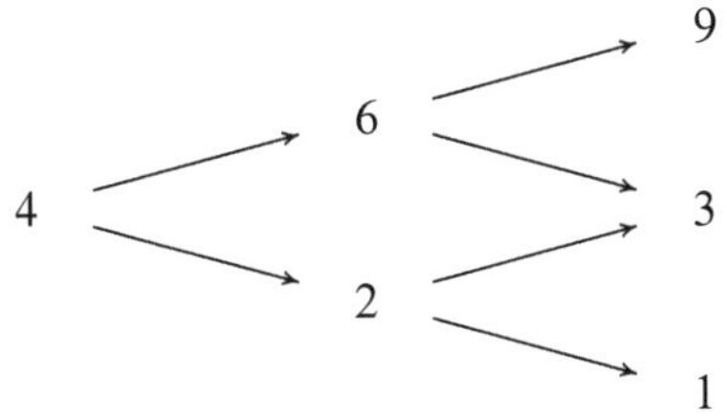

(i) Assume that the interest rate is $\rho = \frac{1}{3}$. For each of the times $r = 0, 1$ determine the price and the composition of the hedging portfolio for a European call option with strike price 2 and expiry time 2. Check that the discounted holding in the bank account and the discounted value of the holding in stock form martingales under the martingale probability.

(ii) Now consider an American put option with strike price $c = 5/2$. Determine the range of values for the interest rate for which the policy to wait until the expiry time to exercise the option is optimal.

Exercise 2.3 Consider the pricing of a claim $C = f(S_n)$ in the standard binomial model. For each of the cases

$$\text{(i)}\quad f(x) = x^2; \quad \text{and} \quad \text{(ii)}\quad f(x) = \ln(x),$$

calculate C_r, the price of the claim at time r in terms of the stock price S_r and derive an expression for Y_r, the amount of stock held in the hedging portfolio.

Exercise 2.4 Consider the hedging of a terminal-value claim $C = f(S_n)$ in the standard binomial model.

(i) Suppose that for any $\lambda \geqslant 1$,

$$\lambda f(x) \leqslant f(\lambda x) \quad \text{for all} \quad x. \tag{2.73}$$

Prove that at each time r, $0 \leqslant r \leqslant n-1$, the holding in the bank account in the hedging strategy is $\leqslant 0$. Check that the condition (2.73) holds in the case of a European call option.

(ii) Suppose that for any $\lambda \geqslant 1$,

$$\lambda f(x) \geqslant f(\lambda x) \quad \text{for all} \quad x. \tag{2.74}$$

Prove that at each time r, $0 \leqslant r \leqslant n-1$, the holding in the bank account in the hedging strategy is $\geqslant 0$. Check that the condition (2.74) holds in the case of a European put option.

Exercise 2.5 Consider the standard binomial model and let $C_r = f_r(S_r)$ represent the price at time r of a claim which pays $C = f(S_n)$ at time n. When f is convex (respectively, concave) show that f_r is convex (respectively, concave) on the possible values that S_r can take on (which are $S_r = S_0 u^i d^{r-i}$, $i = 0, 1, \ldots, r$).

Show that when f is convex (respectively, concave) then the amount of stock held in the hedging portfolio increases (respectively, decreases) between the times r and $r+1$ ($< n$) if the stock price increases between r and $r+1$.
[Here, 'increase' (or 'decrease') should be interpreted in the weak sense, unless the function f is *strictly* convex (or *strictly* concave).]

Exercise 2.6 In the utility maximization problem for the multi-period binomial model, determine the optimal wealth at time n for an investor with initial wealth w_0 in the cases when his utility function is

(i) $v(x) = \gamma x^{1/\gamma}$, where $\gamma > 1$; and

(ii) $v(x) = \left(1 - e^{-ax}\right)/a$, where $a > 0$.

Exercise 2.7 Suppose that over two periods a stock price moves on the binomial tree shown.

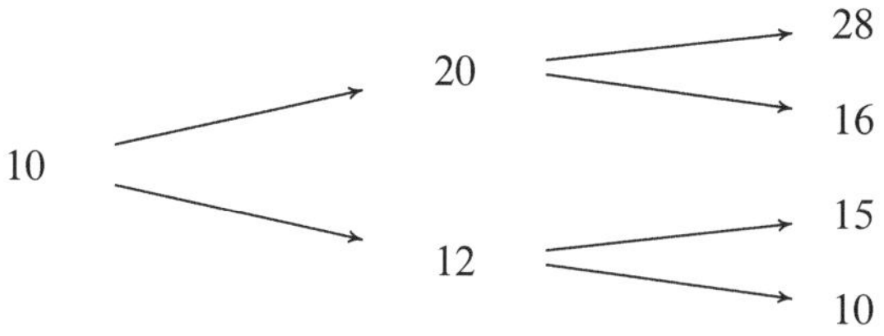

(i) Suppose that the interest rate ρ is constant over the two periods; determine the range of values of ρ for which there is no arbitrage in the model.

(ii) Now suppose that the interest rate is random, taking the value $\rho = \frac{2}{5}$ in the first period, while for the second period it takes the value $\rho = \frac{1}{5}$ when the stock price is 20 and the value $\rho = \frac{1}{6}$ when the stock price is 12. Evaluate the martingale probabilities and determine the price at time 0:

(a) for a European call option at strike price 12 expiring at time 2 and calculate the replicating portfolio at each stage;

(b) for an American claim expiring at time 2 paying $f(S_r) = |S_r - 12|$, when exercised at time $r = 0, 1, 2$, and establish when the option would be exercised.

Exercise 2.8 *Static hedging.* In the standard binomial model, consider the terminal-value claim $C = f(S_n)$ where $f(\cdot)$ is a convex function. By using linear programming, or otherwise, determine the portfolio of minimal initial cost at time 0 which, when held unchanged, is worth at least C at time n, for all values of S_n. Explain why the initial cost of this portfolio is an upper bound for the price of the claim at time 0.

Exercise 2.9 *Bid/Offer spread.* In practice, at any time there are two prices for a stock, the **bid price** at which an investor may sell one unit of the stock, and the **offer price** at which an investor may purchase one unit of the stock; the offer price is higher than the bid price and the difference between the two is known as the **bid/offer spread.** The single price quoted is typically the mid point of these two prices. For the model in Exercise 2.2 (i), consider the pricing of the European call option with strike price 2 and expiry time 2. For simplicity suppose that the bid/offer spreads are zero at times 0 and 2, but that at time 1 the spread when the quoted stock price is 6 is 2ϵ, $0 \leqslant \epsilon < 6$, while when the quoted price is 2 the spread is 2δ, $0 \leqslant \delta < 2$; that means that when the quoted stock price is 6, an investor may sell stock at $6 - \epsilon$ and may buy stock at $6 + \epsilon$, and similarly when the quoted stock price is 2.

Determine the dynamic hedging portfolio that replicates the payoff of the call option exactly at time 2. Compare the initial cost of this exact hedging portfolio with that of the portfolio which at time 0 holds $\frac{7}{8}$ units of stock and borrows $\frac{63}{128}$ from the bank and which is not adjusted between times 0 and 2; note that at time 2 this latter portfolio is worth at least as much as the payoff of the option for all values of the stock price S_2. Show that the cost of the exact hedging portfolio is higher whenever the spreads are such that

$$48\delta\epsilon + 15\delta - 5\epsilon - 60 > 0. \tag{2.75}$$

This example demonstrates how the method of pricing an option by determining the initial cost of a replicating portfolio breaks down in the presence of transaction costs.

Chapter 3

A GENERAL DISCRETE-TIME MODEL

3.1 One-period model

3.1.1 Introduction

Consider an economy operating over one period from time 0 to time 1. Suppose that there are s risky assets, $i = 1, \ldots, s$; the prices of these at time 0 are given by a deterministic vector $\boldsymbol{S}_0 = (S_{1,0}, \ldots, S_{s,0})^\top \in \mathbb{R}^s$ and the prices at time 1 are determined by a random vector $\boldsymbol{S}_1 = (S_{1,1}, \ldots, S_{s,1})^\top$ taking values in $\mathbb{R}^s$. In addition there is a riskless asset, 0, which provides a deterministic return $r_1 > 0$ between time 0 and time 1; the initial price of the riskless asset may be taken as 1 and here $r_1 - 1$ is the fixed interest rate, with the price of the riskless asset at time 1 being r_1. Underlying the model is a probability space $(\Omega, \mathcal{F}, \mathbb{P})$ on which the random vector $\boldsymbol{S}_1$ is defined. The set Ω, which represents the set of possible states of nature $\omega \in \Omega$, is equipped with a σ-field $\mathcal{F}$ of measurable events, or subsets, of Ω and $\mathbb{P}$ is a given probability. In this section assume that $\mathbb{E}\,\|\boldsymbol{S}_1\|^2 < \infty$, that is $\mathbb{E}\,(S_{i,1})^2 < \infty$ for each $i = 1, \ldots, s$, and without any loss of generality assume that the covariance matrix

$$
\begin{aligned}
V &= \mathbb{C}ov\,(\boldsymbol{S}_1) = \mathbb{E}\left[(\boldsymbol{S}_1 - \mathbb{E}\,\boldsymbol{S}_1)(\boldsymbol{S}_1 - \mathbb{E}\,\boldsymbol{S}_1)^\top\right] \\
&= \mathbb{E}\left[\boldsymbol{S}_1 \boldsymbol{S}_1^\top\right] - (\mathbb{E}\,\boldsymbol{S}_1)(\mathbb{E}\,\boldsymbol{S}_1)^\top
\end{aligned}
$$

is positive definite. The (i, j) element of the matrix V is the covariance between the prices (at time 1) of assets i, j $(= 1, \ldots, s)$. The assumption that V is positive definite means in effect that there is only one riskless asset; it is not possible to form a new asset for which the variance of the price at time 1 is zero by taking some linear combination of the s risky assets. As pointed out in Section 1.3, this is not a serious restriction since if there are two, or more, riskless assets then investors would all choose the one with the highest return so the others could all be discarded and the risky assets relabelled, if necessary, to give the situation described here.

A **contingent claim**, C, is a random variable defined on the underlying probability space, and it may be regarded as the payoff at time 1 of some contract; the value of C is not observed until time 1. The archetypal example to keep in mind is a call option at some strike price c on one of the risky assets, asset 1 say; this would pay $(S_{1,1} - c)_+$ at time 1. We will confine the discussion to the class of contingent claims with finite second moment which will be denoted $\mathcal{C} = \{C : \mathbb{E}\,C^2 < \infty\}$.

Consider an individual with the liability to pay the contingent claim C at time 1 and suppose that he wishes to hedge this liability by forming a portfolio, at time 0, holding $\boldsymbol{x} = (x_1, \ldots, x_s)^\top \in \mathbb{R}^s$ in the risky assets and y in the riskless asset. Here, $\boldsymbol{x}$ and y are deterministic and we assume that x_i and y are the amounts of asset i and the riskless asset, respectively, held in the portfolio not the proportions of wealth invested as was the case in Chapter 1. The initial worth of this portfolio is $\boldsymbol{x}^\top \boldsymbol{S}_0 + y$ while its value at time 1 is $\boldsymbol{x}^\top \boldsymbol{S}_1 + y r_1$. Suppose that the individual chooses $\boldsymbol{x}$ and y so as to approximate C optimally in the least-squares sense; that is, the portfolio is chosen to minimize

$$\mathbb{E}\left(C - \boldsymbol{x}^\top \boldsymbol{S}_1 - y r_1\right)^2. \tag{3.1}$$

Set to zero the gradient with respect to $\boldsymbol{x}$ and the derivative with respect to y of the expression in (3.1) to obtain

$$\mathbb{E}\left(C \boldsymbol{S}_1\right) - \mathbb{E}\left[\left(\boldsymbol{x}^\top \boldsymbol{S}_1\right) \boldsymbol{S}_1\right] - y r_1 \mathbb{E}\, \boldsymbol{S}_1 = 0 \tag{3.2}$$

and

$$\mathbb{E}\, C - \mathbb{E}\left(\boldsymbol{x}^\top \boldsymbol{S}_1\right) - y r_1 = 0. \tag{3.3}$$

Note that

$$\mathbb{E}\left[\left(\boldsymbol{x}^\top \boldsymbol{S}_1\right) \boldsymbol{S}_1\right] = V \boldsymbol{x} + \mathbb{E}\left(\boldsymbol{x}^\top \boldsymbol{S}_1\right) \mathbb{E}\, \boldsymbol{S}_1. \tag{3.4}$$

Substitute for $\mathbb{E}\left[\left(\boldsymbol{x}^\top \boldsymbol{S}_1\right) \boldsymbol{S}_1\right]$ from (3.2) into (3.4), to get

$$\begin{aligned} V \boldsymbol{x} &= \mathbb{E}\left(C \boldsymbol{S}_1\right) - y r_1 \mathbb{E}\, \boldsymbol{S}_1 - \mathbb{E}\left(\boldsymbol{x}^\top \boldsymbol{S}_1\right) \mathbb{E}\, \boldsymbol{S}_1 \\ &= \mathbb{E}\left(C \boldsymbol{S}_1\right) - \mathbb{E}\left(C\right) \mathbb{E}\, \boldsymbol{S}_1, \end{aligned}$$

from (3.3), and then solve for the minimizing $\overline{\boldsymbol{x}}$ and $\overline{y}$ to see that

$$\overline{\boldsymbol{x}} = V^{-1} \mathbb{E}\left[C\left(\boldsymbol{S}_1 - \mathbb{E}\, \boldsymbol{S}_1\right)\right] \quad \text{where} \quad \overline{y} = \frac{1}{r_1}\left[\mathbb{E}\, C - \mathbb{E}\left(\overline{\boldsymbol{x}}^\top \boldsymbol{S}_1\right)\right]. \tag{3.5}$$

The cost of setting up this optimal hedging portfolio at time 0 is

$$\overline{\boldsymbol{x}}^\top \boldsymbol{S}_0 + \overline{y} = \frac{1}{r_1}\left[\mathbb{E}\, C - \left(\mathbb{E}\, \boldsymbol{S}_1 - r_1 \boldsymbol{S}_0\right)^\top V^{-1} \mathbb{E}\left[C\left(\boldsymbol{S}_1 - \mathbb{E}\, \boldsymbol{S}_1\right)\right]\right], \tag{3.6}$$

while the **intrinsic risk** of the claim C is defined to be $\mathcal{R}(C) = \mathbb{E}(U^2)$ where $U = C - \overline{\boldsymbol{x}}^\top \boldsymbol{S}_1 - \overline{y} r_1$, the difference between the payoff of the claim and the value of the optimal portfolio at time 1. It follows that

$$\begin{aligned} U &= C - \overline{\boldsymbol{x}}^\top \boldsymbol{S}_1 - \overline{y} r_1 \\ &= C - \mathbb{E}\, C - \overline{\boldsymbol{x}}^\top \left(\boldsymbol{S}_1 - \mathbb{E}\, \boldsymbol{S}_1\right) + \left(\mathbb{E}\, C - \overline{\boldsymbol{x}}^\top \mathbb{E}\, \boldsymbol{S}_1 - \overline{y} r_1\right) \\ &= C - \mathbb{E}\, C - \left(\boldsymbol{S}_1 - \mathbb{E}\, \boldsymbol{S}_1\right)^\top V^{-1} \mathbb{E}\left[C\left(\boldsymbol{S}_1 - \mathbb{E}\, \boldsymbol{S}_1\right)\right], \end{aligned} \tag{3.7}$$

after using (3.3) and (3.5). Note that $\mathbb{E}\, U = 0$ and that the intrinsic risk is

$$\mathcal{R}(C) = \mathbb{E}(U^2) = \mathbb{V}ar(C) - \mathbb{E}\left[C\left(\boldsymbol{S}_1 - \mathbb{E}\, \boldsymbol{S}_1\right)\right]^\top V^{-1} \mathbb{E}\left[C\left(\boldsymbol{S}_1 - \mathbb{E}\, \boldsymbol{S}_1\right)\right].$$

We will say that a claim C may be **hedged exactly** when its associated intrinsic risk is zero; that is, $\mathcal{R}(C) = 0$ or, equivalently, $U \equiv 0$; a contingent claim that may be hedged exactly is also called **attainable** or **redundant**. From (3.7), the claim C may be hedged exactly if and only if it satisfies

$$C = \mathbb{E}C + (\boldsymbol{S}_1 - \mathbb{E}\boldsymbol{S}_1)^\top V^{-1}\mathbb{E}\left[C(\boldsymbol{S}_1 - \mathbb{E}\boldsymbol{S}_1)\right], \tag{3.8}$$

so that it is a particular linear combination of the components of the random vector $\boldsymbol{S}_1$, plus a constant. That is, the claim may be hedged exactly when a portfolio of the assets $0, 1, \ldots, s$, may be formed at time 0, the value of which matches exactly the payoff of the claim at time 1.

To conform with future notation, let $B_1 = 1/r_1$ represent the one-period discount factor which reduces time-1 prices to time-0 values; B_1 is the price at time 0 of a riskless bond which pays exactly the amount 1 at time 1. Consider again the one-period model and say that a (signed) measure $\mathbb{Q}$ is a **dominated martingale measure** if it is dominated by $\mathbb{P}$ (see Appendix A.1.3), has total mass 1 and

$$\boldsymbol{S}_0 = \mathbb{E}_{\mathbb{Q}}(B_1\boldsymbol{S}_1). \tag{3.9}$$

The terminology here comes from the fact that the equation (3.9) expresses the fact that the pair $(\boldsymbol{S}_0, B_1\boldsymbol{S}_1)$ would satisfy the martingale property under the measure $\mathbb{Q}$; here the filtration would be $(\mathcal{F}_0, \mathcal{F}_1)$ where $\mathcal{F}_0$ is the trivial σ-field because the vector $\boldsymbol{S}_0$ is constant and $\mathcal{F}_1 = \sigma(\boldsymbol{S}_1)$ is the σ-field generated by the asset prices at time 1. Note that $B_1\boldsymbol{S}_1$ are the time-1 risky asset prices discounted to time 0 by the discount factor B_1 and that (3.9) expresses the fact that

$$S_{i,0} = \mathbb{E}_{\mathbb{Q}}(B_1 S_{i,1}) \quad \text{for} \quad i = 1, \ldots, s.$$

When we let $L = d\mathbb{Q}/d\mathbb{P}$ be the Radon–Nikodym derivative of $\mathbb{Q}$ with respect to $\mathbb{P}$, then it satisfies $\mathbb{E}L = 1$ and $\boldsymbol{S}_0 = \mathbb{E}(B_1 L\boldsymbol{S}_1)$. We need to be sure that the expectation $\mathbb{E}(L\boldsymbol{S}_1)$ is well defined; a condition sufficient to ensure that it is well defined is that the measure $\mathbb{Q}$ satisfies

$$\mathbb{E}\left(\frac{d\mathbb{Q}}{d\mathbb{P}}\right)^2 < \infty, \tag{3.10}$$

and we will assume this condition holds for all the measures that we consider below. This condition, together with the assumption above that $\mathbb{E}\|\boldsymbol{S}_1\|^2 < \infty$, ensures that $\mathbb{E}\|L\boldsymbol{S}_1\| < \infty$ (by the Cauchy–Schwarz inequality); remember that, for any random vector $\boldsymbol{X} = (X_1, \ldots, X_s)^\top$, $\mathbb{E}\|\boldsymbol{X}\| < \infty$ if and only if $\mathbb{E}|X_i| < \infty$, for each $i = 1, \ldots, s$.

Say that a probability $\mathbb{Q}$, equivalent to $\mathbb{P}$, is an **equivalent martingale probability** when (3.9) holds; thus an equivalent martingale probability is a dominated martingale measure for which $d\mathbb{Q}/d\mathbb{P} > 0$ with $\mathbb{P}$-probability (and hence also $\mathbb{Q}$-probability) one.

We obtain the most important example of a dominated martingale measure by setting

$$L = 1 - (\mathbb{E}\boldsymbol{S}_1 - r_1\boldsymbol{S}_0)^\top V^{-1}(\boldsymbol{S}_1 - \mathbb{E}\boldsymbol{S}_1); \tag{3.11}$$

notice that

$$\begin{aligned}\mathbb{E}L &= \mathbb{E}\left[1-(\mathbb{E}S_1-r_1S_0)^{\top}V^{-1}(S_1-\mathbb{E}S_1)\right]\\ &= 1-(\mathbb{E}S_1-r_1S_0)^{\top}V^{-1}\mathbb{E}(S_1-\mathbb{E}S_1)=1,\end{aligned}$$

since $\mathbb{E}(S_1-\mathbb{E}S_1)=0$ and define $\mathbb{Q}$ by $d\mathbb{Q}/d\mathbb{P}=L$ (see Appendix A.1.3). Now we may calculate that

$$\begin{aligned}\mathbb{E}_{\mathbb{Q}}(B_1S_1) &= \mathbb{E}(B_1LS_1)\\ &= \mathbb{E}(B_1S_1)-\mathbb{E}\left[(B_1\mathbb{E}S_1-S_0)^{\top}V^{-1}(S_1-\mathbb{E}S_1)S_1\right]\end{aligned}$$

after substituting for L from (3.11), and this expression

$$\begin{aligned}&= \mathbb{E}(B_1S_1)-\mathbb{E}\left[(B_1\mathbb{E}S_1-S_0)^{\top}V^{-1}(S_1-\mathbb{E}S_1)(S_1-\mathbb{E}S_1)\right]\\ &= \mathbb{E}(B_1S_1)-\mathbb{E}\left[(S_1-\mathbb{E}S_1)(S_1-\mathbb{E}S_1)^{\top}\right]V^{-1}(B_1\mathbb{E}S_1-S_0);\end{aligned}$$

recall that the covariance matrix $V=\mathbb{E}\left[(S_1-\mathbb{E}S_1)(S_1-\mathbb{E}S_1)^{\top}\right]$ to see that this

$$= \mathbb{E}(B_1S_1)-VV^{-1}(B_1\mathbb{E}S_1-S_0)=S_0.$$

This shows that the measure $\mathbb{Q}$ defined by (3.11) is a dominated martingale measure because it satisfies (3.10); it is known as the **minimal martingale measure** (see Exercise 3.2 for an explanation of the minimality property).

The minimal martingale measure is a probability only when the random variable L defined in (3.11) satisfies $L\geqslant 0$ and it is an equivalent martingale probability when $L>0$ (that is when $\mathbb{P}(L>0)=1$); we will investigate conditions under which this can be guaranteed in the next section. The importance of the minimal martingale measure resides in the following result in Theorem 3.1 which establishes that for any contingent claim the initial cost of its optimal hedging portfolio is computed by first discounting the payoff C of the claim by the discount factor B_1 and then taking the 'expected' value using the minimal martingale measure; this procedure works for any martingale measure when the claim may be hedged exactly.

Theorem 3.1 *The value at time 0 of the optimal hedging portfolio for any contingent claim $C\in\mathscr{C}$ is $\mathbb{E}_{\mathbb{Q}}(B_1C)$, where $\mathbb{Q}$ is the minimal martingale measure.*

*When the claim C may be hedged exactly then this value is $\mathbb{E}_{\mathbb{Q}}(B_1C)$ for **any** dominated martingale measure $\mathbb{Q}$.*

Proof. For the first part compute $\mathbb{E}_{\mathbb{Q}}(B_1C)=\mathbb{E}(B_1LC)$ where L is given in (3.11) and check that it gives the expression in (3.6). When C may be hedged exactly, $U=0$ where U is given in (3.7) and C is given in (3.8), so that for any martingale measure $\mathbb{Q}$,

$$\mathbb{E}_{\mathbb{Q}}(B_1 C) = \frac{1}{r_1}\mathbb{E}_{\mathbb{Q}}(C) = \frac{1}{r_1}\mathbb{E}_{\mathbb{Q}}\left[\mathbb{E}C + (S_1 - \mathbb{E}S_1)^\top V^{-1}\mathbb{E}\left[C(S_1 - \mathbb{E}S_1)\right]\right]$$
$$= \frac{1}{r_1}\left[\mathbb{E}C + \left(\mathbb{E}_{\mathbb{Q}}S_1 - \mathbb{E}S_1\right)^\top V^{-1}\mathbb{E}\left[C(S_1 - \mathbb{E}S_1)\right]\right]$$
$$= \frac{1}{r_1}\left[\mathbb{E}C + (r_1 S_0 - \mathbb{E}S_1)^\top V^{-1}\mathbb{E}\left[C(S_1 - \mathbb{E}S_1)\right]\right],$$

which is the right-hand side in (3.6). □

Let $\mathcal{A} = \{A \in \mathcal{C} : A = \boldsymbol{x}^\top S_1 + y r_1,\ \boldsymbol{x} \in \mathbb{R}^s,\ y \in \mathbb{R}\}$ be the set of **attainable** contingent claims; any claim $A = \boldsymbol{x}^\top S_1 + y r_1 \in \mathcal{A}$ may be hedged exactly by forming the portfolio consisting of y in the riskless asset and $\boldsymbol{x}$ in the s risky assets, and as usual it is assumed that the relations in the definition of $\mathcal{A}$ hold with probability one. As defined previously, the attainable claims (those in $\mathcal{A}$) are precisely those for which the intrinsic risk is zero. If it is assumed that an investor is indifferent between two assets for which the payoff at time 1 is identical then Theorem 3.1 is saying that the 'fair' price at time 0 of an attainable claim $C \in \mathcal{A}$ is $\mathbb{E}_{\mathbb{Q}}(B_1 C)$, where $\mathbb{Q}$ is any dominated martingale measure, since the values of the hedging portfolio and of the claim C are identical at time 1 they should have the same price at time 0 (in particular, $\mathbb{Q}$ may be the minimal martingale measure). The requirement that the model contains no arbitrage opportunities is a condition which is sufficient to ensure that investors put the same value at time 0 on two assets having the same payoff at time 1; the notion of arbitrage in the context of the present model is defined and discussed in detail in the next section.

We will refer to $\mathcal{U} = \{U \in \mathcal{C} : \mathbb{E}U = 0,\ \mathbb{E}(U S_1) = 0\}$ as the set of **unattainable** claims. Notice that $\mathcal{A}$ and $\mathcal{U}$ are linear subspaces of $\mathcal{C}$ since they are closed under addition and multiplication by scalars and it follows immediately that $\mathcal{A} \cap \mathcal{U} = \{0\}$ and $\mathbb{E}(AU) = 0$ whenever $A \in \mathcal{A}$ and $U \in \mathcal{U}$. The reader may check easily that the argument that gave (3.6) and (3.8) shows that any $C \in \mathcal{C}$ may be decomposed as

$$C = A + U, \quad \text{with} \quad A \in \mathcal{A},\ U \in \mathcal{U};$$

this decomposition is essentially unique because if $C = A_1 + U_1 = A_2 + U_2$ where $A_i \in \mathcal{A}$, and $U_i \in \mathcal{U}$, $i = 1, 2$, then $A_1 - A_2 = U_2 - U_1 \in \mathcal{A} \cap \mathcal{U}$, since $\mathcal{A}$ and $\mathcal{U}$ are linear subspaces, which implies that $U_1 = U_2$ and $A_1 = A_2$. We are representing an element $C \in \mathcal{C}$ as the sum of its projection onto the subspace of random variables spanned by S_1 and a constant and a component orthogonal to that subspace in the sense that it is uncorrelated with S_1. Formally $\mathcal{C}$ is represented as $\mathcal{C} = \mathcal{A} \oplus \mathcal{U}$ where $\mathcal{A} \cap \mathcal{U} = \{0\}$.

We will say that the one-period model is a **complete market** if all contingent claims may be hedged exactly, that is $\mathcal{U} = \{0\}$ or $\mathcal{C} \equiv \mathcal{A}$, or equivalently, all claims have zero intrinsic risk. The most important case where the model is complete is the binomial model (over one period) discussed in Section 2.1. The binomial model should be contrasted with the next example.

Example 3.1 *Trinomial model.* This is similar to the set-up of the binomial model except that we now assume that the underlying probability space $\Omega = \{\omega_1, \omega_2, \omega_3\}$, so that there are three possible states of nature, but there are still just two assets. Asset 0 is again the riskless bank account, as in the binomial model, where 1 unit of wealth at time 0 held in the bank account becomes $r_1 = 1 + \rho$ with certainty at time 1 where ρ is the interest rate for the period; asset 1 is a stock for which the price at time 1 is $S_1(\omega_i) = u_i S_0$, $i = 1, 2, 3$, with S_0 being the price of the stock at time 0. We will assume that the values u_1, u_2 and u_3 are distinct. To hedge a claim C exactly would now require the solution of the three linearly independent equations

$$x u_i S_0 + y r_1 = C(\omega_i), \quad i = 1, 2, 3,$$

for the two quantities x and y, which shows that not all claims can be hedged and thus the trinomial model is incomplete. □

It is clear from the discussion of the binomial and trinomial models that when the probability space Ω has n points then n linearly independent assets would be required for completeness; in the present discrete-time framework with a finite number of assets, for the market to be complete it is necessary for Ω to be finite. A formal characterization of completeness which carries through to more general situations is the following.

Theorem 3.2 *The market is complete if and only if there is a unique dominated martingale measure.*

Proof. The existence of at least one dominated martingale measure is assured by the construction of the minimal martingale measure above. Suppose that the market is complete, so that $\mathcal{U} = \{0\}$, and that $\mathbb{Q}$, $\mathbb{Q}'$ are dominated martingale measures with $L = d\mathbb{Q}/d\mathbb{P}$, $L' = d\mathbb{Q}'/d\mathbb{P}$; recall that we are restricting to measures satisfying (3.10) so that $\mathbb{E}(L^2) < \infty$ and $\mathbb{E}(L')^2 < \infty$. Then

$$\mathbb{E} L = \mathbb{E} L' = 1 \quad \text{and} \quad S_0 = \mathbb{E}(B_1 L S_1) = \mathbb{E}\left(B_1 L' S_1\right)$$

which implies that $\mathbb{E}(L - L') = 0$ and $\mathbb{E}((L - L') S_1) = 0$ and shows that the difference $L - L' \in \mathcal{U} = \{0\}$, whence $L = L'$ and so $\mathbb{Q} = \mathbb{Q}'$.

Conversely, when the market is not complete then there exists $U \in \mathcal{U}$ with $U \not\equiv 0$. If $L = d\mathbb{Q}/d\mathbb{P}$ is the Radon–Nikodym derivative of a dominated martingale measure then, since $\mathbb{E} U = 0$ and $\mathbb{E}(U S_1) = 0$, it follows that $L + U$ gives a distinct dominated martingale measure $\mathbb{Q}'$ through setting $d\mathbb{Q}'/d\mathbb{P} = L + U$ because

$$\mathbb{E}(L + U) = 1 \quad \text{and} \quad \mathbb{E}_{\mathbb{Q}'}(B_1 S_1) = \mathbb{E}(B_1 (L + U) S_1) = S_0,$$

which completes the argument. □

Note that for the binomial model we have already confirmed that there is a unique dominated martingale measure $\mathbb{Q}(\{\omega_i\}) = q_i$, $i = 1, 2$ by the calculations that gave the values of q_1 and q_2 given in (2.4) on page 27 and the relations (2.5) and (2.7); also for the binomial model, we have checked the second statement in Theorem 3.1 in (2.6).

3.1.2 Arbitrage

We introduced the notion of arbitrage in the context of the binomial model; here, we will extend the definition to the more general model considered in this chapter. Recall that, in common terminology an arbitrage opportunity occurs when prices in the market are such that some portfolio of assets may be bought at one time and then at a later time sold at then prevailing prices so that a profit ensues without any risk in the transaction. Excluding arbitrage opportunities from a mathematical model is a natural assumption to make and we will characterize in mathematical terms such arbitrage-free models in which there is no possibility of riskless profits.

For the one-period model described in the previous section, an **arbitrage** is a pair $(\boldsymbol{x}, y)$, with $\boldsymbol{x} \in \mathbb{R}^s$, $y \in \mathbb{R}$ satisfying

$$\boldsymbol{x}^\top \boldsymbol{S}_0 + y \leqslant 0, \quad \text{and} \quad \boldsymbol{x}^\top \boldsymbol{S}_1 + y r_1 \geqslant 0 \tag{3.12}$$

with the requirement that $\mathbb{P}(\boldsymbol{x}^\top \boldsymbol{S}_1 + y r_1 > 0) > 0$. Note that the second relation in (3.12) is to be interpreted as the condition that $\mathbb{P}(\boldsymbol{x}^\top \boldsymbol{S}_1 + y r_1 \geqslant 0) = 1$.

An arbitrage, $(\boldsymbol{x}, y)$, occurs when it is possible to form a portfolio at time 0 holding $\boldsymbol{x}$ in the risky assets and y in the riskless asset so that the net value of this portfolio is non-positive at time 0 and non-negative, but non-zero with positive probability, at time 1. Notice that the possibility of forming a portfolio $(\boldsymbol{x}, y) \neq 0$ which at time 1 is worth $\boldsymbol{x}^\top \boldsymbol{S}_1 + y r_1 \equiv 0$ is precluded by the assumption in this chapter that there is only one riskless asset. Consequently, the condition in the definition of an arbitrage that $\mathbb{P}(\boldsymbol{x}^\top \boldsymbol{S}_1 + y r_1 > 0) > 0$ could be replaced by the requirement that $(\boldsymbol{x}, y) \neq 0$.

An equivalent statement of this definition is that an arbitrage is a pair $(\boldsymbol{x}, y)$ satisfying the inequalities in (3.12) where not both $\boldsymbol{x}^\top \boldsymbol{S}_0 + y = 0$ and $\boldsymbol{x}^\top \boldsymbol{S}_1 + y r_1 \equiv 0$. This formulation is the one we will use when generalizing to the multi-period model of Section 3.2. See also Exercise 3.6 for a further equivalent statement in the one-period case.

Example 3.2 *Binomial model.* While the definition of an arbitrage in this section when applied to the binomial model is slightly different from that presented in (2.8) of Section 2.1.3, it should be noted that the existence of an arbitrage in either sense implies the existence of an arbitrage in the other sense. It is clear that when (x, y) is an arbitrage in the sense of (2.8) then (x, y) is an arbitrage in the sense of (3.12); conversely, when (x, y) is an arbitrage in the sense of (3.12) with

$$xS_0 + y = c \leqslant 0 \quad \text{and} \quad xS_1 + y(1+\rho) \geqslant 0$$

then $x' = x$ and $y' = y - c$ satisfy

$$x'S_1 + y'(1+\rho) = xS_1 + (y-c)(1+\rho) \geqslant xS_1 + y(1+\rho) \geqslant 0,$$

which shows that (x', y') is an arbitrage in the sense of (2.8). ∎

The characterization of the property that no arbitrage is possible in the model comes from the Separating Hyperplane Theorem (see Theorem A.1 on page 204) which states that for a non-empty convex subset $\mathcal{Z} \subset \mathbb{R}^n$ and a point $\boldsymbol{y} \notin \mathcal{Z}$, there exists a hyperplane $\mathcal{H} = \{\boldsymbol{z} : \ \boldsymbol{x}^\top \boldsymbol{z} = \beta\}$, say, which separates $\boldsymbol{y}$ and $\mathcal{Z}$ in that the inequalities $\boldsymbol{x}^\top \boldsymbol{y} \leqslant \beta \leqslant \boldsymbol{x}^\top \boldsymbol{z}$ hold for all $\boldsymbol{z} \in \mathcal{Z}$; furthermore, the hyperplane $\mathcal{H}$ may be chosen so that not both $\boldsymbol{y}$ and $\mathcal{Z}$ are contained in $\mathcal{H}$. We first establish a preliminary result.

Lemma 3.1 *Let $\boldsymbol{A}_0$ be a fixed vector in $\mathbb{R}^r$ and $\boldsymbol{A}_1$ a random vector taking values in $\mathbb{R}^r$. Exactly one of the following alternatives* (a) *or* (b) *holds.*

(a) *There exists a vector $\boldsymbol{x} \in \mathbb{R}^r$ satisfying either*

$$\begin{aligned} &\text{(i)} \quad \boldsymbol{x}^\top \boldsymbol{A}_0 \leqslant 0, \quad \boldsymbol{x}^\top \boldsymbol{A}_1 \geqslant 0 \ \text{ and } \ \mathbb{P}\left(\boldsymbol{x}^\top \boldsymbol{A}_1 > 0\right) > 0 \\ \text{or} \ &\text{(ii)} \quad \boldsymbol{x}^\top \boldsymbol{A}_0 < 0 \ \text{ and } \ \boldsymbol{x}^\top \boldsymbol{A}_1 \geqslant 0. \end{aligned} \tag{3.13}$$

(b) *There exists a positive random variable ν, $\mathbb{P}(\nu > 0) = 1$, with $\mathbb{E}\,\|\nu \boldsymbol{A}_1\| < \infty$ such that $\boldsymbol{A}_0 = \mathbb{E}\,(\nu \boldsymbol{A}_1)$.*

Proof. Both (a) and (b) cannot hold, for if they do then we have a contradiction in each sub-case of (3.13),

$$\begin{aligned} &\text{(i)} \ \ 0 \geqslant \boldsymbol{x}^\top \boldsymbol{A}_0 = \boldsymbol{x}^\top \mathbb{E}\,(\nu \boldsymbol{A}_1) = \mathbb{E}\left[\nu\left(\boldsymbol{x}^\top \boldsymbol{A}_1\right)\right] > 0, \\ &\text{(ii)} \ \ 0 > \boldsymbol{x}^\top \boldsymbol{A}_0 = \boldsymbol{x}^\top \mathbb{E}\,(\nu \boldsymbol{A}_1) = \mathbb{E}\left[\nu\left(\boldsymbol{x}^\top \boldsymbol{A}_1\right)\right] \geqslant 0. \end{aligned}$$

Let $\mathcal{Z}$ be the set in $\mathbb{R}^r$ given by

$$\mathcal{Z} = \{\boldsymbol{z} : \boldsymbol{z} = \mathbb{E}\,(\nu \boldsymbol{A}_1) \ \text{for some } \nu, \ \mathbb{P}(\nu > 0) = 1, \ \text{with } \mathbb{E}\,\|\nu \boldsymbol{A}_1\| < \infty\}.$$

Set $\overline{\nu} = 1/\left(1 + \max_i |(\boldsymbol{A}_1)_i|\right)$ so that $1 \geqslant \overline{\nu} > 0$, then by taking $\nu = \overline{\nu}$ it may be seen that $\mathcal{Z}$ is non-empty since $\mathbb{E}\,\|\overline{\nu} \boldsymbol{A}_1\| < \infty$, since $|\left(\overline{\nu} \boldsymbol{A}_1\right)_i| \leqslant 1$, for each i; furthermore, it is straightforward to check that $\mathcal{Z}$ is convex. Now suppose that (b) does not hold. Then $\mathcal{Z}$ is a non-empty convex set not containing the point $\boldsymbol{A}_0$, hence, by the Separating Hyperplane Theorem, there exists a hyperplane $\mathcal{H} = \{\boldsymbol{z} : \boldsymbol{x}^\top \boldsymbol{z} = \beta\}$ that separates $\mathcal{Z}$ and $\boldsymbol{A}_0$ but does not contain both, so that

$$\boldsymbol{x}^\top \boldsymbol{A}_0 \leqslant \beta \leqslant \boldsymbol{x}^\top \mathbb{E}\,(\nu \boldsymbol{A}_1) \quad \text{for all} \quad \nu, \tag{3.14}$$

with $\mathbb{P}(\nu > 0) = 1$ and $\mathbb{E}\,\|\nu \boldsymbol{A}_1\| < \infty$. By taking $\nu = \epsilon \overline{\nu}$ where $\epsilon > 0$ and letting $\epsilon \downarrow 0$ shows that $\beta \leqslant 0$. To see that $\boldsymbol{x}^\top \boldsymbol{A}_1 \geqslant 0$, suppose that $\mathbb{P}(\boldsymbol{x}^\top \boldsymbol{A}_1 < 0) > 0$, so that

$$\mathbb{E}\left(\frac{\overline{\nu}\,(\boldsymbol{x}^\top \boldsymbol{A}_1)_-}{1 + (\boldsymbol{x}^\top \boldsymbol{A}_1)_-}\right) > 0.$$

For $\lambda > 0$ define the positive random variable ν_λ by setting

$$\nu_\lambda = \left(\frac{\lambda \overline{\nu}}{1 + (\boldsymbol{x}^\top \boldsymbol{A}_1)_-} \right) I_{(\boldsymbol{x}^\top \boldsymbol{A}_1 < 0)} + \left(\frac{\overline{\nu}}{1 + (\boldsymbol{x}^\top \boldsymbol{A}_1)_+} \right) I_{(\boldsymbol{x}^\top \boldsymbol{A}_1 \geqslant 0)}.$$

The inclusion of $\overline{\nu}$ in the definition of ν_λ is to ensure that $\mathbb{E}\,\|\nu_\lambda \boldsymbol{A}_1\| < \infty$, since, for all i, $\nu_\lambda |\,(\boldsymbol{A}_1)_i\,| \leqslant \max(\lambda, 1)$. Then

$$\begin{aligned} \boldsymbol{x}^\top \mathbb{E}\,(\nu_\lambda \boldsymbol{A}_1) &= \mathbb{E}\left[\nu_\lambda \left(\boldsymbol{x}^\top \boldsymbol{A}_1\right)\right] \\ &= -\lambda \mathbb{E}\left(\frac{\overline{\nu}\,(\boldsymbol{x}^\top \boldsymbol{A}_1)_-}{1 + (\boldsymbol{x}^\top \boldsymbol{A}_1)_-} \right) + \mathbb{E}\left(\frac{\overline{\nu}\,(\boldsymbol{x}^\top \boldsymbol{A}_1)_+}{1 + (\boldsymbol{x}^\top \boldsymbol{A}_1)_+} \right). \end{aligned}$$

Letting λ grow large would show that the inequality $\beta \leqslant \boldsymbol{x}^\top \mathbb{E}\,(\nu \boldsymbol{A}_1)$ in (3.14) is violated for $\nu = \nu_\lambda$ when λ is sufficiently large; this gives a contradiction, hence $\boldsymbol{x}^\top \boldsymbol{A}_1 \geqslant 0$.

Finally, it is not possible that both $\boldsymbol{x}^\top \boldsymbol{A}_0 = 0$ and $\boldsymbol{x}^\top \boldsymbol{A}_1 \equiv 0$ hold, for if they did both $\mathcal{Z}$ and $\boldsymbol{A}_0$ would lie in the hyperplane $\mathcal{H}$. □

Remark Notice that no assumptions on the finiteness of the mean of $\boldsymbol{A}_1$ are required in Lemma 3.1. If the assumption that $\mathbb{E}\,\|\boldsymbol{A}_1\| < \infty$ is added then the proof may be streamlined slightly. In this case, to show in the proof that $\mathcal{Z}$ is non-empty take $\nu \equiv 1$ while ν_λ may be defined by

$$\nu_\lambda = \lambda I_{(\boldsymbol{x}^\top \boldsymbol{A}_1 < 0)} + I_{(\boldsymbol{x}^\top \boldsymbol{A}_1 \geqslant 0)},$$

to give

$$\mathbb{E}\left[\nu_\lambda \left(\boldsymbol{x}^\top \boldsymbol{A}_1\right)\right] = -\lambda \mathbb{E}\left(\boldsymbol{x}^\top \boldsymbol{A}_1\right)_- + \mathbb{E}\left(\boldsymbol{x}^\top \boldsymbol{A}_1\right)_+,$$

with the remainder of the argument as before. □

It is now possible to establish that the lack of arbitrage in the model corresponds to the existence of an equivalent martingale probability.

Theorem 3.3 *For the one-period model in which there is only one riskless asset, there is no arbitrage if and only if there exists an equivalent martingale probability.*

Proof. In Lemma 3.1 take $r = s + 1$,

$$\boldsymbol{A}_0 = \begin{pmatrix} \boldsymbol{S}_0 \\ 1 \end{pmatrix} \quad \text{and} \quad \boldsymbol{A}_1 = \begin{pmatrix} \boldsymbol{S}_1 \\ r_1 \end{pmatrix}.$$

Because there does not exist a non-trivial $(\boldsymbol{x}, y)$ with $\boldsymbol{x}^\top \boldsymbol{S}_1 + y r_1 \equiv 0$, then the statement that there is no arbitrage is equivalent to the non-occurrence of case (a) in the Lemma. This is then equivalent to the existence of a random variable ν, $\mathbb{P}(\nu > 0) = 1$, with

$$\mathbb{E}\,\nu < \infty, \quad \mathbb{E}\,\|\nu S_1\| < \infty \quad \text{and} \quad \begin{pmatrix} S_0 \\ 1 \end{pmatrix} = \mathbb{E}\begin{pmatrix} \nu S_1 \\ \nu r_1 \end{pmatrix}. \tag{3.15}$$

Relate ν to an equivalent martingale probability $\mathbb{Q}$, with $d\mathbb{Q}/d\mathbb{P} = L$, by setting $L = r_1 \nu$. When ν satisfies (3.15) then $\mathbb{Q}$ is an equivalent martingale probability, since $\mathbb{E}\,L = \mathbb{E}\,(r_1\nu) = 1$ and

$$S_0 = \mathbb{E}\,(\nu S_1) = \mathbb{E}\,(B_1 L S_1) = \mathbb{E}_{\mathbb{Q}}\,(B_1 S_1), \tag{3.16}$$

while conversely when $\mathbb{Q}$ is an equivalent martingale probability then from (3.16) it may be seen that $\nu = (1/r_1)\, d\mathbb{Q}/d\mathbb{P}$ satisfies (3.15); the probability $\mathbb{Q}$ is equivalent to the probability $\mathbb{P}$ because we have $\mathbb{P}(L > 0) = 1$. ∎

Recall the comment immediately preceding Theorem 3.2 that in this model the sample space Ω must be finite for the market to be complete. We may refine the conclusion of Theorem 3.2 in the case when there is no arbitrage.

Corollary 3.1 *Suppose that the sample space Ω is finite and that there is no arbitrage. The market is complete if and only if there is a unique equivalent martingale probability.*

Proof. From Theorem 3.2, it is only necessary to establish that when the market is not complete there is more than one equivalent martingale probability; the existence of at least one, say $\mathbb{Q}$ with $d\mathbb{Q}/d\mathbb{P} = L$, is assured by the lack of arbitrage from Theorem 3.3; necessarily, $\mathbb{P}(L > 0) = 1$. By the lack of completeness, there exists $U \in \mathcal{U}, U \not\equiv 0$; since Ω is finite, letting $L^{\epsilon} = L + \epsilon U$, it follows that $\mathbb{P}(L^{\epsilon} > 0) = 1$ for $\epsilon \neq 0$, with $|\epsilon|$ sufficiently small. Since $\mathbb{E}\,U = 0$ and $\mathbb{E}\,(U S_1) = 0$, it follows that $\mathbb{E}\,L^{\epsilon} = 1$ and $\mathbb{E}\,(B_1 L^{\epsilon} S_1) = S_0$ so that $\mathbb{Q}^{\epsilon}$ defines a distinct equivalent martingale probability when we set $d\mathbb{Q}^{\epsilon}/d\mathbb{P} = L^{\epsilon}$. ∎

Remark For the binomial model, observe that the criterion for the exclusion of an arbitrage of Theorem 3.3 is the same as the condition $u > 1 + \rho > d$ set out in Theorem 2.1. This is because the existence of an equivalent martingale probability for the binomial model is equivalent to the requirement that q_1 and q_2 given in (2.4) satisfy $q_1 > 0$ and $q_2 > 0$. ∎

3.2 Multi-period model

3.2.1 Introduction

Here we consider a market at times $0, 1, \ldots, n$ and suppose that there are s risky assets for which the prices are specified by $\boldsymbol{S}_0, \boldsymbol{S}_1, \ldots, \boldsymbol{S}_n$. The random vector $\boldsymbol{S}_j = (S_{1,j}, \ldots, S_{s,j})^\top$, which is defined on some underlying probability space Ω, is such that $S_{i,j}$ is the price of asset i $(i = 1, \ldots, s)$ at time j $(j = 0, 1, \ldots, n)$.

We are going to model a market evolving in time so we need to represent the information available to investors at each time point. Mathematically, this is done by specifying an expanding sequence of σ-fields $\mathcal{F}_0 \subseteq \mathcal{F}_1 \subseteq \cdots \subseteq \mathcal{F}_n$ in Ω; such a sequence is known as a **filtration**. The σ-field $\mathcal{F}_j$ represents the information available at time j and intuitively it may be thought of as being specified by a partition of events (subsets) of Ω; indeed, when Ω is finite this is precisely the situation but for more general Ω the intuition that this provides will be adequate normally. The evolution of the system is governed by the actual state of nature $\omega \in \Omega$, which is not observed; at time j we have gathered information which enables us to narrow down which is the reigning $\omega \in \Omega$ by observing which events have occurred and which have not occurred. Then $\mathcal{F}_j$ is the collection of events whose occurrence or non-occurrence is known at time j, so knowing $\mathcal{F}_j$ is telling us in which events in the partition the actual ω lies. At time $j + 1$, typically we have more information than at time j, so we may think of $\mathcal{F}_{j+1}$ as being a finer subdivision or partition of Ω, and so on. We observe the prices $\boldsymbol{S}_j$ at time j so the components of the random vector must be random variables which are effectively constant on the events in the partition determining $\mathcal{F}_j$ (for, if not, they would be giving further information about the 'true' ω). We say that a random variable is an $\mathcal{F}_j$ random variable if its value is known after observing $\mathcal{F}_j$ and when the sequence $\{\boldsymbol{X}_j\}$ is such that $\boldsymbol{X}_j$ is an $\mathcal{F}_j$ random vector, for each j, we say that the sequence is **adapted** to the filtration $\{\mathcal{F}_j\}$.

In addition to the s risky assets we will assume that there is a further asset, asset 0 say, for which the prices at times $0, 1, \ldots, n$ are specified by random variables $R_0 \equiv 1, R_1, R_2, \ldots, R_n$ where each $R_j > 0$; we will assume that R_j is an $\mathcal{F}_{j-1}$ random variable so that the price of asset 0 at time j is known at time $j - 1$. This means that for the time period from $j - 1$ to j this asset is riskless in that 1 unit invested in asset 0 at time $j - 1$ returns the amount R_j / R_{j-1} at time j, and this amount is fixed at time $j - 1$. This formulation allows for asset 0 to be a bank account on which there is a random interest rate $R_j / R_{j-1} - 1$ for the time period from $j - 1$ to j, but this interest rate is determined at the start of the period. Note that the ratio R_{j-1} / R_j may be thought of as being a one-period random discount factor, discounting prices at time j back to time $j - 1$, while

$$B_j = \prod_{i=1}^{j} \frac{R_{i-1}}{R_i} = \frac{1}{R_j} \tag{3.17}$$

is the discount factor from time j back to time 0. The quantity B_j / B_k for $j > k$

would be the discount factor from time j back to time k. In the case where the riskless asset 0 corresponds to a bank account with a fixed interest rate ρ per period then the discount factor $B_j = \alpha^j$, where $\alpha = 1/(1+\rho)$ as in the context of the binomial model.

Note that $\mathcal{F}_j$ will often be determined by $\{S_0, \ldots, S_j, R_0, \ldots, R_{j+1}\}$ (in this situation, formally we say $\mathcal{F}_j$ is **generated** by $\{S_0, \ldots, S_j, R_0, \ldots, R_{j+1}\}$ and write $\mathcal{F}_j = \sigma\{S_0, \ldots, S_j, R_0, \ldots, R_{j+1}\}$), but we do not need to restrict ourselves to that case as we may be able to observe other random variables (other than the asset prices) which are providing information about the underlying $\omega \in \Omega$.

A **trading strategy** $\boldsymbol{T} = ((\boldsymbol{X}_0, Y_0), (\boldsymbol{X}_1, Y_1), \ldots, (\boldsymbol{X}_n, Y_n))$ consists of an adapted sequence of random vectors $\boldsymbol{X} = \{\boldsymbol{X}_j\}$ and an adapted sequence of random variables $\boldsymbol{Y} = \{Y_j\}$; here $\boldsymbol{X}_j = \left(X_{1,j}, \ldots, X_{s,j}\right)^\top$, with $X_{i,j}$ representing the amount of asset i and Y_j the amount of asset 0 held from time j to time $j+1$ using the strategy $\boldsymbol{T}$. We will assume that $\boldsymbol{X}_n \equiv 0$ and $Y_n \equiv 0$, so that the model terminates at time n. The requirement that the strategy be adapted means that we may wait until we have observed the prices $\boldsymbol{S}_j$ and R_{j+1} at time j before assembling the portfolio to hold for the period from j to $j+1$. As a shorthand write the trading strategy $\boldsymbol{T} = (\boldsymbol{X}, \boldsymbol{Y})$.

Associated with any strategy $\boldsymbol{T}$ is a **dividend sequence** $\boldsymbol{D}^{\boldsymbol{T}} = \left(D_1^{\boldsymbol{T}}, \ldots, D_n^{\boldsymbol{T}}\right)$, given by

$$D_j^{\boldsymbol{T}} = \left(\boldsymbol{X}_{j-1} - \boldsymbol{X}_j\right)^\top \boldsymbol{S}_j + (Y_{j-1} - Y_j)R_j \quad \text{for} \quad j = 1, \ldots, n. \tag{3.18}$$

The $\mathcal{F}_j$-random variable $D_j^{\boldsymbol{T}}$ is the amount 'consumed' at time j using the strategy $\boldsymbol{T}$ and it is the difference between the amount the portfolio is worth at time j through investing from time $j-1$ and the amount re-invested to be carried over to time $j+1$.

In line with the terminology of the previous section, a **one-period arbitrage** at time $j = 1, \ldots, n$ is a pair $(\boldsymbol{X}_{j-1}, Y_{j-1})$, where $\boldsymbol{X}_{j-1}$ is an $\mathcal{F}_{j-1}$random vector and Y_{j-1} is an $\mathcal{F}_{j-1}$ random variable, with

$$\boldsymbol{X}_{j-1}^\top \boldsymbol{S}_{j-1} + Y_{j-1}R_{j-1} \leqslant 0 \quad \text{and} \quad \boldsymbol{X}_{j-1}^\top \boldsymbol{S}_j + Y_{j-1}R_j \geqslant 0, \tag{3.19}$$

with at least one of these inequalities being strict in the sense that

$$\begin{aligned} \text{either} \quad & \mathbb{P}\left(\boldsymbol{X}_{j-1}^\top \boldsymbol{S}_{j-1} + Y_{j-1}R_{j-1} < 0\right) > 0 \\ \text{or} \quad & \mathbb{P}\left(\boldsymbol{X}_{j-1}^\top \boldsymbol{S}_j + Y_{j-1}R_j > 0\right) > 0. \end{aligned}$$

An **arbitrage** is a trading strategy $\boldsymbol{T} = ((\boldsymbol{X}_0, Y_0), (\boldsymbol{X}_1, Y_1), \ldots, (\boldsymbol{X}_n, Y_n))$ with

$$\boldsymbol{X}_0^\top \boldsymbol{S}_0 + Y_0 R_0 \leqslant 0 \quad \text{and} \quad \boldsymbol{D}^{\boldsymbol{T}} \geqslant 0,$$

with at least one of these inequalities being strict; this last statement is to mean that at least one of $\mathbb{P}(\boldsymbol{X}_0^\top \boldsymbol{S}_0 + Y_0 R_0 > 0) < 0$ or $\mathbb{P}(\boldsymbol{D}^{\boldsymbol{T}} \neq 0) > 0$ holds.

First note the trivial observation that if there exists a one-period arbitrage at $j \geqslant 1$, then an arbitrage exists. For, suppose that $(\boldsymbol{X}_{j-1}, Y_{j-1})$ satisfies (3.19), then define $(\boldsymbol{X}_k, Y_k) \equiv 0$ for $k \neq j-1$; it follows from (3.18) that $D_k^{\boldsymbol{T}} \equiv 0$ for $k \neq j-1$ and $k \neq j$, while we have

$$D_{j-1}^{\boldsymbol{T}} = -\left(\boldsymbol{X}_{j-1}^\top \boldsymbol{S}_{j-1} + Y_{j-1}R_{j-1}\right) \geqslant 0 \quad \text{and} \quad D_j^{\boldsymbol{T}} = \boldsymbol{X}_{j-1}^\top \boldsymbol{S}_j + Y_{j-1}R_j \geqslant 0$$

with at least one of the inequalities $D_{j-1}^{T} \geqslant 0$ and $D_j^{T} \geqslant 0$ holding strictly with positive probability; that is, $\boldsymbol{D^T} \geqslant 0$ with $\boldsymbol{D^T} \neq 0$. The reverse implication, that the existence of an arbitrage implies the existence of a one-period arbitrage, is a consequence of Theorem 3.4 and Corollary 3.3 below.

Assumptions For the remainder of this chapter we will assume that the risky asset prices are non-negative, $\boldsymbol{S}_j \geqslant 0$, and that $\mathbb{E}\,\|\boldsymbol{S}_j\| < \infty$, for each $j = 0, \ldots, n$; that is, each component of $\boldsymbol{S}_j$ has finite expectation. Further assume that $\mathbb{E}\, R_j < \infty$ for each j and that R_j and the components of $\boldsymbol{S}_j$ take just countably many values. This last assumption is to avoid any discussion of measurability technicalities. We may characterize the lack of a one-period arbitrage using a generalization of Lemma 3.1. As we will carry over results from the one-period case, assume further that in each time period there is essentially only the one riskless asset, asset 0, so that the covariance matrix of $\boldsymbol{S}_j$ conditional on $\mathcal{F}_{j-1}$ has full rank with probability 1; this means that it is not possible to form a riskless asset at time $j-1$ from a portfolio of the risky assets.

Lemma 3.2 *There is no one-period arbitrage at time j, $1 \leqslant j \leqslant n$, if and only if there exists an $\mathcal{F}_j$-random variable ν_j satisfying*

$$\mathbb{P}\left(\nu_j > 0\right) = 1, \ \mathbb{E}\left(\nu_j \mid \mathcal{F}_{j-1}\right) = B_j / B_{j-1} \ \text{ and } \ \boldsymbol{S}_{j-1} = \mathbb{E}\left(\nu_j \boldsymbol{S}_j \mid \mathcal{F}_{j-1}\right).$$

Proof. Conditional on $\mathcal{F}_{j-1}$, we may treat $\boldsymbol{S}_{j-1}$ as a constant and apply the corresponding argument as in the proof of Theorem 3.3 directly. We would have $R_{j-1} = \mathbb{E}\left(\nu_j R_j \mid \mathcal{F}_{j-1}\right)$ but we use the fact that R_j is known given $\mathcal{F}_{j-1}$ so it may be taken outside the conditional expectation and recall that $B_j = 1/R_j$ from the relation (3.17). ∎

Corollary 3.2 *For each i, $1 \leqslant i \leqslant n$, there is no one-period arbitrage at all times j, $i \leqslant j \leqslant n$, if and only if there exist $\mathcal{F}_j$-random variables ν_j satisfying $\mathbb{P}\left(\nu_j > 0\right) = 1$, $\mathbb{E}\left(\nu_j \mid \mathcal{F}_{j-1}\right) = B_j / B_{j-1}$ and*

$$\boldsymbol{S}_{j-1} = \mathbb{E}\left[\left(\prod_{k=j}^{n} \nu_k\right) \boldsymbol{S}_n \,\middle|\, \mathcal{F}_{j-1}\right], \ \text{for } j = i, \ldots, n. \tag{3.20}$$

Proof. The proof is by backwards induction on i, $i = n, n-1, \ldots, 1$. For $i = n$ it is just a special case of Lemma 3.2. Assume the result for $i+1$, then by Lemma 3.2 there is no one-period arbitrage at i if and only if there exists an $\mathcal{F}_i$-random variable ν_i with $\mathbb{P}\left(\nu_i > 0\right) = 1$,

$$\mathbb{E}\left(\nu_j \mid \mathcal{F}_{j-1}\right) = B_j / B_{j-1} \ \text{ and } \ \boldsymbol{S}_{i-1} = \mathbb{E}\left(\nu_i \boldsymbol{S}_i \mid \mathcal{F}_{i-1}\right). \tag{3.21}$$

But the inductive hypothesis will imply from (3.20) that

$$S_i = \mathbb{E}\left[\left(\prod_{k=i+1}^{n} \nu_k\right) S_n \,\middle|\, \mathcal{F}_i\right],$$

and so

$$S_{i-1} = \mathbb{E}\left[\nu_i \,\mathbb{E}\left[\left(\prod_{k=i+1}^{n} \nu_k\right) S_n \,\middle|\, \mathcal{F}_i\right] \,\middle|\, \mathcal{F}_{i-1}\right],$$

but, since ν_i may be treated as a constant given $\mathcal{F}_i$ it may be taken inside the inner conditional expectation, this gives

$$S_{i-1} = \mathbb{E}\left[\mathbb{E}\left[\left(\prod_{k=i}^{n} \nu_k\right) S_n \,\middle|\, \mathcal{F}_i\right] \,\middle|\, \mathcal{F}_{i-1}\right]$$

and, by the tower property of conditional expectations, this in turn

$$= \mathbb{E}\left[\left(\prod_{k=i}^{n} \nu_k\right) S_n \,\middle|\, \mathcal{F}_{i-1}\right],$$

which completes the proof. □

Now say that an adapted sequence $\boldsymbol{\nu} = (\nu_1, \ldots, \nu_n)$ of positive random variables, $\mathbb{P}\left(\nu_j > 0\right) = 1$, is a **deflating** sequence when it satisfies the conditions in the statement of Corollary 3.2, that is $\mathbb{E}\left(\nu_j \mid \mathcal{F}_{j-1}\right) = B_j / B_{j-1}$ and

$$S_{j-1} = \mathbb{E}\left[\left(\prod_{k=j}^{n} \nu_k\right) S_n \,\middle|\, \mathcal{F}_{j-1}\right], \text{ for } j = 1, \ldots, n.$$

It is an immediate consequence of the definition that when $\boldsymbol{\nu}$ is a deflating sequence then necessarily

$$\mathbb{E}\left\|\left(\prod_{k=1}^{j} \nu_k\right) S_j\right\| < \infty \quad \text{for each} \quad j = 1, \ldots, n.$$

Corollary 3.2 shows that there is no one-period arbitrage at any time if and only if a deflating sequence exists.

3.2.2 Pricing claims

In line with the terminology in Chapter 2, we refer to $\{B_j S_j, \mathcal{F}_j : 0 \leqslant j \leqslant n\}$, as the **discounted price process**.

Theorem 3.4 *The following three statements are equivalent.*

(a) *At each time* $j = 1, \ldots, n$ *there is no one-period arbitrage.*

(b) *There exists a deflating sequence.*

(c) *There exists an equivalent probability* $\mathbb{Q}$ *such that the discounted price process is a martingale under* $\mathbb{Q}$*; that is*

$$\mathbb{E}_{\mathbb{Q}} \|B_j \boldsymbol{S}_j\| < \infty \quad \text{and} \quad B_{j-1}\boldsymbol{S}_{j-1} = \mathbb{E}_{\mathbb{Q}}\left(B_j \boldsymbol{S}_j \mid \mathcal{F}_{j-1}\right) \quad \text{for} \quad 1 \leqslant j \leqslant n.$$

Proof. Corollary 3.2 demonstrates the equivalence of (a) and (b). Now suppose that (b) holds, and let $\boldsymbol{\nu} = (\nu_1, \ldots, \nu_n)$ be the deflating sequence, so that for each j, $1 \leqslant j < n$,

$$\boldsymbol{S}_{j-1} = \mathbb{E}\left(\nu_j \boldsymbol{S}_j \mid \mathcal{F}_{j-1}\right) = \mathbb{E}\left[\left(\prod_{k=j}^{n} \nu_k\right)\boldsymbol{S}_n \,\middle|\, \mathcal{F}_{j-1}\right].$$

Recall that B_j and B_{j-1} are known at time $j-1$, that is given $\mathcal{F}_{j-1}$, and that we have $\mathbb{E}\left(\nu_j \mid \mathcal{F}_{j-1}\right) = B_j / B_{j-1}$. Define positive random variables $L_j = B_j^{-1} \prod_{r=1}^{j} \nu_r$ for $j \geqslant 1$ with $L_0 = 1$ and put $L = L_n$. Then it follows that $L_{j-1} = \mathbb{E}\left(L_j \mid \mathcal{F}_{j-1}\right)$, so that $\{L_j, \mathcal{F}_j : 0 \leqslant j \leqslant n\}$ is a martingale (under the original probability $\mathbb{P}$); this implies that $L_j = \mathbb{E}\left(L \mid \mathcal{F}_j\right)$, and since $B_0 = 1$ we have $\mathbb{E}\, L_j = 1$, for each j. Define the equivalent probability $\mathbb{Q}$ by setting $\mathbb{Q}(A) = \mathbb{E}\,(L I_A)$ so that $L = d\mathbb{Q}/d\mathbb{P}$ is the Radon–Nikodym derivative of $\mathbb{Q}$ with respect to $\mathbb{P}$. Since $\nu_j = \left(L_j B_j\right) / \left(L_{j-1} B_{j-1}\right)$, we have

$$B_{j-1}\boldsymbol{S}_{j-1} = \mathbb{E}\left(\nu_j B_{j-1} \boldsymbol{S}_j \,\middle|\, \mathcal{F}_{j-1}\right) = \mathbb{E}\left(L_j B_j \boldsymbol{S}_j \,\middle|\, \mathcal{F}_{j-1}\right) / L_{j-1}.$$

But use the tower property of conditional expectations, the fact that B_j and $\boldsymbol{S}_j$ are known given $\mathcal{F}_j$ and (A.5) to see that

$$\begin{aligned}
\mathbb{E}_{\mathbb{Q}}\left(B_j \boldsymbol{S}_j \,\middle|\, \mathcal{F}_{j-1}\right) &= \frac{\mathbb{E}\left(L B_j \boldsymbol{S}_j \,\middle|\, \mathcal{F}_{j-1}\right)}{\mathbb{E}\left(L \mid \mathcal{F}_{j-1}\right)} \\
&= \frac{\mathbb{E}\left(\mathbb{E}\left(L B_j \boldsymbol{S}_j \mid \mathcal{F}_j\right) \,\middle|\, \mathcal{F}_{j-1}\right)}{L_{j-1}} = \frac{\mathbb{E}\left(L_j B_j \boldsymbol{S}_j \mid \mathcal{F}_{j-1}\right)}{L_{j-1}},
\end{aligned}$$

which shows that (c) holds.

Conversely, when (c) holds the argument may be reversed by taking $L = d\mathbb{Q}/d\mathbb{P}$ with $L_j = \mathbb{E}\left(L \mid \mathcal{F}_j\right)$ and defining $\nu_j = \left(L_j B_j\right) / \left(L_{j-1} B_{j-1}\right)$. □

When the discounted price process $\{B_j \boldsymbol{S}_j, \mathcal{F}_j : 0 \leqslant j \leqslant n\}$ is a martingale under a probability $\mathbb{Q}$, equivalent to $\mathbb{P}$, then $\mathbb{Q}$ is said to be an **equivalent martingale probability**; there is a one-to-one correspondence between such probabilities and deflating sequences as the proof of Theorem 3.4 demonstrates. Note the argument implicit in the proof that shows that when X is an $\mathcal{F}_j$-random variable then $\mathbb{E}_{\mathbb{Q}}(X) = \mathbb{E}(L_j X)$, where $L_j = \mathbb{E}\left(d\mathbb{Q}/d\mathbb{P} \mid \mathcal{F}_j\right)$.

For the remainder of this section, in order to ensure that all expectations (and conditional expectations) are well defined in the following, we confine attention to the case where the prices for the assets are such that

$$\mathbb{E}\,\|(B_j/B_r)\boldsymbol{S}_j\|^2 < \infty \quad \text{and} \quad \mathbb{E}\,(B_j/B_r)^2 < \infty \quad \text{for all} \quad j, r, \tag{3.22}$$

and consider only trading strategies $\boldsymbol{T} = (\boldsymbol{X}, \boldsymbol{Y})$ restricted to the set

$$\begin{aligned}\mathcal{T} = \{&\boldsymbol{T} : \boldsymbol{T} = ((\boldsymbol{X}_0, Y_0), (\boldsymbol{X}_1, Y_1), \ldots, (\boldsymbol{X}_n, Y_n)),\\ &\text{where } \mathbb{E}\,\|\boldsymbol{X}_j\|^2 < \infty, \mathbb{E}\,(Y_j^2) < \infty \text{ for each } j \text{ and } \boldsymbol{X}_n \equiv 0, Y_n \equiv 0\}.\end{aligned}$$

Theorem 3.5 *Suppose that $\mathbb{Q}$ is an equivalent martingale probability. Then for any trading strategy, $\boldsymbol{T} \in \mathcal{T}$, its value V_r at time r satisfies*

$$V_r = \boldsymbol{X}_r^\top \boldsymbol{S}_r + Y_r R_r = \mathbb{E}_{\mathbb{Q}}\left(\sum_{j=r+1}^{n} (B_j/B_r) D_j^{\boldsymbol{T}} \;\middle|\; \mathcal{F}_r \right), \tag{3.23}$$

for each $0 \leqslant r \leqslant n-1$.

Proof. First, recall that $B_j R_j = 1$ so that

$$\begin{aligned}\sum_{j=r+1}^{n} B_j D_j^{\boldsymbol{T}} &= \sum_{j=r+1}^{n} B_j \left[(\boldsymbol{X}_{j-1} - \boldsymbol{X}_j)^\top \boldsymbol{S}_j + (Y_{j-1} - Y_j) R_j \right]\\ &= \sum_{j=r+1}^{n} B_j (\boldsymbol{X}_{j-1} - \boldsymbol{X}_j)^\top \boldsymbol{S}_j + \sum_{j=r+1}^{n} (Y_{j-1} - Y_j)\\ &= \sum_{j=r+1}^{n} B_j (\boldsymbol{X}_{j-1} - \boldsymbol{X}_j)^\top \boldsymbol{S}_j + Y_r,\end{aligned} \tag{3.24}$$

because the summation of $Y_{j-1} - Y_j$ is a telescoping sum and $Y_n = 0$. Now use the tower property of conditional expectations, the fact that $\boldsymbol{X}_{j-1}$ is known at time $j-1$ and the martingale property, to see that for $j > r$

$$\begin{aligned}&\mathbb{E}_{\mathbb{Q}}\left[B_j (\boldsymbol{X}_{j-1} - \boldsymbol{X}_j)^\top \boldsymbol{S}_j \mid \mathcal{F}_r \right]\\ &\quad = \mathbb{E}_{\mathbb{Q}}\left[\mathbb{E}_{\mathbb{Q}}\left(B_j \boldsymbol{X}_{j-1}^\top \boldsymbol{S}_j \mid \mathcal{F}_{j-1} \right) \mid \mathcal{F}_r \right] - \mathbb{E}_{\mathbb{Q}}\left(B_j \boldsymbol{X}_j^\top \boldsymbol{S}_j \mid \mathcal{F}_r \right)\\ &\quad = \mathbb{E}_{\mathbb{Q}}\left[\boldsymbol{X}_{j-1}^\top \mathbb{E}_{\mathbb{Q}}\left(B_j \boldsymbol{S}_j \mid \mathcal{F}_{j-1} \right) \mid \mathcal{F}_r \right] - \mathbb{E}_{\mathbb{Q}}\left(B_j \boldsymbol{X}_j^\top \boldsymbol{S}_j \mid \mathcal{F}_r \right)\\ &\quad = \mathbb{E}_{\mathbb{Q}}\left(B_{j-1} \boldsymbol{X}_{j-1}^\top \boldsymbol{S}_{j-1} \mid \mathcal{F}_r \right) - \mathbb{E}_{\mathbb{Q}}\left(B_j \boldsymbol{X}_j^\top \boldsymbol{S}_j \mid \mathcal{F}_r \right).\end{aligned}$$

When the right-hand side is summed on j this provides a telescoping sum, and when we recall that $X_n \equiv 0$, we have

$$\begin{aligned}
\mathbb{E}_{\mathbb{Q}}\left(\sum_{j=r+1}^{n} B_j \left(X_{j-1} - X_j\right)^{\top} S_j \;\middle|\; \mathcal{F}_r\right)
&= \sum_{j=r+1}^{n} \left[\mathbb{E}_{\mathbb{Q}}\left(B_{j-1} X_{j-1}^{\top} S_{j-1} \mid \mathcal{F}_r\right) - \mathbb{E}_{\mathbb{Q}}\left(B_j X_j^{\top} S_j \mid \mathcal{F}_r\right)\right] \\
&= \mathbb{E}_{\mathbb{Q}}\left(B_r X_r^{\top} S_r \mid \mathcal{F}_r\right) = B_r X_r^{\top} S_r. \qquad (3.25)
\end{aligned}$$

Now, because $B_r X_r^{\top} S_r$ and Y_r are known given $\mathcal{F}_r$, from (3.24) and (3.25) we have that

$$\mathbb{E}_{\mathbb{Q}}\left(\sum_{j=r+1}^{n} B_j D_j^{\boldsymbol{T}} \;\middle|\; \mathcal{F}_r\right) = B_r (X_r^{\top} S_r + Y_r R_r)$$

and we may divide through by B_r and take it inside the conditional expectation to give the result. □

Remark The random variable $D_j^{\boldsymbol{T}}$ is the dividend paid by the trading strategy $\boldsymbol{T}$ at time j and $B_j D_j^{\boldsymbol{T}} / B_r$ is its value discounted to time r. Theorem 3.5 is central to the valuation of assets as it shows that if an equivalent martingale probability exists the value of any portfolio generated by a trading strategy at time r is just the conditional expected value of its future discounted dividends after r under the martingale probability; it is immediate that this is the same for all martingale probabilities. The restriction of T to $\mathcal{T}$ and the requirement that the conditions in (3.22) hold is to ensure that in the proof the expectation of $(B_j / B_r) \left(X_{j-1} - X_j\right)^{\top} S_j$ is defined (which would follow using the Cauchy–Schwarz inequality), and hence also its conditional expectation with respect to $\mathcal{F}_r$ is defined.

The relationship (3.23) in Theorem 3.5 helps to give an intuitive explanation of the commonly observed phenomenon that when interest rates go down asset prices rise, and vice versa. Suppose that there is a fixed interest rate ρ per period which will give the discount factor as $B_j = 1/(1+\rho)^j$, so that the value V_r of the trading strategy as given by (3.23) is

$$V_r = \mathbb{E}_{\mathbb{Q}}\left(\sum_{j=r+1}^{n} \left(\frac{1}{1+\rho}\right)^{j-r} D_j^{\boldsymbol{T}} \;\middle|\; \mathcal{F}_r\right); \qquad (3.26)$$

the right-hand side of (3.26) will decrease when ρ increases (or alternatively, will increase as ρ decreases). □

Corollary 3.3 *If there exists an equivalent martingale probability then there is no arbitrage* $\boldsymbol{T} \in \mathcal{T}$.

Proof. Suppose that $\boldsymbol{T} \in \mathcal{T}$ is an arbitrage and that $\mathbb{Q}$ is an equivalent martingale probability. Then $X_0^{\top} S_0 + Y_0 R_0 \leqslant 0$ and $D^{\boldsymbol{T}} \geqslant 0$ with at least one of these relations being strict, and by Theorem 3.5,

$$X_0^\top S_0 + Y_0 R_0 = \mathbb{E}_{\mathcal{Q}}\left(\sum_{j=1}^{n} B_j D_j^{T} \;\middle|\; \mathcal{F}_0\right).$$

When L and $\mathcal{Q}$ are related as in the proof of Theorem 3.4, taking expectations with respect to $\mathcal{Q}$ gives

$$0 \geqslant \mathbb{E}_{\mathcal{Q}}\left(X_0^\top S_0 + Y_0 R_0\right) = \mathbb{E}_{\mathcal{Q}}\left(\sum_{j=1}^{n} B_j D_j^{T}\right) = \mathbb{E}\left(L_n \sum_{j=1}^{n} B_j D_j^{T}\right) \geqslant 0,$$

with at least one of these inequalities being strict, which gives a contradiction. □

A **contingent claim** is any adapted sequence $C = (C_1, \ldots, C_n)$ of random variables; C_j is the payoff of the claim at time j. We restrict attention to claims in $\mathcal{C} = \{C : \mathbb{E}\left|C_j\right|^2 < \infty \text{ for each } j\}$. Say that a contingent claim is **attainable** (or may be **hedged**) if there exists a trading strategy $T \in \mathcal{T}$ with $D_j^{T} = C_j$ for each $j = 1, \ldots, n$. As in the one-period case an attainable claim has no inherent risk since its payoff may be replicated exactly by trading in the market.

Example 3.3 Consider the case where the contingent claim is a European call option on asset 1 at strike price c expiring at time n. Then $C_1 = \cdots = C_{n-1} = 0$ and $C_n = (S_{1,n} - c)_+$. □

Theorem 3.6 *Suppose that there exists an equivalent martingale probability. The unique time-0 price of any attainable contingent claim $C \in \mathcal{C}$ is*

$$\mathbb{E}_{\mathcal{Q}}\left(\sum_{j=1}^{n} B_j C_j \;\middle|\; \mathcal{F}_0\right),$$

where $\mathcal{Q}$ is any equivalent martingale probability.

Proof. Suppose that $T \in \mathcal{T}$ is any trading strategy such that $D_j^{T} = C_j$ for each j. The lack of arbitrage implies that the time-0 price of the claim must be the same as the time-0 value, $X_0^\top S_0 + Y_0 R_0$, of the trading strategy and the result follows from Theorem 3.5. □

3.3 Exercises

Exercise 3.1 Consider the case of the single-period model in which there is just one risky asset with price S_1 at time 1.

(i) Show that when the intrinsic risk $\mathcal{R}(C) = 0$ for all claims C then the underlying probability space has effectively at most two points (so that the model is the binomial model).

(ii) Suppose that C_1, C_2 are two claims such that (C_1, C_2, S_1) have a joint normal distribution. Show that

$$\mathcal{R}(C_1 + C_2) = \mathcal{R}(C_1) + \mathcal{R}(C_2) + 2\,\mathbb{C}ov\left(C_1, C_2 \mid S_1\right);$$

hence $\mathcal{R}(C_1 + C_2) - \mathcal{R}(C_1) - \mathcal{R}(C_2) \geqslant 0$, or $\leqslant 0$, according as the random variables C_1 and C_2 are positively, or negatively, correlated conditional on S_1.

Exercise 3.2 In the context of the one-period model, show that the minimal martingale measure minimizes the expression $\mathbb{E}\left[(d\mathbb{Q}/d\mathbb{P})^2\right]$ over all dominated martingale measures $\mathbb{Q}$.

Exercise 3.3 Let $X_1, \ldots, X_n$ be independent, identically distributed (i.i.d.) random variables defined on some probability space with probability $\mathbb{P}$, each having the $N(\mu, \sigma^2)$-distribution under $\mathbb{P}$. Set $Y_n = X_1 + \cdots + X_n$. Show that, for any real θ,

$$\frac{d\mathbb{Q}}{d\mathbb{P}} = e^{\theta Y_n - n\theta\mu - n\theta^2\sigma^2/2}$$

defines a new probability, equivalent to $\mathbb{P}$, under which $X_1, \ldots, X_n$ are i.i.d. with the $N(\mu + \theta\sigma^2, \sigma^2)$-distribution. In particular, deduce that

$$\mathbb{E}\left[e^{\theta X_1} f(X_1)\right] = e^{\theta\mu + \theta^2\sigma^2/2}\,\mathbb{E}\left[f\left(X_1 + \theta\sigma^2\right)\right],$$

for any appropriate function f for which either expectation is defined (also see (A.15) on page 199).

Exercise 3.4 In the one-period model with just one risky asset, suppose that the price at time 1 of that asset has a log-normal distribution (so that $S_1 = S_0 e^X$, where the random variable X has the $N\left(\mu, \sigma^2\right)$-distribution).

(i) Starting from the definition of an unattainable claim, that is an element of the set $\mathcal{U}$, show directly that the model is not complete by constructing an example of a non-trivial unattainable claim $U = f(S_1)$.

(ii) Is there arbitrage in this model? Justify your answer.

Exercise 3.5 Let $\boldsymbol{A}_0$ be a fixed vector in $\mathbb{R}^r$ and $\boldsymbol{A}_1$ a random vector taking values in $\mathbb{R}^r$. Prove that exactly one of the following alternatives (a) or (b) holds.

(a) There exists a vector $\boldsymbol{x} \in \mathbb{R}^r$ satisfying

$$\boldsymbol{x}^\top(\boldsymbol{A}_1 - \boldsymbol{A}_0) \geqslant 0 \quad \text{with} \quad \mathbb{P}\left(\boldsymbol{x}^\top(\boldsymbol{A}_1 - \boldsymbol{A}_0) > 0\right) > 0.$$

(b) There exists a positive random variable ν, satisfying

$$\mathbb{P}(\nu > 0) = 1, \quad \mathbb{E}\,\|\nu \boldsymbol{A}_1\| < \infty, \quad \mathbb{E}\nu = 1 \quad \text{and} \quad \boldsymbol{A}_0 = \mathbb{E}\,(\nu \boldsymbol{A}_1).$$

Exercise 3.6 For the one-period model prove that the following two statements are equivalent.

(a) There exists an arbitrage.

(b) There exists $\boldsymbol{x} \in \mathbb{R}^s$ satisfying

$$\boldsymbol{x}^\top(\boldsymbol{S}_1 - r_1 \boldsymbol{S}_0) \geqslant 0 \quad \text{and} \quad \mathbb{P}\left(\boldsymbol{x}^\top(\boldsymbol{S}_1 - r_1 \boldsymbol{S}_0) > 0\right) > 0.$$

Hence use Exercise 3.5 to give an alternative proof of Theorem 3.3.

Exercise 3.7 In the context of the one-period model of Section 3.1, suppose that an investor has a strictly increasing, concave, differentiable utility function v and initial wealth w. He acts to maximize the expected utility of his final wealth and achieves the optimal final wealth $\overline{W}$. Show that there is an equivalent martingale probability, which may be expressed in terms of $\overline{W}$ and v, and hence that there is no arbitrage in the model.

Chapter 4

BROWNIAN MOTION

4.1 Introduction

A stochastic process $\{W_t,\ t \geqslant 0\}$ in continuous time taking real values is a **Brownian motion** (or a **Wiener process**) if, for some real constant σ,

(a) for each $s \geqslant 0$, and $t > 0$, the random variable $W_{t+s} - W_s$ has the normal distribution with mean 0 and variance $\sigma^2 t$;

(b) for each $n \geqslant 1$ and any times $0 = t_0 \leqslant t_1 \leqslant \cdots \leqslant t_n$ the random variables $\{W_{t_r} - W_{t_{r-1}},\ 1 \leqslant r \leqslant n\}$ are independent;

(c) $W_0 \equiv 0$; and

(d) W_t is continuous in $t \geqslant 0$.

Let us consider the conditions (a)- (d) in turn. The condition (a) is self-explanatory, it just specifies the distribution of the displacement of the process between two time points. Here, σ^2 is known as the **variance parameter** of the process; in the context of finance σ is referred to as the **volatility** of the process (see Chapter 5). By the scaling property of the normal distribution, when a is a real constant then we may see that $\{aW_t,\ t \geqslant 0\}$ is a Brownian motion with variance parameter $a^2\sigma^2$, so it is immediate that the process $\{W_t/\sigma,\ t \geqslant 0\}$ is a Brownian motion with variance parameter 1; the process with $\sigma^2 = 1$ is called a **standard Brownian motion**. Since a simple scaling enables any calculation to be reduced to one involving a standard Brownian motion we will assume from now on, unless we indicate to the contrary, that we are dealing with the case of the standard Brownian motion.

Condition (b) is referred to by saying that the process has **independent increments**. It is stronger than the Markov property. It implies that the position of the process at time t_n, say, depends on what has happened up to time $t_{n-1} < t_n$ only through the position at time t_{n-1} (which is the Markov property) and **moreover**, the displacement $W_{t_n} - W_{t_{n-1}}$ between t_{n-1} and t_n, is independent of the position at time t_{n-1}. This enables us to write down the transition probabilities explicitly as

$$\begin{aligned} \mathbb{P}\big(W_{t_n} \leqslant x_n \mid W_{t_i} = x_i,\ 0 \leqslant i \leqslant n-1\big) \\ &= \mathbb{P}\big(W_{t_n} - W_{t_{n-1}} \leqslant x_n - x_{n-1}\big) \\ &= \int_{-\infty}^{x_n - x_{n-1}} \phi(u, t_n - t_{n-1})\,du \end{aligned} \tag{4.1}$$

where $\phi(x,t) = e^{-x^2/2t}/\sqrt{2\pi t}$, is the probability density function of the normal distribution with mean 0 and variance t. We may also write down the joint probability density function of $W_{t_1}, \ldots, W_{t_n}$ as

$$f(x_1, \ldots, x_n) = \prod_1^n \phi(x_i - x_{i-1}, t_i - t_{i-1}). \tag{4.2}$$

Notice that condition (b) is consistent with condition (a) by the property of the normal distribution that the sum of independent random variables each having a normal distribution again has a normal distribution. Conditions (a) and (b) also imply that the process is spatially homogeneous so that the distribution of the increment $W_{t+s} - W_s$ does not depend on the position, W_s, at time s for $s, t > 0$. For any $t > s > 0$, since $\mathbb{E}\, W_s = \mathbb{E}\, W_t = 0$ and $W_t - W_s$ is independent of W_s, it follows that the covariance of W_s and W_t is

$$\begin{aligned} \mathbb{C}ov\ (W_s, W_t) &= \mathbb{E}\ (W_s W_t) = \mathbb{E}\ [W_s(W_t - W_s + W_s)] \\ &= \mathbb{E}\ [W_s(W_t - W_s)] + \mathbb{E}\,(W_s^2) = \mathbb{E}\,(W_s)\mathbb{E}\,(W_t - W_s) + s = s; \end{aligned}$$

we then have for any $s, t > 0$ that the covariance is given by

$$\mathbb{C}ov\ (W_s, W_t) = s \wedge t, \tag{4.3}$$

where $s \wedge t = \min(s, t)$. Since the multivariate normal distribution is determined by its means and covariances and normally-distributed random variables are independent if and only if their covariances are zero, it is immediate that, (when $\sigma^2 = 1$) (a) (b) and (c) are equivalent to requiring that for any $n \geqslant 1$ and $t_1, \ldots, t_n$, the joint distribution of $W_{t_1}, \ldots, W_{t_n}$ is normal with zero means and covariances specified by (4.3). The joint distributions of $W_{t_1}, \ldots, W_{t_n}$ for each $n \geqslant 1$ and all $t_1, \ldots, t_n$ are known as the **finite-dimensional distributions** of the process.

Condition (c) is just a convention which is useful for our purposes. If we replace (c) by $W_0 = x$, we say that the process is a Brownian motion started at x. By the spatial homogeneity referred to above, if (c) holds then the process $\{x + W_t, t \geqslant 0\}$ is a Brownian motion started at x.

Turn now to condition (d). Recall that the random variables W_t, $t \geqslant 0$ are all defined on some underlying probability (sample) space Ω. So for each $t \geqslant 0$, W_t is a function, $W_t(\omega)$ say, of the points $\omega \in \Omega$. Normally we suppress this dependence on ω in the notation. However, if we take a fixed $\omega \in \Omega$ and consider $W_t(\omega)$ as a function of $t \geqslant 0$ we obtain what is called a **sample path** (or **trajectory**) of the process. The condition (d) requires that all the sample paths of the processes are continuous. We could replace (d) by the requirement that the set of ω for which the corresponding sample path is continuous has probability 1. Since our discussion of Brownian motion will place reliance on arguments involving sample paths of this process, this property of continuity of sample paths will be very important for our purposes.

It should be pointed out that although we have taken condition (d) as part of the definition of a Brownian motion, in a certain sense (d) is a consequence of (a) and (b).

In fact, when $\{X_t,\ t \geqslant 0\}$ is a stochastic process satisfying (a) - (c) then there exists a stochastic process $\{W_t,\ t \geqslant 0\}$ satisfying (a) - (d) and which is indistinguishable from $\{X_t,\ t \geqslant 0\}$ in that $\mathbb{P}(X_t = W_t, \text{for all } t \geqslant 0) = 1$.

It should not be assumed that because the sample paths of Brownian motion are continuous that they are 'nice' in any other sense. Among properties possessed by the paths are that, with probability one, the paths are nowhere differentiable (see also Section 4.5).

4.2 Hitting-time distributions

4.2.1 The reflection principle

We begin our study of Brownian motion by deriving some properties of the process using elementary arguments. We will see in Section 4.2.3 that many of these results may be obtained more easily using the machinery of martingale theory, but it is instructive to get a feel for working with Brownian motion from first principles. By the symmetry of the normal distribution it is immediate that when $\{W_t,\ t \geqslant 0\}$ is a standard Brownian motion then $\{-W_t,\ t \geqslant 0\}$ is again a standard Brownian motion. Also when $s \geqslant 0$ is any fixed time $\{W_{t+s} - W_s,\ t \geqslant 0\}$ is a standard Brownian motion. What is also true is that for certain random times T, called stopping times of the process, $\{W_{t+T} - W_T,\ t \geqslant 0\}$ is again a standard Brownian motion and is independent of the process $\{W_s,\ 0 \leqslant s \leqslant T\}$.

A **stopping time** T for the process $\{W_t,\ t \geqslant 0\}$ is a random time such that for each $t \geqslant 0$, the event $\{T \leqslant t\}$ depends only on the history of the process up to and including time t, that is, $\mathcal{F}_t = \sigma\,(W_s,\ 0 \leqslant s \leqslant t)$. We shall encounter stopping times only in the context of **hitting times**. For example, for fixed a the hitting time of the level a is defined by

$$T_a = \inf\{t \geqslant 0 : W_t = a\}, \tag{4.4}$$

and we take $T_a = \infty$ if a is never reached. It is seen easily that T_a is a stopping time since, by the continuity of paths,

$$\{T_a \leqslant t\} = \{W_s = a, \quad \text{for some } s,\ 0 \leqslant s \leqslant t\},$$

which only depends on $\{W_s,\ 0 \leqslant s \leqslant t\}$. Notice that when $T_a < \infty$ then $W_{T_a} = a$, this is again because the sample paths of W are continuous; we will see below that $\mathbb{P}(T_a < \infty) = 1$.

An example of a random time which is not a stopping time is the last time the process is at some level, say 0. Let

$$L_0 = \sup\{t \geqslant 0 : W_t = 0\}.$$

Then $\{L_0 \leqslant t\} = \{W_s \neq 0, \text{ for all } s > t\}$, and to determine whether this event occurs or not requires knowledge of the whole path of W_s for $s > t$. Regarding t as the 'present', knowing whether the time L_0 has occurred before t requires knowledge of the whole 'future' after t.

Let T be a stopping time and define a new process

$$\widetilde{W}_t = \begin{cases} W_t & \text{if } t \leqslant T \\ 2W_T - W_t & \text{if } t > T \end{cases}$$

then it follows from the observation above, which we will not prove, that the process $\{\widetilde{W}_t,\ t \geqslant 0\}$ is also a standard Brownian motion; in the special case when T is a fixed (deterministic) time then this is immediate. This fact is known as the **reflection principle**. We will apply it in the case $T = T_a$, when the statement is intuitively clear; in this situation for $t > T_a$, $\widetilde{W}_t = 2a - W_t$ since $W_{T_a} = a$, so that after the level a has been hit, we obtain the process $\widetilde{W}_t$ by reflecting W_t in the level a, which is illustrated in Figure 4.1.

Example 4.1 *Joint distribution of standard Brownian motion and its maximum.* As an example of the application of the reflection principle, we will derive the joint distribution of W_t and $M_t = \max_{0 \leqslant s \leqslant t} W_s$, where M_t is the highest level reached by the Brownian motion in the time interval $[0, t]$. Notice that $M_t \geqslant 0$ and is non-decreasing in t and when T_a is defined by (4.4) for $a > 0$ then $\{M_t \geqslant a\} = \{T_a \leqslant t\}$. Taking $T = T_a$, we have for $a \geqslant 0$, $a \geqslant x$, and all $t \geqslant 0$,

$$\begin{aligned} \mathbb{P}\left(M_t \geqslant a,\ W_t \leqslant x\right) &= \mathbb{P}\left(T_a \leqslant t, W_t \leqslant x\right) = \mathbb{P}\left(T_a \leqslant t,\ 2a - x \leqslant \widetilde{W}_t\right) \\ &= \mathbb{P}\left(2a - x \leqslant \widetilde{W}_t\right) = 1 - \Phi\left(\frac{2a - x}{\sqrt{t}}\right), \end{aligned} \tag{4.5}$$

where $\Phi(x) = \int_{-\infty}^{x} \phi(u)du = \int_{-\infty}^{x} e^{-u^2/2}du/\sqrt{2\pi}$ is the standard normal distribution function, where $\phi(u) = \phi(u, 1)$. For $x \geqslant a$ note that, by the continuity of the paths of Brownian motion, $W_t \geqslant x$ implies that $M_t \geqslant a$, from which we see that

$$\mathbb{P}\left(M_t \geqslant a,\ W_t \geqslant x\right) = \mathbb{P}\left(W_t \geqslant x\right) = 1 - \Phi\left(\frac{x}{\sqrt{t}}\right). \tag{4.6}$$

We will derive some expressions involving the joint distribution of W_t and M_t which will be useful in the discussion of lookback and barrier options in the context of the Black–Scholes model of the next chapter. The first-time reader may move to Example 4.2.

Firstly, with the same reflection argument as the above, for any real ν and $a \geqslant 0$,

$$\begin{aligned} \mathbb{E}\left[e^{\nu W_t} I_{(M_t \geqslant a)}\right] &= \mathbb{E}\left[e^{\nu W_t} I_{(M_t \geqslant a, W_t \geqslant a)}\right] + \mathbb{E}\left[e^{\nu W_t} I_{(M_t \geqslant a, W_t < a)}\right] \\ &= \mathbb{E}\left[e^{\nu W_t} I_{(W_t \geqslant a)}\right] + \mathbb{E}\left[e^{\nu(2a - W_t)} I_{(W_t > a)}\right] \\ &= e^{\nu^2 t/2}\, \mathbb{P}(W_t + \nu t \geqslant a) + e^{2a\nu + \nu^2 t/2}\, \mathbb{P}(W_t - \nu t > a), \end{aligned}$$

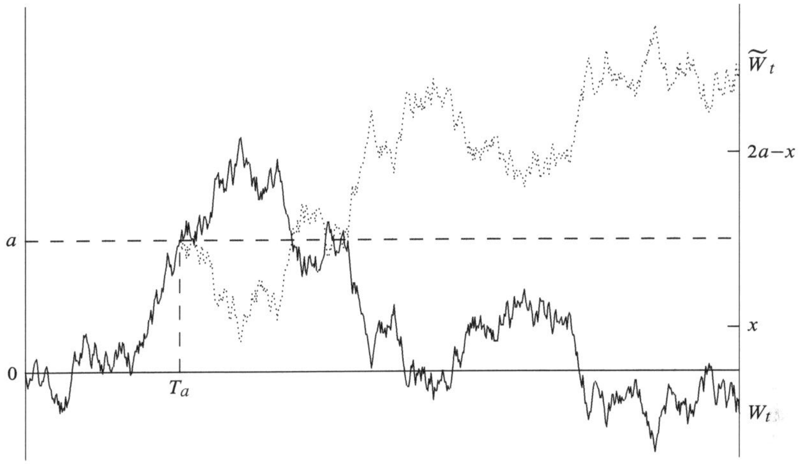

Figure 4.1: The reflection principle for $T = T_a$

where the last equality uses the fact that, since W_t has the $N(0,t)$-distribution, for any set $A \subseteq \mathbb{R}$,

$$\mathbb{E}\left[e^{\nu W_t} I_{(W_t \in A)}\right] = e^{\nu^2 t/2} \mathbb{E}\left[I_{(W_t + \nu t \in A)}\right] = e^{\nu^2 t/2} \mathbb{P}(W_t + \nu t \in A).$$

This is an application of the result that for a random variable Z having the $N(\beta, \tau^2)$-distribution,

$$\mathbb{E}\left[e^Z f(Z)\right] = e^{\beta + \tau^2/2} \mathbb{E} f\left(Z + \tau^2\right)$$

for any function f for which the expectations are defined (see (A.15) on page 199). We then have

$$\mathbb{E}\left[e^{\nu W_t} I_{(M_t \geqslant a)}\right] = e^{\nu^2 t/2}\left[\Phi\left(\frac{\nu t - a}{\sqrt{t}}\right) + e^{2a\nu}\Phi\left(\frac{-\nu t - a}{\sqrt{t}}\right)\right]. \tag{4.7}$$

To derive $\mathbb{E}\left[e^{\theta M_t + \nu W_t} I_{(M_t \geqslant a)}\right]$, for real $\theta \neq 0$, note that

$$\frac{1}{\theta}\left(e^{\theta M_t} - e^{\theta a}\right) I_{(M_t \geqslant a)} = \int_a^\infty e^{\theta u} I_{(M_t \geqslant u)} du,$$

whence

$$\begin{aligned} \mathbb{E}\left[\frac{1}{\theta}\left(e^{\theta M_t} - e^{\theta a}\right) e^{\nu W_t} I_{(M_t \geqslant a)}\right] &= \mathbb{E}\left[\int_a^\infty e^{\theta u + \nu W_t} I_{(M_t \geqslant u)} du\right] \\ &= \int_a^\infty e^{\theta u} \mathbb{E}\left[e^{\nu W_t} I_{(M_t \geqslant u)}\right] du. \end{aligned} \tag{4.8}$$

To evaluate the right-hand side of (4.8), using the expression for $\mathbb{E}\left[e^{\nu W_t} I_{(M_t \geqslant u)}\right]$ obtainable from (4.7), we need the following identities, for $c > 0$ and any real b,

$$\int_a^\infty e^{\theta u} \Phi(b - cu)\, du = \frac{1}{\theta}\left[e^{\theta(\theta + 2bc)/(2c^2)} \Phi(b - ac + \theta/c) - e^{\theta a} \Phi(b - ac)\right],$$

when $\theta \neq 0$, and

$$\int_a^\infty \Phi(b - cu)\, du = \frac{1}{c}\left[(b - ac)\Phi(b - ac) + \phi(b - ac)\right].$$

After calculation, from (4.8) when $\theta + 2\nu \neq 0$ we obtain

$$\begin{aligned} \mathbb{E}\left[e^{\theta M_t + \nu W_t} I_{(M_t \geqslant a)}\right] = {} & \frac{2(\theta + \nu)}{\theta + 2\nu} e^{(\theta+\nu)^2 t/2} \Phi\left(\frac{(\theta + \nu)t - a}{\sqrt{t}}\right) \\ & + \frac{2\nu}{\theta + 2\nu} e^{(\theta + 2\nu)a + \nu^2 t/2} \Phi\left(\frac{-\nu t - a}{\sqrt{t}}\right), \end{aligned} \tag{4.9}$$

while in the case $\theta = -2\nu$,

$$\begin{aligned} & \mathbb{E}\left[e^{\nu W_t - 2\nu M_t} I_{(M_t \geqslant a)}\right] \\ & = 2e^{\nu^2 t/2}\left[\{1 + \nu(\nu t + a)\}\Phi\left(\frac{-\nu t - a}{\sqrt{t}}\right) - \nu\sqrt{t}\phi\left(\frac{\nu t + a}{\sqrt{t}}\right)\right]. \end{aligned} \tag{4.10}$$

It is worth recording the case when $a = 0$, which gives

$$\begin{aligned} & \mathbb{E}\left[e^{\theta M_t + \nu W_t}\right] \\ & = \frac{2(\theta + \nu)}{\theta + 2\nu} e^{(\theta+\nu)^2 t/2} \Phi\left((\theta + \nu)\sqrt{t}\right) + \frac{2\nu}{\theta + 2\nu} e^{\nu^2 t/2} \Phi\left(-\nu\sqrt{t}\right) \end{aligned} \tag{4.11}$$

when $\theta + 2\nu \neq 0$, and when $\theta = -2\nu$,

$$\mathbb{E}\left[e^{\nu W_t - 2\nu M_t}\right] = 2e^{\nu^2 t/2}\left[\{1 + \nu^2 t\}\Phi\left(-\nu\sqrt{t}\right) - \nu\sqrt{t}\phi\left(\nu\sqrt{t}\right)\right]. \tag{4.12}$$

Note that of course these formulae could have been derived alternatively by obtaining the joint probability density function of M_t and W_t by differentiating the expressions in (4.5) and (4.6) with respect to a and x and evaluating the appropriate integrals. □

Example 4.2 *The distribution of the hitting time of a level.* By using reflection about the level a after the time T_a we have, for $x \leqslant a$,

$$\mathbb{P}(M_t \geqslant a,\ W_t \geqslant 2a - x) = \mathbb{P}(M_t \geqslant a,\ W_t \leqslant x)$$

from which, using (4.5), we see that

$$\begin{aligned} \mathbb{P}(T_a \leqslant t) = \mathbb{P}(M_t \geqslant a) & = \mathbb{P}(M_t \geqslant a,\ W_t > a) + \mathbb{P}(M_t \geqslant a,\ W_t \leqslant a) \\ & = 2\mathbb{P}(M_t \geqslant a,\ W_t \leqslant a), \text{ by the observation above} \\ & = 2\left[1 - \Phi\left(a/\sqrt{t}\right)\right], \text{ after setting } x = a \text{ in (4.5).} \end{aligned} \tag{4.13}$$

Letting $t \uparrow \infty$, we have $\mathbb{P}(T_a < \infty) = 1$ since $\Phi(0) = 1/2$. Differentiating the expression in (4.13) with respect to t gives the probability density function of T_a as

$$f_{T_a}(t) = ae^{-a^2/2t}/\sqrt{2\pi t^3}, \quad t > 0, \tag{4.14}$$

and we can check directly that $\mathbb{E}\, T_a = \infty$, since this density is $O(t^{-3/2})$ as $t \to \infty$. Thus we can see that a Brownian motion displays behaviour analogous to the simple symmetric random walk or, more generally, an irreducible null-recurrent Markov chain; the probability of reaching any other state from 0 is 1 but the expected time to reach it is infinite. Again, by direct computation from (4.14) we may derive the moment-generating function of T_a as

$$\mathbb{E}\left(e^{-\theta T_a}\right) = e^{-a\sqrt{2\theta}}, \quad \text{for} \quad \theta > 0 \quad \text{and} \quad a \geqslant 0. \tag{4.15}$$

To see this, write

$$\begin{aligned}\mathbb{E}\left(e^{-\theta T_a}\right) &= \int_0^\infty \frac{a}{\sqrt{2\pi t^3}} e^{-(\theta t + a^2/2t)} dt \\ &= e^{-a\sqrt{2\theta}} \int_0^\infty \frac{a}{\sqrt{2\pi t^3}} e^{-(\frac{a}{\sqrt{t}} - \sqrt{2\theta t})^2/2} dt.\end{aligned}$$

To calculate the integral in this expression, first make the substitution $t = a^2/\,(2\theta u)$ to obtain

$$\begin{aligned}I_1 &= \int_0^\infty \frac{a}{\sqrt{2\pi t^3}} e^{-(\frac{a}{\sqrt{t}} - \sqrt{2\theta t})^2/2} dt \\ &= \int_0^\infty \sqrt{\frac{2\theta}{2\pi u}} e^{-(\frac{a}{\sqrt{u}} - \sqrt{2\theta u})^2/2} du = I_2 = \frac{1}{2}\left(I_1 + I_2\right) \\ &= \frac{1}{2}\int_0^\infty \frac{1}{\sqrt{2\pi}}\left(\frac{a}{t^{3/2}} + \sqrt{\frac{2\theta}{t}}\right) e^{-(\frac{a}{\sqrt{t}} - \sqrt{2\theta t})^2/2} dt\end{aligned}$$

and then setting $v = (a/\sqrt{t}) - \sqrt{2\theta t}$, this

$$= \int_{-\infty}^\infty \frac{1}{\sqrt{2\pi}} e^{-v^2/2} dv = 1.$$

Use the symmetry of the Brownian motion about 0 to see that $\mathbb{E}\left(e^{-\theta T_a}\right) = e^{-|a|\sqrt{2\theta}}$ for $-\infty < a < \infty$ and $\theta \geqslant 0$. □

Example 4.3 *The ruin problem.* Now we will compute the probability that the standard Brownian motion exits from the interval $[a, b]$, $a \leqslant 0$, $b \geqslant 0$, for the first time through the right-hand end point b, say. This is known as the ruin problem because if the Brownian motion represents your fortune evolving while playing some game, and you start with fortune x and wish to compute the probability that your fortune reaches some level c before you go bankrupt, this would be the situation where $b = c - x$ and $a = -x$. We wish to compute $\mathbb{P}(T_b < T_a)$; by the spatial homogeneity of Brownian motion this is equivalent to the problem of computing

$$q(c, x) = \mathbb{P}(T_c < T_0 \mid W_0 = x), \quad 0 \leqslant x \leqslant c,$$

and then setting $x = -a$ and $c = b + x$. By symmetry $q(c, c/2) = 1/2$ and so for any integers $n \geqslant m \geqslant 1$, by considering which of $(m+1)c/n$ and $(m-1)c/n$ is hit first, we have

$$q(c, mc/n) = \tfrac{1}{2}q(c, (m-1)c/n) + \tfrac{1}{2}q(c, (m+1)c/n)$$

with $q(c, c) = 1$ and $q(c, 0) = 0$. Fix n and solve for $u_m = q(c, mc/n)$ from the second-order recurrence relation

$$u_m = \tfrac{1}{2}u_{m-1} + \tfrac{1}{2}u_{m+1}, \text{ for } 1 \leqslant m < n, \text{ with } u_n = 1, \, u_0 = 0. \tag{4.16}$$

The auxiliary equation of the recurrence relation is

$$x^2 - 2x + 1 = 0$$

which has 1 as a repeated root, so that the general solution of (4.16) is $u_m = A + Bm$; use the boundary conditions at $m = n$ and $m = 0$ to see that

$$u_m = q(c, mc/n) = m/n.$$

But $q(c, x)$ is monotonic in x, since for $0 \leqslant x < y \leqslant c$

$$q(c, x) = q(y, x)q(c, y) \leqslant q(c, y),$$

because, starting from x to hit the level c the process must first hit y and then starting from y it must hit c; we have shown that $q(c, x) = x/c$ when x/c is rational, hence $q(c, x) = x/c$, for $0 \leqslant x \leqslant c$. Thus we have

$$\mathbb{P}(T_b < T_a) = |a|/(|a| + b), \quad \text{for } a \leqslant 0 \leqslant b,$$

is the probability that the standard Brownian motion hits the level b before it hits the level a. □

Example 4.4 *Hitting a sloping line.* Finally in this section, let us consider the hitting time of the line $a + bt$ and set

$$T_{a,b} = \inf\{t \geqslant 0 : W_t = a + bt\}, \tag{4.17}$$

where again $T_{a,b} = \infty$ if no such time exists. We will use (4.15) to compute $\mathbb{E}\left(e^{-\theta T_{a,b}}\right)$ and $\mathbb{P}\left(T_{a,b} < \infty\right)$. Fix $\theta > 0$ and for $a > 0$, $b \geqslant 0$ set

$$\psi(a, b) = \mathbb{E}\left(e^{-\theta T_{a,b}}\right) = \mathbb{E}\left(e^{-\theta T_{a,b}} I_{(T_{a,b} < \infty)}\right).$$

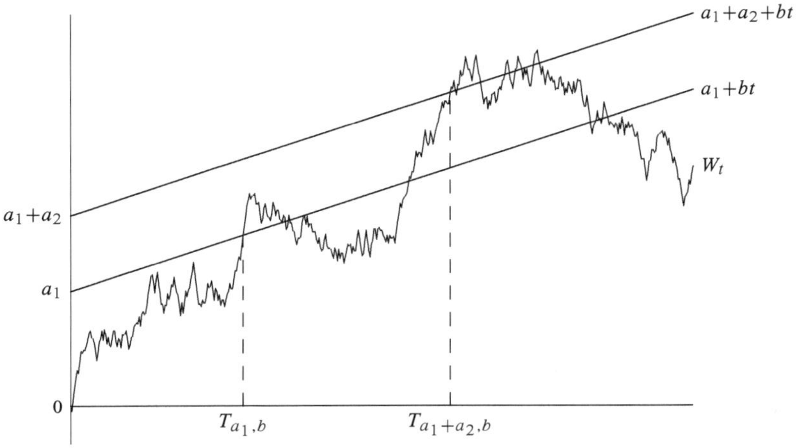

Figure 4.2: The decomposition (4.18) for the hitting time of $a + bt$

There are two steps in the argument given in the relations (4.18) and (4.19) below which are illustrated in Figure 4.2 and Figure 4.3, respectively. Firstly, for $a_1, a_2 > 0$

$$T_{a_1+a_2,b} = T_{a_1,b} + \left(T_{a_1+a_2,b} - T_{a_1,b}\right); \tag{4.18}$$

furthermore, we can see that $T_{a_1+a_2,b} - T_{a_1,b}$ is independent of $T_{a_1,b}$ and that it has the same distribution as $T_{a_2,b}$, so that

$$\psi(a_1 + a_2, b) = \psi(a_1, b)\psi(a_2, b)$$

which implies that

$$\psi(a, b) = e^{-\kappa(b)a}, \quad \text{for some function } \kappa(b).$$

Since $b \geqslant 0$ the process must hit the level a before it can hit the line $a + bt$ so, by conditioning on T_a, we have

$$\begin{aligned}
\psi(a, b) &= \int_0^\infty f_{T_a}(t)\mathbb{E}\left(e^{-\theta T_{a,b}} \mid T_a = t\right) dt \\
&= \int_0^\infty f_{T_a}(t)e^{-\theta t}\,\mathbb{E}\left(e^{-\theta T_{bt,b}}\right) dt \\
&= \int_0^\infty f_{T_a}(t)e^{-[\theta+\kappa(b)b]t} dt = \mathbb{E}\left(e^{-[\theta+\kappa(b)b]T_a}\right) \\
&= e^{-a\sqrt{2(\theta+\kappa(b)b)}}.
\end{aligned} \tag{4.19}$$

Equating the two expressions for $\psi(a, b)$ gives

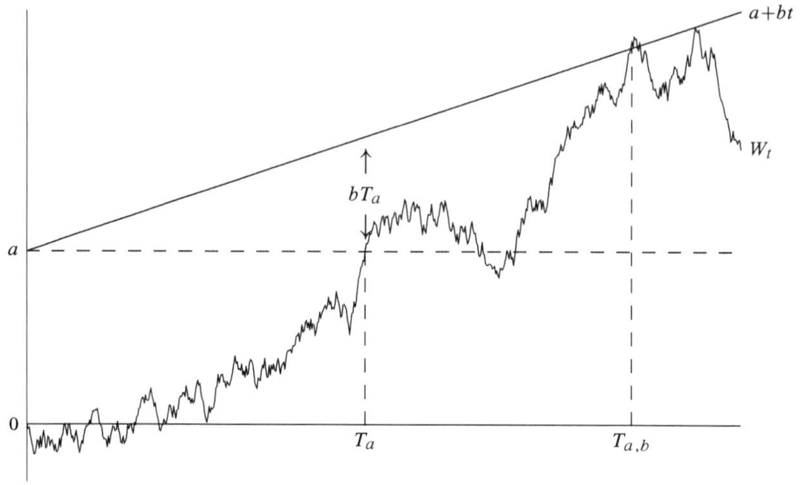

Figure 4.3: The calculation (4.19) for the hitting time of $a + bt$

$$\kappa^2(b) = 2\theta + 2\kappa(b)b \quad \text{or} \quad \kappa(b) = b \pm \sqrt{b^2 + 2\theta}.$$

Since for $\theta > 0$ we must have $\psi(a, b) \leqslant 1$ or $\kappa(b) \geqslant 0$ we may exclude the negative root and so

$$\mathbb{E}\left(e^{-\theta T_{a,b}}\right) = e^{-a\left(b+\sqrt{b^2+2\theta}\right)}. \tag{4.20}$$

Letting $\theta \downarrow 0$ in (4.20) gives for $a > 0$, $b \geqslant 0$,

$$\mathbb{P}\left(T_{a,b} < \infty\right) = e^{-2ab}. \tag{4.21}$$

In principle, we may obtain the distribution of $T_{a,b}$ by inverting (4.20), though later we will give a direct argument using (4.21). We may also see that the moment-generating function of $T_{a,b}$ is given by the expression in (4.20) for $b < 0$ and $a > 0$. In this case, since the line $a + bt$ must be hit before the level a, we may decompose the hitting time of a as $T_a = T_{a,b} + \left(T_a - T_{a,b}\right)$ where, conditional on $T_{a,b}$, the random variable $\left(T_a - T_{a,b}\right)$ has the same distribution as the time of hitting the level $-bT_{a,b}$. Denoting the probability density function of $T_{a,b}$ by $f_{T_{a,b}}$,

$$\begin{aligned} e^{-a\sqrt{2\nu}} = \mathbb{E}\left(e^{-\nu T_a}\right) &= \int_0^\infty \mathbb{E}\left(e^{-\nu T_a} \mid T_{a,b} = t\right) f_{T_{a,b}}(t)dt \\ &= \int_0^\infty e^{-\nu t}\, \mathbb{E}\left(e^{-\nu T_{bt}}\right) f_{T_{a,b}}(t)dt = \mathbb{E}\left[e^{-(\nu-\sqrt{2\nu}b)T_{a,b}}\right]. \end{aligned}$$

Put $\theta = \nu - \sqrt{2\nu}b$ which implies that $\sqrt{2\nu} = b + \sqrt{b^2 + 2\theta}$ and shows that (4.20) holds for all b, $-\infty < b < \infty$. Of course, for $b \leqslant 0$, $\mathbb{P}\left(T_{a,b} < \infty\right) = 1$ since in this case $T_{a,b} \leqslant T_a$ and $\mathbb{P}(T_a < \infty) = 1$. ∎

For a real constant μ, we refer to the process $W_t^{\mu} = W_t + \mu t$ as a Brownian motion with **drift** μ. In the notation here, $T_{a,b}$ is the first hitting time of the level a by a Brownian motion with drift $-b$. Another way of writing (4.21) is, for $a > 0$, $b > 0$,

$$\mathbb{P}\left(\sup_{0\leqslant t<\infty} (W_t - bt) \geqslant a\right) = e^{-2ab},$$

showing that in this case, $\sup_{0\leqslant t<\infty} (W_t - bt)$ has the exponential distribution with mean $1/(2b)$.

4.2.2 Transformations of Brownian motion

One useful tool when working with Brownian motion is the fact that it is invariant under certain transformations. When $\{W_t,\ t \geqslant 0\}$ is a standard Brownian motion then so are the following:

1. $\{cW_{t/c^2},\ t \geqslant 0\}$, for any real $c \neq 0$;
2. $\{tW_{1/t},\ t \geqslant 0\}$, where $tW_{1/t}$ is taken to be 0 when $t = 0$; and
3. $\{W_s - W_{s-t},\ 0 \leqslant t \leqslant s\}$, for any fixed $s \geqslant 0$.

The proofs of each of 1, 2 and 3 are similar. For example, in the case of 2 it is clear that $tW_{1/t}$ has continuous sample paths for $t > 0$; to see that this process is continuous at $t = 0$, observe from (4.21) that for $a > 0$ and $b > 0$,

$$\mathbb{P}\left(\limsup_{t\to\infty} (W_t/t) \leqslant b\right) \geqslant \mathbb{P}(W_t \leqslant a + bt,\ \text{for all } t \geqslant 0) = 1 - e^{-2ab}.$$

Now let $a \to \infty$ to see that $\mathbb{P}(\limsup_{t\to\infty} (W_t/t) \leqslant b) = 1$, for all $b > 0$. Let $b \downarrow 0$ to obtain $\mathbb{P}(\limsup_{t\to\infty} (W_t/t) \leqslant 0) = 1$; symmetry of the process $\{W_t\}$ about 0 gives $\mathbb{P}(\liminf_{t\to\infty} (W_t/t) \geqslant 0) = 1$ from which we see that

$$\mathbb{P}\left(\lim_{t\to\infty} (W_t/t) = 0\right) = 1, \quad \text{or equivalently} \quad \mathbb{P}\left(\lim_{t\downarrow 0} \left(tW_{1/t}\right) = 0\right) = 1.$$

Furthermore, for any choice of times $t_1, \ldots, t_n$, the collection of random variables $\{t_1 W_{1/t_1}, \ldots, t_n W_{1/t_n}\}$ has a joint normal distribution with zero means, so by the remarks following (4.3) the assertion is true provided the covariance takes the required form; but,

$$\begin{aligned} \mathcal{C}ov\left(sW_{1/s}, tW_{1/t}\right) &= \mathbb{E}\left(sW_{1/s}tW_{1/t}\right) = st\ \mathbb{E}\left(W_{1/s}W_{1/t}\right) \\ &= st\ \min(1/s, 1/t) = \min(s,t). \end{aligned}$$

The transformation 2 enables us to relate the properties of W_t as $t \to \infty$ to those of W_t as $t \to 0$, or vice versa.

Example 4.5 *Hitting a sloping line.* We determine the distribution of the time of hitting the line $a + bt$, that is, $\mathbb{P}\left(T_{a,b} \leqslant t\right)$ for $a > 0$, $-\infty < b < \infty$, which was considered in Example 4.4. Using 2 above, compute

$$\begin{aligned}\mathbb{P}\left(T_{a,b} \leqslant t\right) &= \mathbb{P}(W_s \geqslant a + bs, \quad \text{for some } s, 0 \leqslant s \leqslant t)\\ &= \mathbb{P}\left(sW_{1/s} \geqslant a + bs, \quad \text{for some } s, 0 < s \leqslant t\right)\\ &= \mathbb{P}(W_s \geqslant as + b, \quad \text{for some } s, 1/t \leqslant s < \infty).\end{aligned}$$

By conditioning on the value taken by $W_{1/t}$ this probability

$$\begin{aligned}&= \int_{-\infty}^{\infty} \mathbb{P}\left(W_s \geqslant as + b, \quad \text{for some } s, 1/t \leqslant s < \infty \mid W_{1/t} = x\right) \phi(x, 1/t)dx\\ &= \int_{-\infty}^{\frac{a}{t}+b} \mathbb{P}\left(T_{b-x+a/t,a} < \infty\right) \phi(x, 1/t)dx + 1 - \Phi\left(\frac{a+bt}{\sqrt{t}}\right).\end{aligned}$$

Now, use (4.21) to see that this gives

$$\begin{aligned}\mathbb{P}\left(T_{a,b} \leqslant t\right) &= \int_{-\infty}^{\frac{a}{t}+b} e^{-2a(a/t+b-x)}\phi(x, 1/t)dt + 1 - \Phi\left(\frac{a+bt}{\sqrt{t}}\right)\\ &= e^{-2ab}\int_{-\infty}^{\frac{a}{t}+b} \sqrt{\frac{t}{2\pi}}\, e^{-(tx-2a)^2/(2t)}dx + 1 - \Phi\left(\frac{a+bt}{\sqrt{t}}\right)\\ &= e^{-2ab}\Phi\left(\frac{bt-a}{\sqrt{t}}\right) + 1 - \Phi\left(\frac{a+bt}{\sqrt{t}}\right), \qquad (4.22)\end{aligned}$$

as the expression for the distribution function of the random variable $T_{a,b}$. □

4.2.3 Computations using martingales

Many of the results derived in Sections 4.2.1 and 4.2.2 are most easily established and extended using the theory of martingales. A stochastic process $\{X_t,\ t \geqslant 0\}$ is a **martingale** relative to the Brownian motion $\{W_t,\ t \geqslant 0\}$ or, more strictly, its filtration $\{\mathcal{F}_t,\ t \geqslant 0\}$ where $\mathcal{F}_t = \sigma\left(W_s, 0 \leqslant s \leqslant t\right)$, if for each $t \geqslant 0$,

(i) X_t is adapted to $\mathcal{F}_t$, so that it is an $\mathcal{F}_t$-random variable,

(ii) $\mathbb{E}\,|X_t| < \infty$, and

(iii) $\mathbb{E}\left(X_t \mid \mathcal{F}_s\right) = X_s$, for each s, $0 \leqslant s \leqslant t$.

The relation (iii) is the martingale property and it embodies the idea of a 'fair' game; when X_t denotes the fortune at time t of a player playing a fair game then his conditional expected fortune at time $t > s$, given the outcomes up to s is just his fortune at time s. In particular (iii) implies that

$$\mathbb{E}\,X_t = \mathbb{E}\,X_0, \quad \text{for each} \quad t \geqslant 0, \qquad (4.23)$$

so that a martingale has constant mean.

Example 4.6 The following processes are martingales:

1. W_t;
2. $W_t^2 - t$;
3. $e^{\lambda W_t - \lambda^2 t/2}$ for any real λ.

To verify the martingale property in each of these cases is straightforward. We will illustrate by considering 3, where $X_t = e^{\lambda W_t - \lambda^2 t/2}$. Recall that when Z is a $N(\mu, \sigma^2)$-random variable, so that it has the normal distribution with mean μ and variance σ^2, then $\mathbb{E}\left(e^{\theta Z}\right) = e^{\theta\mu + \theta^2\sigma^2/2}$ for all real θ. Since for $t > s$,

$$X_t = X_s e^{\lambda(W_t - W_s) - \lambda^2(t-s)/2},$$

and $W_t - W_s$ is independent of $\mathcal{F}_s$ and has the $N(0, t-s)$-distribution, we have

$$\mathbb{E}\left(X_t \mid \mathcal{F}_s\right) = X_s \mathbb{E}\left(e^{\lambda(W_t - W_s) - \lambda^2(t-s)/2}\right) = X_s.$$

A process of the form $e^{aW_t + bt}$ is known as **exponential Brownian motion** (or **geometric Brownian motion**). A similar argument to the above shows that such a process is a martingale if and only if $b = -a^2/2$. □

The property of martingales that we will use in this section is that for certain stopping times T, the relation (4.23) continues to hold if t is replaced by T. The following result, which we will not prove, is a particular case of the Optional Sampling Theorem (see Appendix A.2 for the statement of this result in discrete time).

Theorem 4.1 *Let $\{X_t,\ t \geqslant 0\}$ be a martingale with continuous sample paths and let T be a stopping time of $\{W_t,\ t \geqslant 0\}$ which is bounded, that is $\mathbb{P}(T \leqslant t) = 1$ for some $t \geqslant 0$. Then $\mathbb{E}\, X_T = \mathbb{E}\, X_0$.*

Remark To see that the conclusion of the Theorem may not be true if the requirement that the stopping time is bounded is lifted, consider the case when the martingale is $X_t = W_t$, a standard Brownian motion, $T = T_a$, the first hitting time of the level a, and then $\mathbb{P}(T_a < \infty) = 1$, with $\mathbb{E}\, W_{T_a} = a$ and $\mathbb{E}\, W_0 = 0$. □

Example 4.7 *The ruin problem.* To illustrate the use of this result let $T = \min(T_a, T_b)$, for $a < 0$, $b > 0$, be the stopping time which is the time of first exit from the interval $[a, b]$. Now T is not a bounded stopping time but for any $t \geqslant 0$, the stopping time $T \wedge t = \min(T, t)$ does have that property. Thus, using the fact that W_t is a martingale, Theorem 4.1 gives

$$\mathbb{E}\, W_{T\wedge t} = \mathbb{E}\, W_0 = 0, \quad \text{for each } t \geqslant 0.$$

But

$$\mathbb{E}\, W_{T\wedge t} = a\,\mathbb{P}(T = T_a \leqslant t) + b\,\mathbb{P}(T = T_b \leqslant t) + \mathbb{E}\,\left(W_t I_{(T>t)}\right)$$

where, as usual, I_A denotes the indicator of an event A. However,

$$\left|\mathbb{E}\,\left(W_t I_{(T>t)}\right)\right| \leqslant \mathbb{E}\,\left(|W_t|\, I_{(T>t)}\right) \leqslant \max(|a|, b)\,\mathbb{P}(T > t) \downarrow 0 \quad \text{as} \quad t \to \infty$$

giving

$$0 = a\,\mathbb{P}(T_a < T_b) + b\,\mathbb{P}(T_b < T_a)\,;$$

use the fact that

$$1 = \mathbb{P}(T_a < T_b) + \mathbb{P}(T_b < T_a)$$

to conclude that

$$\mathbb{P}(T_b < T_a) = |a|/(|a| + b),$$

as obtained in Example 4.3. □

Example 4.8 *The expected time to exit an interval.* Using the fact that $W_t^2 - t$ is a martingale we may see that a similar argument to that in the previous example when $T = T_a \wedge T_b$ gives

$$\begin{aligned} 0 &= \mathbb{E}\, W_{T\wedge t}^2 - \mathbb{E}\,(T \wedge t) \\ &= a^2\,\mathbb{P}(T = T_a \leqslant t) + b^2\,\mathbb{P}(T = T_b \leqslant t) + \mathbb{E}\,\left(W_t^2 I_{(T>t)}\right) - \mathbb{E}\,(T \wedge t). \end{aligned}$$

As before,

$$\mathbb{E}\,\left(W_t^2 I_{(T>t)}\right) \leqslant \max(a^2, b^2)\,\mathbb{P}(T > t) \downarrow 0$$

and since $\mathbb{E}\,(T \wedge t) \uparrow \mathbb{E}\,T$ as $t \to \infty$ we obtain

$$\begin{aligned} \mathbb{E}\,\min(T_a, T_b) &= a^2\,\mathbb{P}(T_a < T_b) + b^2\,\mathbb{P}(T_b < T_a) \\ &= |a| b \quad \text{for} \quad a < 0 < b, \end{aligned}$$

when we use the result from Example 4.7, giving an expression for the expected time that the standard Brownian motion takes to first exit from the interval $[a, b]$. □

Example 4.9 *Hitting a sloping line.* Finally, we give an alternative derivation of the formula (4.20) by using the martingale $X_t = e^{\lambda W_t - \lambda^2 t/2}$ with $\lambda = b + \sqrt{b^2 + 2\theta}$, $\theta > 0$. When $T_{a,b} < \infty$, $W_{T_{a,b}} = a + bT_{a,b}$, and we have

$$\lambda W_{T_{a,b}} - \frac{1}{2}\lambda^2 T_{a,b} = \lambda a + \frac{1}{2}\left(2\lambda b - \lambda^2\right) T_{a,b} = \lambda a - \theta T_{a,b},$$

which implies that

$$1 = \mathbb{E}X_0 = \mathbb{E}X_{T_{a,b}\wedge t} = e^{\lambda a}\mathbb{E}\left(e^{-\theta T_{a,b}} I_{(T_{a,b}\leqslant t)}\right) + \mathbb{E}\left(X_t I_{(T_{a,b}>t)}\right).$$

However, if $T_{a,b} > t$, then $W_t \leqslant a + bt$ and since $\lambda > 0$,

$$X_t \leqslant e^{\lambda W_t - \lambda^2 t/2} \leqslant e^{\lambda a + (b\lambda - \lambda^2/2)t} = e^{\lambda a - \theta t},$$

which tends to 0 as $t \to \infty$, since $\theta > 0$, and so $1 = e^{\lambda a}\mathbb{E}e^{-\theta T_{a,b}}$ giving (4.20). Notice that when we solve $b\lambda - \lambda^2/2 = -\theta$ we must take the positive root for λ to give the inequality for X_t on the set $T_{a,b} > t$. □

4.3 Girsanov's Theorem

Girsanov's Theorem is an important tool which enables calculations for Brownian motion with drift ν to be reduced to calculations for the case $\nu = 0$ and which also plays a central role in finance. We will illustrate the simplest case. Recall that $W_t^\nu = W_t + \nu t$ represents Brownian motion with drift ν. We use notation such as $\mathbb{P}(W_t \in dx)$ to represent $\mathbb{P}(W_t \in (x, x + dx))$ for small dx and so it is just the probability density function of W_t, evaluated at x, times dx; the multivariate analogue is defined in the obvious way. Take times $0 = t_0 < t_1 < \cdots < t_n = t$ and $x_0, \ldots, x_n$ with $x_0 = 0$ and $x_n = x$ and remember that the random variable $W_{t_i}^\nu - W_{t_{i-1}}^\nu$ has the normal distribution with mean $\nu\,(t_i - t_{i-1})$ and variance $t_i - t_{i-1}$. Using the independent increments property we have

$$\mathbb{P}\left(W_{t_i}^\nu \in dx_i, i = 1, \ldots, n\right) = \left[\prod_{i=1}^{n}\left(\frac{e^{-[x_i - x_{i-1} - \nu(t_i - t_{i-1})]^2/2(t_i - t_{i-1})}}{\sqrt{2\pi\,(t_i - t_{i-1})}}\right)\right] dx_1 \ldots dx_n;$$

since $\sum_{i=1}^{n}(x_i - x_{i-1}) = x_n = x$ and $\sum_{i=1}^{n}(t_i - t_{i-1}) = t_n = t$, we see that this expression

$$= \left(e^{\nu x - \nu^2 t/2}\right)\left[\prod_{i=1}^{n}\left(\frac{e^{-(x_i - x_{i-1})^2/2(t_i - t_{i-1})}}{\sqrt{2\pi\,(t_i - t_{i-1})}}\right)\right] dx_1 \ldots dx_n$$

$$= \left(e^{\nu x - \nu^2 t/2}\right)\mathbb{P}\left(W_{t_i} \in dx_i,\ i = 1, \ldots, n\right). \tag{4.24}$$

Set $L_t = e^{\nu W_t - \nu^2 t/2}$, then it follows from the previous section that L_t is a martingale and $\mathbb{E}L_t \equiv 1$. The relation (4.24) implies immediately that

$$\mathbb{E}f\left(W_{t_1}^\nu, \ldots, W_{t_n}^\nu\right) = \mathbb{E}\left[L_t f\left(W_{t_1}, \ldots, W_{t_n}\right)\right],$$

for any function $f : \mathbb{R}^n \to \mathbb{R}$ for which either, and hence both, expectation is defined. This shows that calculations involving Brownian motion with drift ν may be converted into calculations for Brownian motion with drift 0. Fix $t > 0$, and define the equivalent probability $\mathbb{Q}$ by setting $\mathbb{Q}(A) = \mathbb{E}(L_t I_A)$ for all events A, then since (4.24) holds for all choices of n, x_i and t_i it is evident that under the probability $\mathbb{Q}$ the process $\{W_s, 0 \leqslant s \leqslant t\}$ becomes a Brownian motion with drift ν. This observation is the idea underlying Girsanov's Theorem which is stated below without proof in a slightly more general form.

Theorem 4.2 (Girsanov) *Let $\{X_s, 0 \leqslant s \leqslant t\}$ be a Brownian motion with drift μ and variance parameter σ^2. Let*

$$L_t = \exp\left[\left(\frac{\nu-\mu}{\sigma^2}\right) X_t - \left(\frac{\nu^2-\mu^2}{2\sigma^2}\right) t\right],$$

and define the equivalent probability $\mathbb{Q}(A) = \mathbb{E}(L_t I_A)$, so that the Radon–Nikodym derivative is $d\mathbb{Q}/d\mathbb{P} = L_t$. Then, under the probability $\mathbb{Q}$, $\{X_s, 0 \leqslant s \leqslant t\}$ is a Brownian motion with drift ν and variance parameter σ^2.

Remark For a discussion and statement of Girsanov's' Theorem in a more general context and its relation to earlier work of Cameron and Martin see Karatzas and Shreve (1988), for example. To see the connection of Theorem 4.2 with the previous derivation, note that since X_t is a Brownian motion with drift μ and variance parameter σ^2 we may represent it as $X_t = \sigma W_t + \mu t$ where W_t is a standard Brownian motion under the original probability $\mathbb{P}$; then

$$\begin{aligned} \ln L_t &= \left(\frac{\nu-\mu}{\sigma^2}\right)(\sigma W_t + \mu t) - \left(\frac{\nu^2-\mu^2}{2\sigma^2}\right) t \\ &= \left(\frac{\nu-\mu}{\sigma}\right) W_t - \frac{1}{2}\left(\frac{\nu-\mu}{\sigma}\right)^2 t, \end{aligned}$$

which implies that $W_t = \overline{W}_t + ((\nu-\mu)/\sigma)\, t$ with $\overline{W}_t$ being a standard Brownian motion under $\mathbb{Q}$, and thus $X_t = \sigma \overline{W}_t + \nu t$ showing that X_t is a Brownian motion with drift ν and variance parameter σ^2 under $\mathbb{Q}$. □

Corollary 4.1 *Let $\{W_s^\nu, 0 \leqslant s \leqslant t\}$, $\{W_s^\mu, 0 \leqslant s \leqslant t\}$ be Brownian motions with drifts ν and μ respectively and each with variance parameter σ^2. Then for all choices of $0 \leqslant t_1 \leqslant \cdots \leqslant t_n \leqslant t$ and functions f,*

$$\mathbb{E} f\left(W_{t_1}^\nu, \ldots, W_{t_n}^\nu\right) = \mathbb{E}\left[L_t f\left(W_{t_1}^\mu, \ldots, W_{t_n}^\mu\right)\right],$$

where $L_t = \exp\left[\left(\frac{\nu-\mu}{\sigma^2}\right) W_t^\mu - \left(\frac{\nu^2-\mu^2}{2\sigma^2}\right) t\right]$.

To demonstrate the use of this idea we derive the joint distribution of Brownian motion with drift and its maximum, which also gives another derivation of the distribution of the time to hit a sloping line.

Example 4.10 *Joint distribution of Brownian motion with drift and its maximum.* Recall that T_a is the first hitting time of the level a by the Brownian motion W_t. For $a > 0$, $x \leqslant a$, choose A to be the event

$$A = \{W_t \leqslant x,\ T_a > t\} = \{W_t \leqslant x,\ M_t < a\}.$$

Let $M_t^\nu = \max_{0 \leqslant s \leqslant t} (W_s + \nu s)$ be the maximum of the Brownian motion with drift ν up to time t. Then it follows from the above argument that when $\mathbb{Q}$ is specified by $d\mathbb{Q}/d\mathbb{P} = \exp\{\nu W_t - \nu^2 t/2\}$, then

$$\mathbb{Q}(A) = \mathbb{P}\left(W_t^\nu \leqslant x,\ M_t^\nu < a\right) = \mathbb{E}\left(e^{\nu W_t - \nu^2 t/2} I_{(W_t \leqslant x,\, M_t < a)}\right).$$

There are at least a couple of ways to compute this last expectation; firstly, from Example 4.1, for $x \leqslant a$,

$$\begin{aligned}\mathbb{P}(W_t \leqslant x,\ T_a > t) &= \mathbb{P}(W_t \leqslant x) - \mathbb{P}(W_t \leqslant x,\ M_t \geqslant a)\\ &= \Phi\left(\frac{x}{\sqrt{t}}\right) - 1 + \Phi\left(\frac{2a - x}{\sqrt{t}}\right).\end{aligned}$$

By differentiating with respect to x, it follows that the probability density of W_t on the event $\{T_a > t\}$ evaluated at y is $\phi(y,t) - \phi(2a - y, t)$, whence

$$\begin{aligned}\mathbb{P}\left(W_t^\nu \leqslant x,\ M_t^\nu < a\right) &= \frac{1}{\sqrt{2\pi t}} \int_{-\infty}^{x} \left(e^{-y^2/2t} - e^{-(2a-y)^2/2t}\right) e^{\nu y - \nu^2 t/2} dy\\ &= \frac{1}{\sqrt{2\pi t}} \int_{-\infty}^{x} \left(e^{-(y-\nu t)^2/2t} - e^{2a\nu} e^{-(y-2a-\nu t)^2/2t}\right) dy\\ &= \Phi\left(\frac{x - \nu t}{\sqrt{t}}\right) - e^{2a\nu}\Phi\left(\frac{x - 2a - \nu t}{\sqrt{t}}\right); \end{aligned} \tag{4.25}$$

it will be useful in the next chapter to note that this expression equals

$$\mathbb{P}\left(W_t^\nu \leqslant x\right) - e^{2a\nu}\,\mathbb{P}\left(W_t^\nu \leqslant x - 2a\right).$$

Alternatively, we could have used the reflection principle directly and argued as in the calculation in Example 4.1,

$$\begin{aligned}\mathbb{E}\left[e^{\nu W_t} I_{(W_t \leqslant x,\, T_a > t)}\right] &= \mathbb{E}\left[e^{\nu W_t} I_{(W_t \leqslant x)}\right] - \mathbb{E}\left[e^{\nu W_t} I_{(W_t \leqslant x,\, M_t \geqslant a)}\right]\\ &= e^{\nu^2 t/2}\,\mathbb{P}(W_t + \nu t \leqslant x) - \mathbb{E}\left[e^{\nu(2a - W_t)} I_{(W_t \geqslant 2a - x)}\right]\\ &= e^{\nu^2 t/2}\left[\mathbb{P}(W_t + \nu t \leqslant x) - e^{2a\nu}\,\mathbb{P}(W_t - \nu t \geqslant 2a - x)\right]\\ &= e^{\nu^2 t/2}\left[\Phi\left(\frac{x - \nu t}{\sqrt{t}}\right) - e^{2a\nu}\Phi\left(\frac{x - 2a - \nu t}{\sqrt{t}}\right)\right],\end{aligned}$$

to give the same result; note that in the second and third equalities we have used Girsanov's Theorem again. We may also use (4.11) and (4.12) to give expressions for the moment-generating function of the maximum until time t of the Brownian motion with drift ν. We have

$$\begin{aligned}\mathbb{E}\left[e^{\theta M_t^\nu}\right] &= \mathbb{E}\left[e^{\theta \sup_{0\leqslant s\leqslant t}(W_s+\nu s)}\right] = \mathbb{E}\left[e^{\nu W_t-\nu^2 t/2}e^{\theta M_t}\right] \\ &= \frac{2(\theta+\nu)}{\theta+2\nu}e^{\theta(\theta+2\nu)t/2}\Phi\left((\theta+\nu)\sqrt{t}\right)+\frac{2\nu}{\theta+2\nu}\Phi\left(-\nu\sqrt{t}\right),\end{aligned} \tag{4.26}$$

in the case when $\theta+2\nu \neq 0$, while when $\theta = -2\nu$, this moment-generating function

$$= 2\left[\left(1+\nu^2 t\right)\Phi\left(-\nu\sqrt{t}\right)-\nu\sqrt{t}\phi\left(\nu\sqrt{t}\right)\right]. \tag{4.27}$$

If, as before, $T_{a,-\nu}$ denotes the first hitting time of the level a by the Brownian motion with drift ν, setting $x = a$ in (4.25) gives

$$\mathbb{P}(T_{a,-\nu} > t) = \mathbb{P}\left(M_t^\nu < a\right) = \Phi\left(\frac{a-\nu t}{\sqrt{t}}\right) - e^{2a\nu}\Phi\left(\frac{-a-\nu t}{\sqrt{t}}\right), \tag{4.28}$$

which agrees with (4.22) when we set $\nu = -b$. □

4.4 Brownian motion as a limit

The reason why Brownian motion occupies a central role in the study of continuous-time stochastic processes is the same as the reason why the normal distribution is of central importance in statistical modelling and that is because both arise naturally as limits of discrete processes. In the case of the normal distribution, it occurs as the limit, via the Central Limit Theorem, of scaled and normed sums of independent and identically distributed random variables. Brownian motion arises from a generalization of that result as the limit of random walks scaled in both time and space. In finance, to appreciate why the Black–Scholes model, which will be discussed in detail in the next chapter, is the natural continuous-time version of the binomial model it is important to understand this limit procedure. Example 4.11 explains how the Black–Scholes model may be approximated by the binomial model.

Let $\{X_i,\, i \geqslant 1\}$ be a sequence of independent, identically distributed random variables with $\mathbb{E}X_i = 0$ and $\mathbb{V}ar X_i = \sigma^2$, $0 < \sigma^2 < \infty$. For each $n \geqslant 1$ define the stochastic process $\{Z_t^n,\, t \geqslant 0\}$ by setting $Z_t^n = \sum_{i=1}^{[nt]} X_i/\sqrt{n}$, where $[nt]$ is the integer part of nt; that is, the largest integer not exceeding nt. When $t < 1/n$ interpret the sum as being 0. Notice that in defining Z_t^n the time scale in the random walk with increments $X_1, X_2, \ldots$ is being speeded up by a factor n while in space the scale is being contracted by the factor $\sqrt{n}$. Then for any choice of $0 \leqslant t_1 \leqslant \cdots \leqslant t_k$ and for any $x_1, \ldots, x_k$,

$$\lim_{n\to\infty} \mathbb{P}\left(Z_{t_i}^n \leqslant x_i,\ 1 \leqslant i \leqslant k\right) = \mathbb{P}\left(W_{t_i} \leqslant x_i,\ 1 \leqslant i \leqslant k\right), \tag{4.29}$$

where $\{W_t,\ t \geqslant 0\}$ is a Brownian motion with variance parameter σ^2. As stated here this result is of the same form as the classical Central Limit Theorem; a stronger mode of convergence (known as weak convergence leading to a Functional Central Limit Theorem) of the random-walk process to the Brownian motion holds which enables one to show that the distribution of an appropriate function of the process $\{Z_t^n,\ t \geqslant 0\}$ converges to the distribution of the corresponding function of $\{W_t,\ t \geqslant 0\}$, for a large class of functions.

In finance, Brownian motion as an approximation to the random walk, or vice versa, is required in a slightly more general form than presented above; this, in turn, is related to a generalization of the classical Central Limit Theorem for sequences of sequences of independent and identically distributed random variables, sometimes referred to as triangular arrays of random variables. Now suppose that for each $n \geqslant 1$, $\left\{X_i^n,\ i \geqslant 1\right\}$ is a sequence of independent, identically distributed random variables with $\mathbb{E} X_i^n = \mu_n$ and $\mathbb{V}ar X_i^n = \sigma_n^2$, with $\sigma_n^2 \to \sigma^2$, $0 < \sigma^2 < \infty$, as $n \to \infty$. Suppose further that the following technical condition holds,

$$\mathbb{E}\left(\left(X_i^n - \mu_n\right)^2 I_{\left(|X_i^n - \mu_n| > \epsilon\sqrt{n}\right)}\right) \to 0 \quad \text{as} \quad n \to \infty \quad \text{for any} \quad \epsilon > 0; \tag{4.30}$$

this is a special case of Lindeberg's condition (see Billingsley (1995)). It may then be shown that, as $n \to \infty$, the sequence of random variables $\sum_{i=1}^n (X_i^n - \mu_n)/\sqrt{n}$ converges in distribution to a random variable with the $N(0, \sigma^2)$ distribution; this conclusion would just be the standard Central Limit Theorem when the distributions of the random variables $\{X_i^n\}$ are the same for each $n \geqslant 1$. When we define the stochastic process $Z_t^n = \sum_{i=1}^{[nt]} (X_i^n - \mu_n)/\sqrt{n}$ for $t \geqslant 0$, with the sum being 0 when $t < 1/n$, then the conclusion of (4.29) holds in this context also. It is easy to see that a sufficient condition for (4.30) to hold is that $\mathbb{E}\,|X_i^n - \mu_n|^{2+\delta} \leqslant c < \infty$, for all i and n, for some $\delta > 0$ and constant c.

Suppose that $\mu_n \to 0$ in such a way that $\sqrt{n}\mu_n \to \mu$ and define the stochastic process $\overline{Z}_t^n = \sum_{i=1}^{[nt]} X_i^n/\sqrt{n}$ for $t \geqslant 0$, with the sum again being 0 when $t < 1/n$. Then it may be shown that for any choice of $0 \leqslant t_1 \leqslant \cdots \leqslant t_k$ and for any $x_1, \ldots, x_k$,

$$\lim_{n\to\infty} \mathbb{P}\left(\overline{Z}_{t_i}^n \leqslant x_i,\ 1 \leqslant i \leqslant k\right) = \mathbb{P}\left(W_{t_i}^\mu \leqslant x_i,\ 1 \leqslant i \leqslant k\right),$$

where $\left\{W_t^\mu,\ t \geqslant 0\right\}$ is a Brownian motion with variance parameter σ^2 and drift μ. It is easy to obtain a feel for why this is true since we may write

$$\overline{Z}_t^n = \frac{\sum_{i=1}^{[nt]} \left(X_i^n - \mu_n\right) + \mu_n[nt]}{\sqrt{n}} = Z_t^n + \mu_n[nt]/\sqrt{n},$$

and then $\mu_n[nt]/\sqrt{n} \to \mu t$ while the remainder of the process, Z_t^n, is approximating Brownian motion with variance parameter σ^2, by the previous observation.

Example 4.11 *Binomial model.* We will see that for appropriate choices of parameters in the binomial model, for a large number of time periods the stock-price process approximates an exponential Brownian motion; that is, a process of the form e^{aW_t+bt+c} for constants a, b and c. Consider a sequence of binomial models indexed by $n = 1, 2, \ldots$, where in the nth model the jumps in the stock-price process are determined by a sequence of independent, identically distributed random variables $\{Y_i^n, i \geqslant 1\}$ where $\mathbb{P}\left(Y_i^n = u_n\right) = p_n = 1 - \mathbb{P}\left(Y_i^n = d_n\right)$. The stock-price process in the nth model is $S_j^n = S_0^n \prod_{i=1}^{j} Y_i^n$. Consider the process $S_{[nt]}^n / S_0^n = \exp\left\{\sum_{i=1}^{[nt]} \ln\left(Y_i^n\right)\right\}$, with continuous parameter $t \geqslant 0$. For suitable choices of the parameters u_n, d_n and p_n the process $\sum_{i=1}^{[nt]} \ln\left(Y_i^n\right)$ will approximate, in the sense described above, a Brownian motion with drift μ and variance parameter σ^2; equivalently, $S_{[nt]}^n$ approximates an exponential Brownian motion initiated at S_0^n. Let $X_i^n = \sqrt{n}\ln\left(Y_i^n\right)$, then choose the parameters u_n, d_n and p_n so that

$$\begin{aligned} \sqrt{n}\mu_n &= \sqrt{n}\mathbb{E}\,X_i^n = n\left(p_n \ln\left(u_n/d_n\right) + \ln d_n\right) \to \mu \\ \text{and}\quad \sigma_n^2 &= \mathbb{V}ar X_i^n = np_n\left(1-p_n\right)\left(\ln\left(u_n/d_n\right)\right)^2 \to \sigma^2 \quad \text{as } n \to \infty. \end{aligned}$$

When we take $u_n = e^{g_n}$ and $d_n = e^{-g_n}$, say, then it may be verified that

$$g_n = \frac{\sigma}{\sqrt{n}} \quad \text{and} \quad p_n = \frac{1}{2}\left(1 + \frac{\mu}{\sigma\sqrt{n}}\right)$$

are appropriate choices. Furthermore, it is straightforward to check that the condition (4.30) holds here. □

4.5 Stochastic calculus

We now consider a very brief introduction to stochastic calculus which underlies the probabilistic approach to derivative pricing in a continuous-time setting. In the following, these ideas arise principally in Sections 5.3.1 and 5.3.2 in the discussion of hedging and self-financing strategies in the context of the Black–Scholes model; the first-time reader may wish to omit the current section at this stage and return to it before tackling those issues.

As has been remarked before, apart from being continuous, the paths of a standard Brownian motion $\{W_t, t \geqslant 0\}$ are highly irregular. It was pointed out that they are not differentiable; in addition, they are of unbounded variation on every finite time interval with probability 1. Recall that a function $f : [0, \infty) \to \mathbb{R}$ is of unbounded variation on the interval $[0, t]$ when

$$\sup_{\mathcal{P}_n} \sum_{i=1}^{n} |f(t_i) - f(t_{i-1})| = \infty,$$

where the supremum is taken over all finite partitions

$$\mathcal{P}_n = \{0 = t_0 < t_1 < \cdots < t_n = t\}$$

of the interval. However, it may be shown that, with probability 1,

$$\lim_{n\to\infty} \sum_{i=0}^{2^n-1} \left(W_{(i+1)t/2^n} - W_{it/2^n}\right)^2 = t \tag{4.31}$$

for any $t > 0$, so that the quadratic variation remains finite when we consider dyadic partitions of the interval $[0, t]$ with subintervals of the form $[(i + 1)t/2^n, it/2^n]$. We will not present a proof of (4.31), but we may observe that when we take the expected value of the random variable in the left-hand side, we have

$$\mathbb{E} \sum_{i=0}^{2^n-1} \left(W_{(i+1)t/2^n} - W_{it/2^n}\right)^2 = \sum_{i=0}^{2^n-1} \left(\frac{(i+1)t}{2^n} - \frac{it}{2^n}\right) = t$$

for all n.

As a consequence of the unbounded variation property, it is not possible to talk about forming an integral $X_t = \int_0^t Z_s dW_s$ in any conventional deterministic sense. One of the important achievements of modern probability has been the development of a stochastic calculus which enables us to define and work with a stochastic process X_t, known as the stochastic integral of a suitable stochastic process $\{Z_t, t \geqslant 0\}$ with respect to the Brownian motion, which has many of the properties that are associated with an ordinary integral and a few more besides.

To begin with, why might we want to study such an integral? To answer this question, suppose that a gambler plays a game where at each time t_i, $i = 0, \ldots, n-1$ of a given deterministic partition $\mathcal{P}_n = \{0 = t_0 < t_1 < \cdots < t_n = t\}$, he may stake the amount Z_{t_i} and at time t_{i+1} be repaid the amount

$$Z_{t_i}\left(W_{t_{i+1}} - W_{t_i}\right)$$

which is proportional to the movement of the Brownian motion in the time interval $[t_i, t_{i+1}]$. If we require that this gambling is non-anticipative then the amount Z_{t_i} would have to be an $\mathcal{F}_{t_i}$-random variable. Remember that $\mathcal{F}_s = \sigma(W_u, 0 \leqslant u \leqslant s)$ is the history of the Brownian motion up to time s. Here, for the moment, assume that $\mathbb{E} Z_{t_i}^2 < \infty$ for each t_i. The total return through using the 'strategy' Z would be the amount

$$X_t(\mathcal{P}_n) = \sum_{i=0}^{n-1} Z_{t_i}\left(W_{t_{i+1}} - W_{t_i}\right).$$

Because $W_{t_{i+1}} - W_{t_i}$ is independent of $\mathcal{F}_{t_i}$, and hence of Z_{t_i}, and has mean 0, it follows that

$$\mathbb{E}\left(X_t\left(\mathcal{P}_n\right)\right) = 0; \tag{4.32}$$

this is because

$$\begin{aligned}\mathbb{E}\left(Z_{t_i}\left(W_{t_{i+1}} - W_{t_i}\right) \mid \mathcal{F}_{t_i}\right) &= Z_{t_i}\mathbb{E}\left(W_{t_{i+1}} - W_{t_i} \mid \mathcal{F}_{t_i}\right) \\ &= Z_{t_i}\mathbb{E}\left(W_{t_{i+1}} - W_{t_i}\right) = 0,\end{aligned} \tag{4.33}$$

which implies that $\mathbb{E}\left(Z_{t_i}\left(W_{t_{i+1}} - W_{t_i}\right)\right) = 0$. Furthermore, if we consider $t_i < t_j$ say, then it is also the case that

$$\mathbb{E}\left(Z_{t_i}\left(W_{t_{i+1}} - W_{t_i}\right) Z_{t_j}\left(W_{t_{j+1}} - W_{t_j}\right)\right) = 0.$$

This is true for a similar reason, since

$$\begin{aligned}&\mathbb{E}\left(Z_{t_i}\left(W_{t_{i+1}} - W_{t_i}\right) Z_{t_j}\left(W_{t_{j+1}} - W_{t_j}\right) \middle| \mathcal{F}_{t_j}\right) \\ &\qquad = Z_{t_i}\left(W_{t_{i+1}} - W_{t_i}\right)\mathbb{E}\left(Z_{t_j}\left(W_{t_{j+1}} - W_{t_j}\right) \middle| \mathcal{F}_{t_j}\right) = 0,\end{aligned}$$

by (4.33). It follows that

$$\begin{aligned}\mathbb{E}\left(X_t\left(\mathcal{P}_n\right)\right)^2 &= \mathbb{E}\left(\sum_{i=0}^{n-1} Z_{t_i}^2\left(W_{t_{i+1}} - W_{t_i}\right)^2\right) \\ &= \mathbb{E}\left(\sum_{i=0}^{n-1} Z_{t_i}^2\mathbb{E}\left(\left(W_{t_{i+1}} - W_{t_i}\right)^2 \mid \mathcal{F}_{t_i}\right)\right) \\ &= \mathbb{E}\left(\sum_{i=0}^{n-1} Z_{t_i}^2\left(t_{i+1} - t_i\right)\right),\end{aligned}$$

using the independence of $W_{t_{i+1}} - W_{t_i}$ and $\mathcal{F}_{t_i}$ and the fact that the variance of $W_{t_{i+1}} - W_{t_i}$ is $t_{i+1} - t_i$.

Now for a stochastic process satisfying

$$\{Z_s, s \geqslant 0\} \quad \text{is adapted, and} \quad \mathbb{E}\left(\int_0^t Z_s^2 ds\right) < \infty, \tag{4.34}$$

write $Z_s\left(\mathcal{P}_n\right)$ for the stochastic process which is constant on the intervals $[t_i, t_{i+1})$ of $\mathcal{P}_n$ and agrees with Z_s on the left end points, so that $Z_s\left(\mathcal{P}_n\right) = Z_{t_i}$ if $s \in [t_i, t_{i+1})$. We may express the variance of $X_t\left(\mathcal{P}_n\right)$ as

$$\mathbb{E}\left(X_t\left(\mathcal{P}_n\right)\right)^2 = \mathbb{E}\left(\int_0^t Z_s^2\left(\mathcal{P}_n\right) ds\right). \tag{4.35}$$

It may be shown that if we take increasingly finer partitions $\mathcal{P}_n$ as $n \to \infty$ in such a way that

$$\mathbb{E} \int_0^t |Z_s - Z_s(\mathcal{P}_n)|^2 \, ds \to 0,$$

then the random variables $X_t(\mathcal{P}_n)$ converge to a limit random variable, denoted X_t, in the sense that $\mathbb{E} |X_t - X_t(\mathcal{P}_n)|^2 \to 0$. Moreover, this limit may be taken so that X_t is continuous in t. The limit stochastic process $\{X_t, t \geqslant 0\}$ is what is known as the stochastic integral of Z_s with respect to W_s and is written as $X_t = \int\limits_0^t Z_s dW_s$. In fact, in the construction something much stronger is true in that

$$\mathbb{P}\left(\sup_{0 \leqslant s \leqslant t} |X_s - X_s(\mathcal{P}_n)| \to 0 \text{ as } n \to \infty \right) = 1.$$

Properties (4.32) and (4.35) are carried over in the limit so that

$$\mathbb{E} X_t = 0 \quad \text{and} \quad \mathbb{E} X_t^2 = \mathbb{E}\left(\int_0^t Z_s^2 ds \right). \tag{4.36}$$

In gambling terms, the stochastic integral X_t would represent the total return by time t from allowing a gambling strategy which may be adjusted continuously so that the amount wagered is being altered instantaneously at each time point. A shorthand differential notation is used, write $dX_t = Z_t dW_t$ to mean that

$$X_t = X_0 + \int_0^t Z_s dW_s,$$

where the stochastic integral is defined as in the construction given above. It may be seen that for any stochastic process satisfying (4.34) the corresponding stochastic integral $\{X_t, t \geqslant 0\}$ is a martingale, which is a strengthened form of the first statement in (4.36).

Example 4.12 *Deterministic integrand.* In the special case where $Z_t = z(t)$ is just a deterministic function of t, then we may compute the distribution of $\int\limits_0^t z(s) dW_s$ explicitly, it has the $N\left(0, \int\limits_0^t z^2(s)ds\right)$-distribution. To see this, note that for the approximating partition $\mathcal{P}_n$, the integral

$$\sum_{i=0}^{n-1} z(t_i)\left(W_{t_{i+1}} - W_{t_i}\right)$$

is a linear combination of independent random variables $W_{t_{i+1}} - W_{t_i}$ having the $N(0, t_{i+1} - t_i)$-distribution and so it has the normal distribution with mean 0 and variance $\sum\limits_{i=0}^{n-1} (z(t_i))^2 (t_{i+1} - t_i)$. Taking the limit as $n \to \infty$ gives the conclusion.

Furthermore, in this case the stochastic integral $X_t = \int_0^t z(s)dW_s$ is a process with independent increments. □

While the construction of the stochastic integral is a non-trivial exercise, once the existence and basic properties of the stochastic integral have been established then working with it is not much more difficult than with an ordinary integral. Stochastic integrals and ordinary integrals of stochastic processes of the form $\int_0^t Y_s ds$, for any adapted stochastic process $\{Y_s, s \geqslant 0\}$ for which the integral is defined, may be mixed to give in differential notation a stochastic differential equation of the form

$$dX_t = Y_t dt + Z_t dW_t;$$

this is shorthand for the statement that $\{X_t, t \geqslant 0\}$ is the process defined by

$$X_t = X_0 + \int_0^t Y_s ds + \int_0^t Z_s dW_s, \quad \text{for} \quad t \geqslant 0,$$

where X_0 is the initial position of the process. When working with stochastic integrals the principal difference from ordinary calculus is in the treatment of quantities of the order of $(dW_t)^2$. In deterministic calculus, for a differentiable function $f(t)$, the differential $df(t)$ is an infinitesimal quantity of the same order of magnitude as dt. When W_t is a standard Brownian motion, the differential dW_t must be regarded as a stochastic infinitesimal quantity of order $\sqrt{dt}$, but $(dW_t)^2$ may be worked with as if it is the *deterministic* quantity dt. This is a consequence of the result in (4.31), but to get a heuristic idea of why this is the case, recall that an increment of Brownian motion $\Delta W_t = W_{t+\Delta t} - W_t$ is a random variable having the normal distribution with mean 0 and variance Δt; then the increment may be represented as $\Delta W_t = U\sqrt{\Delta t}$, where U has the standard normal distribution with mean 0 and variance 1. By Chebychev's inequality

$$\mathbb{P}\left(\left|(\Delta W_t)^2 - \Delta t\right| > \epsilon\right) = \mathbb{P}\left(\Delta t\left|U^2 - 1\right| > \epsilon\right) \leqslant (\Delta t)^2\, \mathbb{E}\left(U^2 - 1\right)^2 / \epsilon^2,$$

so that for any ϵ which is of larger order than Δt the right-hand side tends to 0 as $\Delta t \to 0$. The most important result in stochastic calculus for our purposes is the following which makes this idea precise and it is known as *Itô's Lemma*.

Theorem 4.3 *Suppose that $\{X_t, t \geqslant 0\}$ is a stochastic process that may be represented as $dX_t = Y_t dt + Z_t dW_t$ and that $f(x,t)$ is a function with continuous second partial derivatives. The stochastic process $f(X_t, t)$ may be represented as*

$$df(X_t, t) = \left(Y_t \frac{\partial f}{\partial x} + \frac{\partial f}{\partial t} + \frac{1}{2} Z_t^2 \frac{\partial^2 f}{\partial x^2}\right) dt + Z_t \frac{\partial f}{\partial x} dW_t,$$

where the partial derivatives are evaluated at (X_t, t).

To appreciate the difference from the deterministic case, when x_t, y_t, z_t and w_t are deterministic and linked by $dx_t = y_t dt + z_t dw_t$, then

$$df(x_t, t) = \frac{\partial f}{\partial x} dx_t + \frac{\partial f}{\partial t} dt = \left(y_t \frac{\partial f}{\partial x} + \frac{\partial f}{\partial t} \right) dt + z_t \frac{\partial f}{\partial x} dw_t.$$

In the stochastic case we pick up an extra term $\frac{1}{2} Z_t^2 \left(\partial^2 f / \partial x^2 \right)$ on the right-hand side. While we will not give a proof of Itô's Lemma here, to get an idea from whence it comes, use Taylor's Theorem to argue that

$$\begin{aligned} \Delta f(X_t, t) &= f(X_{t+\Delta t}, t + \Delta t) - f(X_t, t) \\ &= \frac{\partial f}{\partial x} \Delta X_t + \frac{\partial f}{\partial t} \Delta t \\ &\quad + \frac{1}{2} \frac{\partial^2 f}{\partial x^2} (\Delta X_t)^2 + \frac{\partial^2 f}{\partial x \partial t} \Delta X_t \Delta t + \frac{1}{2} \frac{\partial^2 f}{\partial t^2} (\Delta t)^2 + \cdots . \end{aligned}$$

But $\Delta X_t = Y_t \Delta t + Z_t \Delta W_t$ and by the explanation above $(\Delta X_t)^2 = Z_t^2 \Delta t + o(\Delta t)$, with terms like $(\Delta X_t)(\Delta t)$ being $o(\Delta t)$, so that they are of smaller order than Δt. If only the terms of order no smaller than Δt are retained, this becomes

$$\Delta f(X_t, t) = \frac{\partial f}{\partial x} \Delta X_t + \left(\frac{\partial f}{\partial t} + \frac{1}{2} Z_t^2 \frac{\partial^2 f}{\partial x^2} \right) \Delta t,$$

which gives the correct expression when we substitute for ΔX_t. A formal proof of Itô's Lemma requires use of Taylor's Theorem along the lines of the above and the definition of the stochastic integral as a limit over approximating Z_t $(\mathcal{P}_n)$.

Example 4.13 *Stochastic differential equation for exponential Brownian motion.* Consider the exponential Brownian motion $S_t = S_0 \exp\{\sigma W_t + \mu t\}$, where W_t is standard Brownian motion and S_0 is a constant. Apply Itô's Lemma with $X_t = W_t$, so that $Y_t \equiv 0$ and $Z_t \equiv 1$, and with $f(x, t) = S_0 \exp\{\sigma x + \mu t\}$ to obtain

$$\begin{aligned} dS_t = df(W_t, t) &= \left(\frac{\partial f}{\partial t} + \frac{1}{2} \frac{\partial^2 f}{\partial x^2} \right) dt + \frac{\partial f}{\partial x} dW_t \\ &= S_t \left[\left(\mu + \frac{1}{2} \sigma^2 \right) dt + \sigma dW_t \right]. \end{aligned}$$

Replace μ by $\mu - \sigma^2/2$ and this shows that the solution of the stochastic differential equation of the form

$$dS_t = S_t \left(\mu dt + \sigma dW_t \right),$$

is given by $S_t = S_0 \exp\left\{\sigma W_t + \left(\mu - \sigma^2/2\right) t\right\}$. □

Example 4.14 *Simple integration by parts.* Consider the case where the process X_t satisfies $dX_t = Y_t dt + Z_t dW_t$ and $g(t)$ is a (deterministic) function of t. We consider computing the stochastic differential of the product $d\left(X_t g(t)\right)$. In Itô's Lemma take $f(x,t) = xg(t)$, so that

$$\frac{\partial f}{\partial x} = g(t), \quad \frac{\partial f}{\partial t} = xg'(t), \quad \text{and} \quad \frac{\partial^2 f}{\partial x^2} = 0.$$

It follows that

$$\begin{aligned} d\left(X_t g(t)\right) &= \left(Y_t g(t) + X_t g'(t)\right) dt + Z_t g(t) dW_t \\ &= X_t g'(t) dt + g(t) dX_t; \end{aligned}$$

in integral form this is

$$X_t g(t) = X_0 g(0) + \int_0^t X_s g'(s) ds + \int_0^t g(s) dX_s,$$

which is the usual integration-by-parts formula. Be careful, when $g(t)$ is a stochastic process, in general we would have an extra term in the integration-by-parts formula (see Exercise 4.8) . □

Example 4.15 *Ornstein–Uhlenbeck process.* The process defined by the stochastic differential equation

$$dX_t = -\alpha X_t dt + \sigma dW_t \quad \text{where} \quad \alpha > 0, \tag{4.37}$$

is known as an Ornstein–Uhlenbeck process and its behaviour is such that it is pulled towards 0, or *mean-reverting*, by a drift proportional to its distance from 0. Think of multiplying the equation (4.37) by $e^{\alpha t}$, or more properly taking $g(t) = e^{\alpha t}$ in the previous example, to see that $d\left(X_t e^{\alpha t}\right) = \sigma e^{\alpha t} dW_t$, which gives

$$X_t = e^{-\alpha t} X_0 + \sigma e^{-\alpha t} \int_0^t e^{\alpha s} dW_s.$$

For the slightly more general case of the process mean-reverting to a level β, which is the Ornstein–Uhlenbeck process satisfying

$$dX_t = -\alpha \left(X_t - \beta\right) dt + \sigma dW_t,$$

using the above calculation, it follows that

$$X_t = \left(1 - e^{-\alpha t}\right) \beta + e^{-\alpha t} X_0 + \sigma e^{-\alpha t} \int_0^t e^{\alpha s} dW_s.$$

By Example 4.12, it is clear that when X_0 is constant, X_t has the normal distribution with mean $\beta\left(1 - e^{-\alpha t}\right) + X_0 e^{-\alpha t}$ and variance

$$\sigma^2 e^{-2\alpha t} \int_0^t e^{2\alpha s} ds = \frac{\sigma^2}{2\alpha}\left(1 - e^{-2\alpha t}\right).$$

Notice that, as $t \to \infty$, then the distribution of X_t tends to the $N\left(\beta, \sigma^2/2\alpha\right)$-distribution. When X_0 has the $N\left(\beta, \sigma^2/2\alpha\right)$-distribution and is independent of the Brownian motion driving the stochastic differential equation then the process $\{X_t,\, t \geqslant 0\}$ is stationary in that X_t has the same distribution, $N\left(\beta, \sigma^2/2\alpha\right)$, for all times t. □

4.6 Exercises

Exercise 4.1 For $a > 0$, let M denote the *last* time that a standard Brownian motion hits the line $at,\, t \geqslant 0$. Show that M has the same distribution as $1/T_a$ and verify that $\mathbb{E}\, M = 1/a^2$.

Exercise 4.2 For a standard Brownian motion let $T = \min(T_a, T_b)$ be the time of first exit from the interval $[a, b], a < 0 < b$, where T_a is the time of first passage to the level a. Show that for $\theta \geqslant 0$,

$$\mathbb{E}\left[e^{-\theta T}\right] = \frac{\cosh\left((a+b)\sqrt{\theta/2}\right)}{\cosh\left((b-a)\sqrt{\theta/2}\right)}.$$

[**Hint:** Either, use the fact that

$$e^{-\theta T_b} = e^{-\theta T} I_{(T_b < T_a)} + e^{-\theta(T_b - T_a + T)} I_{(T_a < T_b)}$$

and observe that $T_b - T_a$ has the same distribution as T_{b-a} conditional on the event $(T_a < T_b)$ or, apply the Optional Sampling Theorem using the martingale $\left\{e^{\lambda W_t - \lambda^2/t}\right\}$, for appropriate choices of λ.]

Exercise 4.3 By repeated use of the reflection principle show that for a standard Brownian motion

$$\mathbb{P}\left(\max_{0 \leqslant s \leqslant t} |W_s| \leqslant x\right) = \mathbb{P}\left(|W_t| \leqslant x\right) + \sum_{r=1}^{\infty} (-1)^r \,\mathbb{P}\left((2r-1)x \leqslant |W_t| \leqslant (2r+1)x\right)$$

for $x > 0$.

Exercise 4.4 For $a > 0$ and $b < a$, let $T_b^a = \inf\{t \geqslant T_a : W_t = b\}$ denote the first time that the standard Brownian motion $\{W_t\}$ hits the level b after it first hits a.

For $t > 0$ and real ν, show that when $x \geqslant b$,

$$\mathbb{E}\left[e^{\nu W_t} I_{(T_b^a \leqslant t,\, W_t \geqslant x)}\right] = e^{-2\nu(a-b)+\nu^2 t/2} \Phi\left(\frac{2(b-a) - x + \nu t}{\sqrt{t}}\right),$$

while for $x \leqslant b$,

$$\mathbb{E}\left[e^{\nu W_t} I_{(T_b^a \leqslant t,\, W_t \leqslant x)}\right] = e^{2a\nu+\nu^2 t/2} \Phi\left(\frac{x - 2a - \nu t}{\sqrt{t}}\right).$$

Use these results and Girsanov's Theorem to derive the distribution of the Brownian motion $\{W_s^\nu\}$ with drift ν at time t on the event where it has hit the level $a > 0$ and then the level b $(< a)$ before time t.

Exercise 4.5 Use the fact that

$$\mathbb{P}\left(\sup_{0 \leqslant s \leqslant t} (W_s + \mu s) \leqslant a\right) = \Phi\left(\frac{a - \mu t}{\sqrt{t}}\right) - e^{2a\mu} \Phi\left(\frac{-a - \mu t}{\sqrt{t}}\right), \quad \text{for } a > 0,$$

to derive an expression for $\mathbb{E}\left[e^{\theta W_t} I_{(T_{a,b} > t)}\right]$, for any real θ, where $T_{a,b}$ is the hitting time of the line $a + bs$. Hence, or otherwise, determine an expression for $\mathbb{E}\left[e^{\theta T_{a,b}} I_{(T_{a,b} \leqslant t)}\right]$, for $2\theta \leqslant b^2$.

Exercise 4.6 Suppose that $\{N_t,\ t \geqslant 0\}$ is a **Poisson process** (defined on some underlying probability space with probability $\mathbb{P}$), with rate $\lambda > 0$; that is, for all choice of times, $0 = t_0 \leqslant t_1 \leqslant \cdots \leqslant t_k$, the random variables $\left\{N_{t_i} - N_{t_{i-1}},\ 1 \leqslant i \leqslant k\right\}$ are independent (so that the process has independent increments) with $N_{t_i} - N_{t_{i-1}}$ having the Poisson distribution with mean $\lambda\,(t_i - t_{i-1})$ for each i and, by convention, $N_0 = 0$. For $\theta > 0$ define $\mathbb{Q}$ by setting the Radon–Nikodym derivative

$$\frac{d\mathbb{Q}}{d\mathbb{P}} = \theta^{N_t} e^{\lambda t(1-\theta)}, \quad \text{for fixed } t > 0.$$

Verify that $\mathbb{Q}$ is a probability and that $\{N_s,\ 0 \leqslant s \leqslant t\}$ is a Poisson process under $\mathbb{Q}$ with rate $\theta\lambda$.

Exercise 4.7 Let $\{W_t,\ t \geqslant 0\}$ be a standard Brownian motion and let $\{N_t,\ t \geqslant 0\}$ be an independent Poisson process of rate λ. For $a > 0$, $-\infty < b < \infty$ and fixed $c \geqslant 0$, let $\tau_{a,b} = \inf\{t \geqslant 0 : W_t = a + bt + cN_t\}$, where the infimum of the empty set is taken to be $+\infty$, as usual. For $\theta > 0$, show that the moment-generating function of $\tau_{a,b}$ is given by $\mathbb{E}\left[e^{-\theta\tau_{a,b}}\right] = e^{-ax}$, where $x = x(b, c, \lambda, \theta)$ is the unique positive root of

$$x = b + \sqrt{b^2 + 2\theta + 2\lambda\left(1 - e^{-cx}\right)}.$$

Consider a model of a stock price subjected to shocks which occur at times governed by a Poisson process, rate λ; each shock produces a proportional reduction $e^{-\nu}$, $\nu > 0$, in the price and otherwise the price evolves according to an independent exponential Brownian motion. At time t, the price may be represented as $S_t = S_0 e^{\sigma W_t + \mu t - \nu N_t}$. Determine the moment-generating function of the first time the price reaches the level $h > S_0$.

Exercise 4.8 Suppose that $\{X_t,\ t \geqslant 0\}$ is a stochastic process that may be represented as $dX_t = Y_t dt + Z_t dW_t$. For (suitably nice) functions $f(x,t)$ and $g(x,t)$ use Itô's Lemma to establish the stochastic integration-by-parts formula

$$d\,(fg) = f dg + g df + Z_t^2 \frac{\partial f}{\partial x}\frac{\partial g}{\partial x}\, dt$$

where f, g and the partial derivatives are evaluated at (X_t, t).

Exercise 4.9 For a standard Brownian motion $\{W_t,\ t \geqslant 0\}$, evaluate the stochastic integral $\int_0^t W_s dW_s$.

Chapter 5

THE BLACK–SCHOLES MODEL

5.1 Introduction

We now consider in detail the Black–Scholes model which is the continuous-time analogue of the binomial model. The model consists of an economy in which there are just two assets, a bank account paying a fixed continuously compounded interest rate ρ per unit time and the second asset for which the price is a stochastic process; we will refer to the latter asset as a stock but it may be any other asset which is traded freely, such as a foreign currency. One unit in the bank at time 0 grows to $e^{\rho t}$ by time t and we will assume that $\rho \geqslant 0$, although from the mathematical viewpoint this is not a requirement for all that follows. The bank account ensures that there is positive riskless borrowing so that at time s a bond paying off one unit at time t may be bought at a positive price $e^{-\rho(t-s)}$; this implies that the discount factor is $e^{-\rho t}$ at time t. The price of the stock at time t is S_t where S_t is determined by the stochastic differential equation

$$dS_t = S_t \left(\mu dt + \sigma dW_t\right),$$

with $\{W_t, t \geqslant 0\}$ being a standard Brownian motion and $\sigma > 0$, μ are constants. In the context of finance the parameter σ is known as the **volatility** of the stock. Recall from Example 4.13 of Chapter 4 that it follows that the stock-price process $\{S_t, t \geqslant 0\}$ is an exponential Brownian motion and it may be represented as

$$S_t = S_0 e^{(\mu - \sigma^2/2)t + \sigma W_t},$$

where S_0 is the initial price of the stock, which it may be assumed is observed at time 0. The information available at time t is the history of the price process, $\mathcal{F}_t = \sigma\left(S_u, 0 \leqslant u \leqslant t\right)$, that is the information obtained by observing the movements of the stock price process up to time t; equivalently, it is $\sigma\left(W_u, 0 \leqslant u \leqslant t\right)$, the information obtained by observing the driving Brownian motion in the stochastic differential equation.

We consider the model over a finite time interval $[0, t_0]$. Initially we will proceed informally and develop the ideas by analogy with the discrete-time situation. We will seek an equivalent martingale measure (or probability) $\mathcal{Q}$; that is, an equivalent probability $\mathcal{Q}$ under which the discounted stock-price process $\{e^{-\rho t} S_t, 0 \leqslant t \leqslant t_0\}$ is a martingale. By Girsanov's Theorem we know how a Brownian motion with one drift may be transformed by a change of probability into a Brownian motion with another drift. When we define the equivalent probability $\mathbb{Q}$ by setting the Radon–Nikodym derivative

$$\frac{d\mathbb{Q}}{d\mathbb{P}} = \exp\left(\frac{\nu}{\sigma} W_{t_0} - \frac{\nu^2}{2\sigma^2} t_0\right),$$

then $\{\sigma W_t, 0 \leqslant t \leqslant t_0\}$, which is a Brownian motion with zero drift and variance parameter σ^2 (under the original probability $\mathbb{P}$), becomes a Brownian motion with drift ν and variance parameter σ^2 under $\mathbb{Q}$. So for calculations with the new probability $\mathbb{Q}$, σW_t may be thought of as being replaced by $\sigma \overline{W}_t + \nu t$ where $\overline{W}_t$ is a standard Brownian motion under $\mathbb{Q}$. It follows that, under $\mathbb{Q}$, the discounted stock-price process $e^{-\rho t} S_t$ may be regarded as being

$$S_0 \exp\left(\left(\nu + \mu - \rho - \frac{\sigma^2}{2}\right) t + \sigma \overline{W}_t\right),$$

which by the observation in Example 4.6 on page 95 is a martingale under $\mathbb{Q}$ if and only if $\nu = \rho - \mu$. This shows that calculations involving the discounted stock-price process when $\mathbb{Q}$ is an equivalent martingale probability are the same as calculations under the original probability after setting $\mu = \rho$. It may be noted that the choice $\nu = \rho - \mu$ gives the unique equivalent martingale measure in this case, although we will not make use of this in the following; the uniqueness is a consequence of the discussion of Proposition 1.6 in Chapter VIII of Revuz and Yor (2004).

5.2 The Black–Scholes formula

5.2.1 Derivation

The classical Black–Scholes formula establishes the price of a **European call option** on the stock at strike price c expiring at time t_0; this is a contract entitling (but not requiring) the holder to buy one unit of the stock at the fixed price c, the **strike price**, at the fixed time t_0, the **expiry time**. In our framework it is a contingent claim C, with the payoff $C = \left(S_{t_0} - c\right)_+$ at time t_0. Remember that it should be distinguished from an **American call option** which entitles (but does not require) the holder to buy one unit of stock at the fixed strike price, c, *at or before* the expiry time t_0; so in the case of the American option the time at which the option is exercised, if ever, is under the control of the holder and may be at any instant up to the expiry time.

As the European call option can be hedged, and a formal description of what that means in the continuous-time context will be developed in Section 5.3, then, by analogy with the binomial model, its unique price at time 0 should be

$$\mathbb{E}_{\mathbb{Q}}\left(e^{-\rho t_0} C\right) = \mathbb{E}_{\mathbb{Q}}\left(e^{-\rho t_0}\left(S_{t_0} - c\right)_+\right), \tag{5.1}$$

where $\mathbb{Q}$ is the equivalent martingale measure. That is, its price is the expected value under the equivalent martingale measure of the discounted payoff. Here, we take $\mathbb{Q}$ to be the measure described in the previous section and so we compute the right-hand side of (5.1) under the assumption that we are taking expectations with the original probability when we have set $\mu = \rho$. First note the elementary calculation contained in the following lemma.

Lemma 5.1 *Suppose that Z is a random variable having the $N(\beta, \tau^2)$-distribution and that a and c are positive constants. Then*

$$\begin{aligned} &\mathbb{E}\left(ae^Z - c\right)_+ \\ &\quad = ae^{(\beta+\tau^2/2)}\Phi\left(\frac{\ln(a/c)+\beta}{\tau}+\tau\right) - c\Phi\left(\frac{\ln(a/c)+\beta}{\tau}\right). \end{aligned} \tag{5.2}$$

Proof. Recall (from (A.15)) that when a random variable Z has the $N(\beta, \tau^2)$-distribution,

$$\mathbb{E}\left[e^Z f(Z)\right] = e^{\beta+\tau^2/2}\mathbb{E}\, f\left(Z+\tau^2\right)$$

for any function f for which the expectations are defined. It follows that the left-hand side of (5.2) is

$$\begin{aligned} \mathbb{E}\left(ae^Z - c\right)_+ &= a\mathbb{E}\left(e^Z I_{(Z>\ln(c/a))}\right) - c\mathbb{P}(Z > \ln(c/a)) \\ &= ae^{\beta+\tau^2/2}\mathbb{P}\left(Z > \ln(c/a) - \tau^2\right) - c\mathbb{P}(Z > \ln(c/a)); \end{aligned}$$

recall the scaling property of the normal distribution which implies that the random variable $(Z-\beta)/\tau$ has the standard normal distribution, $N(0,1)$, then we see that this expression is

$$ae^{(\beta+\tau^2/2)}\left(1-\Phi\left(\frac{\ln(c/a)-\beta}{\tau}-\tau\right)\right) - c\left(1-\Phi\left(\frac{\ln(c/a)-\beta}{\tau}\right)\right).$$

Use the fact that the standard normal distribution function satisfies $\Phi(-x) = 1-\Phi(x)$, since the distribution is symmetric about 0, to obtain the right-hand side of (5.2). □

We are now in a position to give the Black–Scholes formula. It should be stressed that a proof of the result will be given later; for the moment the development of the formula presented here is just by analogy with the discrete-time case since we have yet to describe the notion of hedging in the continuous-time framework.

Theorem 5.1 (Black–Scholes) *The time-0 price of a European call option at strike price c with expiry time t_0 is*

$$S_0\Phi\left(\frac{\ln(S_0/c)+(\rho+\sigma^2/2)t_0}{\sigma\sqrt{t_0}}\right) - ce^{-\rho t_0}\Phi\left(\frac{\ln(S_0/c)+(\rho-\sigma^2/2)t_0}{\sigma\sqrt{t_0}}\right). \tag{5.3}$$

To see why (5.3) accords with (5.1) use the fact that, under the equivalent martingale probability $\mathbb{Q}$, the random variable S_{t_0} has the same distribution as the random variable $S_0 e^{(\rho-\sigma^2/2)t_0+\sigma W_{t_0}}$ has under the original probability, so that S_{t_0} may be expressed as $S_{t_0} = S_0 e^Z$ where Z has the $N((\rho-\sigma^2/2)t_0, \sigma^2 t_0)$-distribution. Then calculate from (5.2) with $\beta = (\rho-\sigma^2/2)t_0$, $\tau = \sigma\sqrt{t_0}$ and $a = S_0$ and multiply by the factor $e^{-\rho t_0}$. There is of course nothing special about the time 0 here so that we may derive the price of the option at time t with essentially the same computation, after replacing the expectation in (5.1) by the conditional expectation given $\mathcal{F}_t$.

Corollary 5.1 *For $0 \leqslant t < t_0$, the time-t price of a European call option at strike price c with expiry time t_0 is*

$$S_t \Phi\left(\frac{\ln(S_t/c) + (\rho+\sigma^2/2)(t_0-t)}{\sigma\sqrt{t_0-t}}\right) \\ - ce^{-\rho(t_0-t)}\Phi\left(\frac{\ln(S_t/c) + (\rho-\sigma^2/2)(t_0-t)}{\sigma\sqrt{t_0-t}}\right).$$

The expression given in Corollary 5.1 will be referred to as the **Black–Scholes formula**. The first thing to observe about this formula is the absence of the drift term, μ, of the stock-price process. The option price at time t depends on the stock-price process only through its current value S_t and through the volatility σ. Given the basic model, two investors may have different ideas about the value of the drift μ but they will agree on the same price for the option. The other parameters appearing in the formula are the time to expiry $t_0 - t$, the strike price c and the interest rate ρ, which is given and assumed to be constant.

5.2.2 Dependence on the parameters: the Greeks

In order to look further at the formula we introduce the following notation. Set

$$d_1(x,t) = \frac{\ln(x/c) + (\rho+\sigma^2/2)(t_0-t)}{\sigma\sqrt{t_0-t}}; \tag{5.4}$$

$$d_2(x,t) = d_1(x,t) - \sigma\sqrt{t_0-t}; \tag{5.5}$$

and define

$$p(x,t) = x\Phi\left(d_1(x,t)\right) - ce^{-\rho(t_0-t)}\Phi\left(d_2(x,t)\right). \tag{5.6}$$

The time-t price of the option given in Corollary 5.1 is $p(S_t,t)$. As observed above, the price formula also depends implicitly on ρ, σ, c and t_0, so it may be more proper to write $p = p(x,t,\rho,\sigma,c,t_0)$, with similar expressions for d_1 and d_2, but normally we will suppress the dependence on these other parameters in the notation.

Notice that the expression given in Corollary 5.1 reduces to $\left(S_{t_0} - c\right)_+$, the value of the option at time t_0, when $S_t \to S_{t_0}$ as $t \uparrow t_0$; this is easy to see at least when

$S_{t_0} \neq c$ using the fact that $\Phi(x) \to 0$ or 1 according as $x \to -\infty$ or $x \to \infty$. The argument is slightly more delicate when $S_{t_0} = c$, but in this case it follows from the fact, established below, that $p(x,t)$ is an increasing function of x and then, for $\epsilon > 0$,

$$\begin{aligned} 0 = \lim_{t \uparrow t_0} p(S_t - \epsilon, t) &\leqslant \liminf_{t \uparrow t_0} p(S_t, t) \\ &\leqslant \limsup_{t \uparrow t_0} p(S_t, t) \leqslant \lim_{t \uparrow t_0} p(S_t + \epsilon, t) = \epsilon; \end{aligned}$$

letting $\epsilon \downarrow 0$ completes the argument.

We see immediately from the formula that, instantaneously at t, an investor would be indifferent between holding the option or holding a portfolio which consists of $\Phi\left(d_1(S_t,t)\right)$ units of stock and which is short $c\Phi\left(d_2(S_t,t)\right)$ in bonds; that is, that amount in bonds is borrowed from the bank. The investor is indifferent between the two holdings at the instant t since the values of the two are the same. This portfolio will be referred to as the **hedging** (or **replicating**) portfolio of the option. How this notion of hedging relates to that in the discrete-time case will be expanded on later. It should be noted that in order to match the option price at each time the hedging portfolio is adjusted continuously as the stock price S_t and t change.

Let $\phi(d) = \Phi'(d) = e^{-d^2/2}/\sqrt{2\pi}$ be the probability density function of the standard normal distribution, and notice that

$$\begin{aligned} d_2^2 &= d_1^2 - 2\left[\ln(x/c) + \left(\rho + \sigma^2/2\right)(t_0 - t)\right] + \sigma^2 (t_0 - t) \\ &= d_1^2 - 2\ln(x/c) - 2\rho\,(t_0 - t), \end{aligned}$$

from which it follows that

$$\begin{aligned} x\phi(d_1) - ce^{-\rho(t_0 - t)}\phi(d_2) &= \frac{1}{\sqrt{2\pi}}\left[xe^{-d_1^2/2} - ce^{-\rho(t_0-t)-d_2^2/2}\right] \\ &= \frac{e^{-d_1^2/2}}{\sqrt{2\pi}}\left[x - ce^{\ln(x/c)}\right] \equiv 0. \end{aligned} \tag{5.7}$$

Since $\partial d_1/\partial x = \partial d_2/\partial x$, from (5.7) we see immediately that

$$\frac{\partial p}{\partial x} = \Phi(d_1) + \left[x\phi(d_1) - ce^{-\rho(t_0-t)}\phi(d_2)\right]\frac{\partial d_1}{\partial x} = \Phi(d_1) > 0.$$

This shows that the price of the option is strictly increasing as a function of the price of the stock, as would be expected. The quantity

$$D_t = \left.\frac{\partial p}{\partial x}\right|_{(S_t,t)} = \Phi\left(d_1\left(S_t,t\right)\right),$$

is known as the **Delta** of the option at time t. The Delta represents the amount of the stock held in the hedging portfolio and of course it determines the sensitivity of the price of the option to changes in the price of the underlying asset, the stock. Derivatives of the option price with respect to various parameters of the formula are known as the 'Greeks' for the rather prosaic reason that Greek letters have traditionally been used to denote these quantities. Now evaluate the second derivative of p with respect to x,

$$\frac{\partial^2 p}{\partial x^2} = \phi(d_1) \frac{\partial d_1}{\partial x} = \frac{\phi(d_1)}{\sigma x \sqrt{t_0 - t}} > 0. \tag{5.8}$$

This implies that, when the other parameters are held fixed, the Black–Scholes formula is a strictly convex function of the stock price. The quantity

$$\Gamma_t = \left. \frac{\partial^2 p}{\partial x^2} \right|_{(S_t, t)} = \frac{\phi(d_1(S_t, t))}{\sigma S_t \sqrt{t_0 - t}},$$

is known as the **Gamma** of the option at time t. The Gamma determines the sensitivity of the Delta to changes in the stock price; that is, the sensitivity of the stock holding in the hedging portfolio to such changes. The larger the Gamma the more sensitive the stock holding in the hedging portfolio is to movements in the stock price. Next consider

$$\frac{\partial p}{\partial t} = x\phi(d_1) \frac{\partial d_1}{\partial t} - ce^{-\rho(t_0 - t)} \left[\rho \Phi(d_2) + \phi(d_2) \frac{\partial d_2}{\partial t} \right].$$

Observe that

$$\frac{\partial d_2}{\partial t} = \frac{\partial d_1}{\partial t} + \frac{\sigma}{2\sqrt{t_0 - t}}, \tag{5.9}$$

so that using (5.7) we see that

$$\frac{\partial p}{\partial t} = -ce^{-\rho(t_0 - t)} \left[\rho \Phi(d_2) + \frac{\sigma \phi(d_2)}{2\sqrt{t_0 - t}} \right] < 0; \tag{5.10}$$

that is, the price of the option decreases with time, all other variables being fixed. The quantity

$$\Theta_t = \left. \frac{\partial p}{\partial t} \right|_{(S_t, t)} = -ce^{-\rho(t_0 - t)} \left[\rho \Phi(d_2(S_t, t)) + \frac{\sigma \phi(d_2(S_t, t))}{2\sqrt{t_0 - t}} \right]$$

is the **Theta** of the option at time t.

Now consider the dependence of the Black–Scholes formula on the two parameters ρ and c, the interest rate and the strike price; because of its importance the dependence on the volatility σ will be treated at some length in the next section. In the case of the interest rate ρ, a similar calculation to those above yields

$$\frac{\partial p}{\partial \rho} = c(t_0 - t) e^{-\rho(t_0 - t)} \Phi(d_2) > 0,$$

demonstrating that the option price increases as the interest rate increases. This last statement should be treated with some caution and the reader should be reminded that it holds when the other arguments in the Black–Scholes formula are being held fixed, in particular the stock price x. In practice, the stock price itself will tend to decrease when the interest rate increases, since the stock price may be regarded as the expectation under a martingale measure of the sum of discounted future dividends, and this would make the option price decrease, which would counteract the fact that $\partial p/\partial \rho > 0$.

Finally, $\partial d_1/\partial c = \partial d_2/\partial c$, and again (5.7) shows that

$$\begin{aligned}\frac{\partial p}{\partial c} &= x\phi\left(d_1\right)\frac{\partial d_1}{\partial c} - e^{-\rho(t_0-t)}\Phi\left(d_2\right) - ce^{-\rho(t_0-t)}\phi\left(d_2\right)\frac{\partial d_2}{\partial c}\\ &= -e^{-\rho(t_0-t)}\Phi\left(d_2\right) < 0,\end{aligned}$$

which quantifies the intuitively obvious fact that the price decreases as the strike price of the call increases; the weaker fact that the price is non-increasing in the strike price c comes from the observation that

$$\left(S_{t_0} - c_1\right)_+ \leqslant \left(S_{t_0} - c_2\right)_+, \quad \text{when} \quad c_1 > c_2,$$

and hence $\mathbb{E}_{\mathbb{Q}}\left[e^{-\rho(t_0-t)}\left(S_{t_0} - c_1\right)_+ \mid \mathcal{F}_t\right] \leqslant \mathbb{E}_{\mathbb{Q}}\left[e^{-\rho(t_0-t)}\left(S_{t_0} - c_2\right)_+ \mid \mathcal{F}_t\right]$. After differentiating p again, we see that the option price p is a strictly convex function of the strike price c, since

$$\frac{\partial^2 p}{\partial c^2} = -e^{-\rho(t_0-t)}\phi\left(d_2\right)\frac{\partial d_2}{\partial c} = \frac{e^{-\rho(t_0-t)}\phi\left(d_2\right)}{\sigma c\sqrt{t_0-t}} > 0.$$

In the same way as above, the weaker observation that p is convex in c may be seen from the fact that $(x)_+$ is a convex function of x and hence the time-t price of the option, $\mathbb{E}_{\mathbb{Q}}\left[e^{-\rho(t_0-t)}\left(S_{t_0} - c\right)_+ \mid \mathcal{F}_t\right]$, is convex in c. To see this, for $0 < \lambda < 1$ and values c_1 and c_2 of the stock price, we have

$$\begin{aligned}\left(S_{t_0} - \lambda c_1 - (1-\lambda)c_2\right)_+ &= \left(\lambda\left(S_{t_0} - c_1\right) + (1-\lambda)\left(S_{t_0} - c_2\right)\right)_+\\ &\leqslant \lambda\left(S_{t_0} - c_1\right)_+ + (1-\lambda)\left(S_{t_0} - c_2\right)_+;\end{aligned} \tag{5.11}$$

now taking conditional expectations through (5.11) it follows that

$$\begin{aligned}&\mathbb{E}_{\mathbb{Q}}\left[\left(S_{t_0} - \lambda c_1 - (1-\lambda)c_2\right)_+ \mid \mathcal{F}_t\right]\\ &\qquad\leqslant \lambda\mathbb{E}_{\mathbb{Q}}\left[\left(S_{t_0} - c_1\right)_+ \mid \mathcal{F}_t\right] + (1-\lambda)\mathbb{E}_{\mathbb{Q}}\left[\left(S_{t_0} - c_2\right)_+ \mid \mathcal{F}_t\right]\end{aligned}$$

and multiplying both sides by the constant $e^{-\rho(t_0-t)}$ gives the required inequality.

5.2.3 Volatility

The volatility σ is the only parameter related directly to the evolution of the stock price which enters into the Black–Scholes formula; as a consequence it is of considerable importance since, apart from the actual stock price S_t itself, the only influence the stock-price process has on the option price is through σ. First observe that

$$\frac{\partial p}{\partial \sigma} = x\phi\left(d_1\right)\frac{\partial d_1}{\partial \sigma} - ce^{-\rho(t_0-t)}\phi\left(d_2\right)\frac{\partial d_2}{\partial \sigma};$$

but

$$\frac{\partial d_1}{\partial \sigma} = \frac{\partial d_2}{\partial \sigma} + \sqrt{t_0 - t} = -\frac{d_2}{\sigma}, \tag{5.12}$$

so that recalling (5.7) again, we find that

$$\frac{\partial p}{\partial \sigma} = ce^{-\rho(t_0-t)}\phi\left(d_2\right)\sqrt{t_0 - t} = x\phi\left(d_1\right)\sqrt{t_0 - t} > 0, \tag{5.13}$$

showing that the price of the option increases as the volatility increases. The quantity

$$\Lambda_t = \left.\frac{\partial p}{\partial \sigma}\right|_{(S_t,t)} = S_t\phi\left(d_1\left(S_t, t\right)\right)\sqrt{t_0 - t}$$

is the **Vega** of the option at time t (also known as the **Lambda** of the option).

The volatility may be estimated by standard statistical techniques from historical data, but more often the traded price of a particular option is used to establish an 'implied' volatility. Suppose that at time $t \leqslant t_0$ the market price of a European call option at strike price c_0 and expiry date t_0 is C_t while the stock price is $S_t = x$ then, keeping x, t, ρ, c_0, and t_0 fixed, the equation

$$p(x, t, \rho, \sigma, c_0, t_0) = C_t \tag{5.14}$$

may be solved to obtain the **implied volatility**, $\sigma = \overline{\sigma}$, as illustrated in Figure 5.2. This resulting value of σ may then be used to value options on the same stock at other strike prices and with other expiry dates in a consistent way.

Note that as $\sigma \downarrow 0$,

$$p(x, t, \rho, \sigma, c_0, t_0) \downarrow \left(x - c_0 e^{-\rho(t_0-t)}\right)_+,$$

while when $\sigma \uparrow \infty$ we have $d_1 \to \infty$ and $d_2 \to -\infty$ so that

$$p(x, t, \rho, \sigma, c_0, t_0) \uparrow x.$$

We note that for a market in which there is no arbitrage it must always be the case that the price, C_t, of the European call option will lie in the range

$$\left(S_t - ce^{-\rho(t_0-t)}\right)_+ \leqslant C_t \leqslant S_t, \tag{5.15}$$

and consequently there will always be a solution $\overline{\sigma}$ for the implied volatility from the equation (5.14). To see that the absence of arbitrage implies (5.15), suppose that $C_t > S_t$, then an arbitrage may be formed by constructing a portfolio at time t at zero net cost by buying one unit of stock, selling one call and putting the difference $C_t - S_t$ in the bank account; at time t_0 this portfolio will be worth

$$S_{t_0} - C_{t_0} + e^{\rho(t_0-t)}\left(C_t - S_t\right) = S_{t_0} - \left(S_{t_0} - c\right)_+ + e^{\rho(t_0-t)}\left(C_t - S_t\right) > 0,$$

so it is an arbitrage. We may argue similarly if we had $C_t < S_t - c^{-\rho(t_0-t)}$, then we would sell the stock and buy the call to form an arbitrage.

The option is often said to be **at the money** at time t when the price $S_t = c$; it is **in the money** when $S_t > c$, so that the American call option would pay out a positive amount if exercised at that instant, and it is **out of the money** when $S_t < c$, so that the American call would pay nothing if exercised. A useful extension of this terminology is to say that the option is **at the money forward** (or **at the discounted money**) at time t when the price $S_t = ce^{-\rho(t_0-t)}$; it is **in the money forward** when $S_t > ce^{-\rho(t_0-t)}$ and it is **out of the money forward** when $S_t < ce^{-\rho(t_0-t)}$.

Often it is observed that, when there are a range of options on the same stock traded in the market, the Black–Scholes formula tends to overprice at-the-money-forward options, that is those for which x is close to $ce^{-\rho(t_0-t)}$, while it under-prices deep in-the-money-forward (when x is much larger than $ce^{-\rho(t_0-t)}$) and deep out-of-the-money-forward options (x much smaller than $ce^{-\rho(t_0-t)}$). This phenomenon is often referred to by saying that there are volatility **'smiles'** in that the implied volatility for strike prices c close to the value $xe^{\rho(t_0-t)}$ are lower than those for c distant from this value; this leads to the implied volatility appearing as a unimodal function of the strike price c. This is illustrated in Figure 5.1 for options at different strike prices $c_1, c_2, \ldots$ and the same expiry date t_0. If the prices in the market did agree exactly with those derived from a Black–Scholes model then the implied volatilities would all be the same and equal to the value of the volatility in the model.

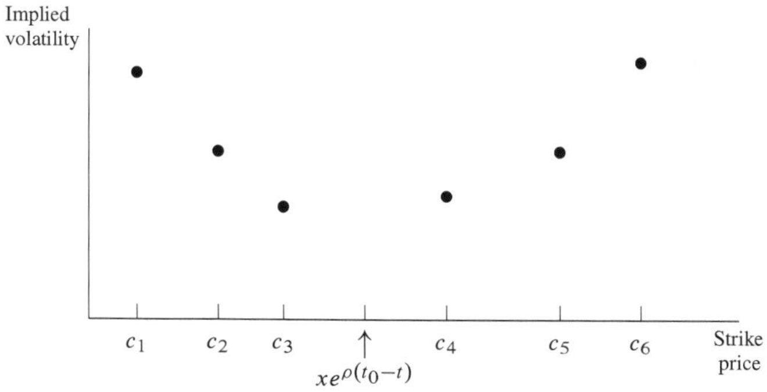

Figure 5.1: A volatility 'smile'

An explanation of why this departure from the model is observed may be found in a closer examination of the dependence of the Black–Scholes formula on the volatility σ. Since $\phi'(d_1) = -d_1\phi(d_1)$, from (5.12) and (5.13) we see that

$$\frac{\partial^2 p}{\partial \sigma^2} = x\phi'(d_1)\frac{\partial d_1}{\partial \sigma}\sqrt{t_0 - t} = xd_1 d_2 \phi(d_1)\sqrt{t_0 - t}/\sigma. \tag{5.16}$$

It may be seen that $d_1 d_2 \geqslant 0$ if and only if $\sigma \leqslant \sigma_0$, where

$$\sigma_0 = \sqrt{\frac{2}{t_0 - t}\left|\ln(x/c) + \rho(t_0 - t)\right|}.$$

This follows by first observing that $d_1 d_2 = \left((d_1+d_2)^2 - (d_1-d_2)^2\right)/4$,

$$d_1 + d_2 = 2\left(\frac{\ln(x/c) + \rho\,(t_0 - t)}{\sigma\sqrt{t_0 - t}}\right) \quad \text{and} \quad d_1 - d_2 = \sigma\sqrt{t_0 - t};$$

then it is clear that

$$d_1 d_2 = \frac{(\ln(x/c) + \rho\,(t_0 - t))^2}{\sigma^2\,(t_0 - t)} - \frac{\sigma^2\,(t_0 - t)}{4} \geqslant 0$$

is equivalent to $\sigma \leqslant \sigma_0$.

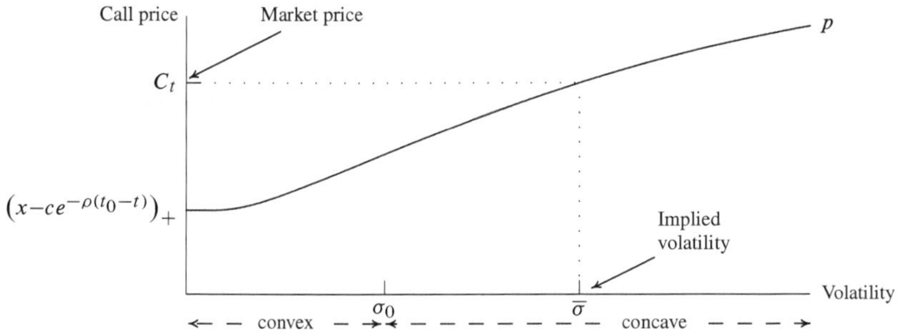

Figure 5.2: Black–Scholes price as a function of the volatility

It is immediate from (5.16) that p is a convex function of σ for σ in the range $0 \leqslant \sigma \leqslant \sigma_0$ while it is concave in σ for $\sigma \geqslant \sigma_0$. Notice that $\sigma_0 = 0$ if and only if $x = ce^{-\rho(t_0-t)}$ while σ_0 is large when the ratio $xe^{\rho(t_0-t)}/c$ is large or small. As a function of σ, essentially p is concave for the whole range of σ for an option which is close to being at the money forward while essentially it is convex for all σ for one which is deep in the money forward or deep out of the money forward.

If now, instead of the volatility being a constant, it is a random variable Σ independent of the Brownian motion driving the stock price then, suppressing in the notation the dependence of p on the other variables, we might expect the true option price to be $\mathbb{E}\,p(\Sigma)$ where we are taking expectations over the values of Σ. We would obtain from Jensen's inequality

$$p(\mathbb{E}\,\Sigma) \geqslant \mathbb{E}\,p(\Sigma) \quad \text{or} \quad p(\mathbb{E}\,\Sigma) \leqslant \mathbb{E}\,p(\Sigma),$$

according as σ_0 is small or σ_0 is large; this demonstrates the tendency of the Black–Scholes formula, which uses a constant value of σ, to overprice or under-price in the respective cases.

5.3 Hedging and the Black–Scholes equation

5.3.1 Self-financing portfolios

Suppose that we have a contingent claim paying C at time t_0; that is, C is an $\mathcal{F}_{t_0}$-random variable. For the European call option we have been studying we would have $C = \left(S_{t_0} - c\right)_+$. By analogy with Section 2.2.2 of Chapter 2, consider a portfolio which at time t holds an amount X_t of stock and Y_t in the bank account, so the value of the portfolio is

$$C_t = X_t S_t + Y_t. \tag{5.17}$$

We will suppose that this portfolio may be adjusted continuously and it is done so non-anticipatively so that $\{X_t, t \geqslant 0\}$ and $\{Y_t, t \geqslant 0\}$ are adapted; that is, for each time t, X_t and Y_t are $\mathcal{F}_t$-random variables, which means that the amounts held in the stock and the bond at time t depend only on information about the stock price up to and including time t. Recall that $e^{-\rho(t_0-t)}$ represents the price at time t of a riskless bond paying 1 with certainty at time t_0; in order to consider the change in the value of the bank account as time varies, write $Y_t = B_t e^{-\rho(t_0-t)}$, so that B_t corresponds to the holding at time t in bonds maturing at time t_0. We may then replace (5.17) by

$$C_t = X_t S_t + B_t e^{-\rho(t_0-t)}. \tag{5.18}$$

Say that the portfolio (X_t, B_t) is **self-financing** when C_t may be represented as

$$dC_t = X_t dS_t + B_t d\left(e^{-\rho(t_0-t)}\right) = X_t dS_t + \rho B_t e^{-\rho(t_0-t)} dt; \tag{5.19}$$

that is,

$$C_t = C_0 + \int_0^t X_u dS_u + \int_0^t \rho B_u e^{-\rho(t_0-u)} du.$$

This means that all the changes in the value of the portfolio after time 0 are due to trading in the stock and the bond and not to the injection (or withdrawal) of funds. The portfolio **replicates** (or **hedges**) the contingent claim C when it is self-financing and $C_{t_0} \equiv C$, so that its value at time t_0 matches exactly the payoff of the claim.

Let $\mathbb{Q}$ be the equivalent martingale probability, that is the equivalent probability under which $\{e^{-\rho t} S_t, 0 \leqslant t \leqslant t_0\}$ is a martingale, then if the claim C can be replicated by a portfolio with value $\{C_t, 0 \leqslant t \leqslant t_0\}$, exactly as in discrete time we see that its unique time-0 price must be C_0, the initial value of the portfolio and furthermore

$$C_0 = E_{\mathbb{Q}}\left(e^{-\rho t_0} C \mid \mathcal{F}_0\right); \tag{5.20}$$

this expression is $E_{\mathbb{Q}}\left(e^{-\rho t_0} C\right)$ if S_0 is regarded as a constant (so that $\mathcal{F}_0$ is the trivial σ-field). Its time-t price will be

$$C_t = E_{\mathbb{Q}}\left(e^{-\rho(t_0-t)} C \mid \mathcal{F}_t\right). \tag{5.21}$$

To see why (5.20) and (5.21) are true, observe that under $\mathbb{Q}$, for a self-financing portfolio the process $\{e^{-\rho t}C_t,\ 0 \leqslant t \leqslant t_0\}$ is a martingale since, using the formula for stochastic integration-by-parts and (5.19),

$$\begin{aligned} d\left(e^{-\rho t}C_t\right) &= e^{-\rho t}dC_t - \rho C_t e^{-\rho t}dt \\ &= e^{-\rho t}\left(X_t dS_t + \rho B_t e^{-\rho(t_0-t)}dt\right) - \rho\left(X_t S_t + B_t e^{-\rho(t_0-t)}\right)e^{-\rho t}dt \\ &= X_t\left(e^{-\rho t}dS_t - \rho S_t e^{-\rho t}dt\right) = X_t d\left(e^{-\rho t}S_t\right). \end{aligned}$$

This means that

$$e^{-\rho t}C_t = C_0 + \int_0^t X_u d\left(e^{-\rho u}S_u\right).$$

But under $\mathbb{Q}$, the process $\left\{e^{-\rho t}S_t\right\}$ is a martingale from which it follows that

$$\int_0^t X_u d\left(e^{-\rho u}S_u\right), \tag{5.22}$$

the stochastic integral with respect to that process, is also a martingale under $\mathbb{Q}$. To understand why this is, argue that

$$\begin{aligned} d\left(e^{-\rho t}S_t\right) &= -\rho e^{-\rho t}S_t dt + e^{-\rho t}dS_t \\ &= -\rho e^{-\rho t}S_t dt + e^{-\rho t}\left[\mu S_t dt + \sigma S_t dW_t\right] = \sigma e^{-\rho t}S_t dW_t, \end{aligned}$$

when $\mu = \rho$; but we know that calculations under $\mathbb{Q}$ correspond to setting $\mu = \rho$ and thus we may think of (5.22) under $\mathbb{Q}$ as being a stochastic integral with respect to the driving Brownian motion $\{W_t\}$,

$$\int_0^t X_u d\left(e^{-\rho u}S_u\right) = \int_0^t \sigma e^{-\rho u}X_u S_u dW_u,$$

and hence a martingale under $\mathbb{Q}$; this presupposes the technical condition (4.34) on page 104 when $Z_t = \sigma e^{-\rho t}X_t S_t$, which we will assume holds. It is now immediate that $\left\{e^{-\rho t}C_t\right\}$ is a martingale under $\mathbb{Q}$ and so

$$e^{-\rho t}C_t = \mathbb{E}_{\mathbb{Q}}\left(e^{-\rho t_0}C_{t_0} \mid \mathcal{F}_t\right) = \mathbb{E}_{\mathbb{Q}}\left(e^{-\rho t_0}C \mid \mathcal{F}_t\right),$$

which is (5.21).

It should be noted that in the general discrete-time model the existence of an equivalent martingale measure was guaranteed by the requirement that arbitrage is excluded from the model. In the present model the existence of the equivalent martingale measure is a consequence of Girsanov's Theorem; while it implies that there are no arbitrage opportunities we will not explore that aspect here. In fact, in the Black–Scholes model any claim paying C at time t_0 for which C is an $\mathcal{F}_{t_0}$-random variable satisfying $\mathbb{E}C^2 < \infty$ may be replicated but again a proof of this is beyond our scope. At the end of this section, we will prove the Black–Scholes formula by showing that the portfolio with

$$X_t = \Phi\left(d_1\left(S_t, t\right)\right) \quad \text{and} \quad B_t = -c\Phi\left(d_2\left(S_t, t\right)\right)$$

is self-financing and hence replicates the European call option, where the notation d_1, d_2 is as in (5.4) and (5.5).

We will consider first the general problem of when a portfolio holding $g(S_t, t)$ in stock and $h(S_t, t)$ in the bond is self-financing, that is the circumstances under which the portfolio with value C_t given by

$$C_t = p(S_t, t) = g(S_t, t)\, S_t + h(S_t, t)\, e^{-\rho(t_0 - t)} \tag{5.23}$$

is self-financing.

Theorem 5.2 *Suppose that $g(x, t)$ and $h(x, t)$ are functions with continuous second partial derivatives. The portfolio with value given in* (5.23) *is self-financing if and only if*

$$x\frac{\partial g}{\partial x} + e^{-\rho(t_0 - t)}\frac{\partial h}{\partial x} = 0, \quad \textit{and} \tag{5.24}$$

$$\frac{1}{2}\sigma^2 x^2 \frac{\partial g}{\partial x} + x\frac{\partial g}{\partial t} + e^{-\rho(t_0 - t)}\frac{\partial h}{\partial t} = 0. \tag{5.25}$$

Proof. With $p(x, t) = xg(x, t) + h(x, t)e^{-\rho(t_0 - t)}$, using Itô's Lemma we see that the portfolio with value given in (5.23) is self-financing if and only if

$$\begin{aligned} dp(S_t, t) &= \left.\frac{\partial p}{\partial x}\right|_{(S_t, t)} dS_t + \left[\frac{\partial p}{\partial t} + \frac{1}{2}\sigma^2 x^2 \frac{\partial^2 p}{\partial x^2}\right]_{(S_t, t)} dt \\ &= g(S_t, t)\, dS_t + \rho h(S_t, t)\, e^{-\rho(t_0 - t)} dt. \end{aligned}$$

Equating the coefficients of dS_t and dt we see that this is true if and only if

$$g = \frac{\partial p}{\partial x} \quad \text{and} \quad \rho h e^{-\rho(t_0 - t)} = \frac{\partial p}{\partial t} + \frac{1}{2}\sigma^2 x^2 \frac{\partial^2 p}{\partial x^2}. \tag{5.26}$$

The first relation in (5.26) gives

$$g = \frac{\partial p}{\partial x} = g + x\frac{\partial g}{\partial x} + e^{-\rho(t_0 - t)}\frac{\partial h}{\partial x},$$

which yields (5.24), and the second relation becomes

$$\frac{1}{2}\sigma^2 x^2 \left(2\frac{\partial g}{\partial x} + x\frac{\partial^2 g}{\partial x^2} + e^{-\rho(t_0 - t)}\frac{\partial^2 h}{\partial x^2}\right) + x\frac{\partial g}{\partial t} + e^{-\rho(t_0 - t)}\frac{\partial h}{\partial t} = 0. \tag{5.27}$$

But differentiating (5.24) with respect to x shows that

$$\frac{\partial g}{\partial x} + x\frac{\partial^2 g}{\partial x^2} + e^{-\rho(t_0 - t)}\frac{\partial^2 h}{\partial x^2} = 0,$$

so that (5.27) may be replaced by (5.25). □

Corollary 5.2 *Suppose that $p(x,t)$ is a function with continuous second partial derivatives. Then $p(S_t,t)$ is the value of a self-financing portfolio of the form* (5.23) *if and only if the function p satisfies*

$$\frac{1}{2}\sigma^2 x^2 \frac{\partial^2 p}{\partial x^2} + \rho x \frac{\partial p}{\partial x} + \frac{\partial p}{\partial t} - \rho p = 0. \tag{5.28}$$

Proof. When $p(S_t,t)$ is the value of such a portfolio, then $p(x,t) = xg(x,t) + h(x,t)e^{-\rho(t_0-t)}$, with g and h satisfying (5.24) and (5.25). From the derivation of (5.24), $g = \partial p/\partial x$ and substituting this and

$$h = e^{\rho(t_0-t)}\left(p - x\frac{\partial p}{\partial x}\right) \tag{5.29}$$

into (5.25) gives (5.28). Conversely, when p satisfies (5.28) then set

$$g = \frac{\partial p}{\partial x} \quad \text{and} \quad h = e^{\rho(t_0-t)}\left(p - x\frac{\partial p}{\partial x}\right);$$

it is immediate that g and h satisfy (5.24) and (5.25), completing the proof. □

Note that when we may represent the value of a self-financing portfolio in the form of (5.23) so that $C_t = p(S_t,t)$ then the proof of Theorem 5.2 shows that the holdings in stock and bonds are

$$X_t = g(S_t,t) = \left.\frac{\partial p}{\partial x}\right|_{(S_t,t)} \quad \text{and} \quad B_t = h(S_t,t) = e^{\rho(t_0-t)}\left[p - \frac{\partial p}{\partial x}\right]_{(S_t,t)}.$$

We may also observe that the holding in bonds $\{B_t\}$ is a martingale under the probability $\mathbb{Q}$; to see this, argue using Itô's Lemma that

$$dB_t = dh(S_t,t) = \left.\frac{\partial h}{\partial x}\right|_{(S_t,t)} dS_t + \left[\frac{\partial h}{\partial t} + \frac{1}{2}\sigma^2 x^2 \frac{\partial^2 h}{\partial x^2}\right]_{(S_t,t)} dt. \tag{5.30}$$

With h given by (5.29), calculate that

$$\begin{aligned}\frac{\partial h}{\partial t} &= e^{\rho(t_0-t)}\left[-\rho p + \rho x\frac{\partial p}{\partial x} + \frac{\partial p}{\partial t} - x\frac{\partial^2 p}{\partial x \partial t}\right] \\ &= -e^{\rho(t_0-t)}\left[\frac{1}{2}\sigma^2 x^2 \frac{\partial^2 p}{\partial x^2} + x\frac{\partial^2 p}{\partial x \partial t}\right],\end{aligned} \tag{5.31}$$

after using the fact that p satisfies (5.28); also calculate that

$$\frac{\partial h}{\partial x} = -e^{\rho(t_0-t)} x \frac{\partial^2 p}{\partial x^2} \quad \text{and} \quad \frac{\partial^2 h}{\partial x^2} = -e^{\rho(t_0-t)}\left[x\frac{\partial^3 p}{\partial x^3} + \frac{\partial^2 p}{\partial x^2}\right]. \tag{5.32}$$

Now substitute from (5.31) and (5.32) into (5.30) and use the relation that

$$\frac{1}{2}\sigma^2 x^3 \frac{\partial^3 p}{\partial x^3} + \sigma^2 x^2 \frac{\partial^2 p}{\partial x^2} + x \frac{\partial^2 p}{\partial x \partial t} = -\rho x^2 \frac{\partial^2 p}{\partial x^2}, \tag{5.33}$$

which is obtained by first differentiating (5.28) with respect to x and then multiplying by x. We see that

$$\begin{aligned} dB_t = dh(S_t, t) &= e^{\rho(t_0 - t)} \left[\rho x^2 \frac{\partial^2 p}{\partial x^2} \right]_{(S_t, t)} dt - e^{\rho(t_0 - t)} \left[x \frac{\partial^2 p}{\partial x^2} \right]_{(S_t, t)} dS_t \\ &= -e^{\rho t_0} \left[x \frac{\partial^2 p}{\partial x^2} \right]_{(S_t, t)} d\left(e^{-\rho t} S_t\right). \end{aligned} \tag{5.34}$$

Now $\{e^{-\rho t} S_t\}$ is a martingale under $\mathbb{Q}$, so it follows from (5.34) that $\{B_t\}$ is a stochastic integral with respect to that process and hence it is also a martingale under $\mathbb{Q}$; in this argument we are assuming implicitly that the function p has continuous third partial derivatives and also that the appropriate form of the integrability condition (4.34) holds. Since the discounted value of the portfolio, $\{e^{-\rho t} C_t\}$, is a martingale under $\mathbb{Q}$ it follows that the discounted value of the holding in stock, $\{e^{-\rho t} X_t S_t\}$, is also a martingale under $\mathbb{Q}$. These observations are analogous to the discussion on page 37 for the binomial model.

Example 5.1 *Holding a fixed proportion of wealth in stock.* Consider an investor who trades in the Black–Scholes model and who constantly rebalances his portfolio so as to maintain a fixed proportion, γ, of his wealth invested in the stock and the remainder invested in the bank account over the time interval $[0, t_0]$. Here, one may wish to consider the situation where γ is in the range $0 < \gamma < 1$ but that is not necessary to what follows; the case $\gamma < 0$ would correspond to being short in stock while $\gamma > 1$ corresponds to being short in the bank account.

Let $p = p(S_t, t)$ represent the value of the portfolio at time t then $p(x, t)$ must satisfy

$$\gamma p = x \frac{\partial p}{\partial x}$$

since $\partial p / \partial x$ units of stock are held in the self-financing portfolio. Rearranging gives

$$\frac{1}{p}\frac{\partial p}{\partial x} = \frac{\gamma}{x} \quad \text{and hence} \quad \ln(p) = \ln(x^\gamma) + \ln(c) \quad \text{or} \quad p = cx^\gamma$$

where $c = c(t)$ depends only on t and not on x. For p to represent the value of a self-financing portfolio it must satisfy (5.28) so that

$$\frac{1}{2}\sigma^2 x^2 \left[\gamma(\gamma - 1) c x^{\gamma - 2}\right] + \rho x \left(c \gamma x^{\gamma - 1}\right) + c' x^\gamma - \rho c x^\gamma = 0,$$

which gives

$$(\gamma - 1)\left(\rho + \gamma \sigma^2 / 2\right) c + c' = 0.$$

Integrating this equation yields

$$c(t) = c(0)e^{(1-\gamma)(\rho+\gamma\sigma^2/2)t},$$

where $c(0)$ is determined from the initial wealth of the investor at time 0, w_0, by setting $c(0) = w_0/S_0^\gamma$. Thus

$$p(x,t) = w_0 e^{(1-\gamma)(\rho+\gamma\sigma^2/2)t}\,(x/S_0)^\gamma,$$

gives the value of the portfolio. □

Now we proceed to give a proof of the Black–Scholes formula by checking that $g(x,t) = \Phi(d_1(x,t))$ and $h(x,t) = -c\Phi(d_2(x,t))$ satisfy (5.24) and (5.25), where d_1 and d_2 are given by (5.4) and (5.5). This, together with the arguments earlier in the chapter, will demonstrate that this choice of g and h defines a replicating portfolio for the option, which is sufficient to establish Theorem 5.1 and Corollary 5.1. It is immediate from the relation (5.7) that g and h satisfy (5.24), while for (5.25)

$$\begin{aligned}\frac{1}{2}\sigma^2 x^2 \frac{\partial g}{\partial x} &+ x\frac{\partial g}{\partial t} + e^{-\rho(t_0-t)}\frac{\partial h}{\partial t}\\ &= \frac{1}{2}\sigma^2 x^2 \phi(d_1)\frac{\partial d_1}{\partial x} + x\phi(d_1)\frac{\partial d_1}{\partial t} - ce^{-\rho(t_0-t)}\phi(d_2)\frac{\partial d_2}{\partial t}\\ &= \frac{1}{2}\sigma^2 x^2 \phi(d_1)\frac{\partial d_1}{\partial x} - \frac{c\sigma e^{-\rho(t_0-t)}}{2\sqrt{t_0-t}}\phi(d_2) = 0,\end{aligned}$$

after using (5.7) and (5.8).

The equation (5.28) is known as the general partial differential equation for contingent claims in the Black–Scholes model or, simply, as the **Black–Scholes equation**. Any contingent claim which may be replicated by a portfolio of the form (5.23) will have a time-t price $p(S_t,t)$ such that the function $p(x,t)$ satisfies (5.28). The boundary conditions imposed on the equation by the particular claim determine the appropriate solution. For example, for the European call of the Black–Scholes formula, $p(x,t)$ is the solution of (5.28) with the boundary condition $p(x,t_0) = (x-c)_+$.

5.3.2 Dividend-paying claims

The discussion and justification of the pricing formulae given in this chapter is based on the idea of a self-financing portfolio for which the value matches the price of the claim at each point in time. By suitably redefining the payoff of a claim if necessary, all contingent claims may be put in this framework, although in its original formulation the price of the claim may not be matched by a self-financing portfolio. Consider for example a claim which pays the holder a 'dividend' at the rate D_t/unit time at time t where $\{D_t,\, t \geqslant 0\}$ is an adapted process and also pays off the sum C, an $\mathcal{F}_{t_0}$ random variable, at time t_0. This may be thought of as being equivalent to a claim paying off a total amount

$$\int_t^{t_0} e^{\rho(t_0-u)} D_u du + C$$

at time t_0 if it is held from time t to t_0; this is because the dividend $D_u du$ paid in the interval $(u, u+du)$, when invested in the bank account until time t_0, yields the amount $e^{\rho(t_0-u)} D_u du$. By the arguments above its time-t price will be

$$V_t = \mathbb{E}_{\mathbb{Q}} \left(\int_t^{t_0} e^{-\rho(u-t)} D_u du + e^{-\rho(t_0-t)} C \;\middle|\; \mathcal{F}_t \right), \tag{5.35}$$

where $\mathbb{Q}$ is the equivalent martingale probability as before. The relation (5.35) should be compared with (2.48) on page 47 where the corresponding situation for the binomial model is considered. We need to modify the requirement that a replicating portfolio be self-financing since funds are being withdrawn at the rate D_t. Suppose that the portfolio has value V_t at time t and it holds X_t in stock and B_t in bonds maturing at t_0, then we have

$$V_t = X_t S_t + B_t e^{-\rho(t_0-t)},$$

as before, but now in place of (5.19) we would require

$$dV_t = X_t dS_t + \rho B_t e^{-\rho(t_0-t)} dt - D_t dt, \tag{5.36}$$

with $V_{t_0} = C$. The additional term $-D_t dt$ on the right-hand side of (5.36) is to take account of the withdrawal of the dividend from the portfolio in the time interval $[t, t+dt]$. Recalling (2.49), set

$$C_t = \int_0^t e^{\rho(t-u)} D_u du + V_t.$$

In a similar way to that in the preceding section, it is easy to check that $\{e^{-\rho t} C_t\}$ is a martingale under $\mathbb{Q}$; this follows from (5.36) because

$$\begin{aligned} d\left(e^{-\rho t} C_t\right) &= e^{-\rho t} D_t dt + e^{-\rho t} dV_t - \rho e^{-\rho t} V_t dt \\ &= e^{-\rho t} X_t dS_t + \rho B_t e^{-\rho t_0} dt - \rho e^{-\rho t} \left(X_t S_t + B_t e^{-\rho(t_0-t)} \right) dt \\ &= X_t d\left(e^{-\rho t} S_t\right), \end{aligned}$$

and $\{e^{-\rho t} S_t\}$ is a martingale under $\mathbb{Q}$ which shows that $\{e^{-\rho t} C_t\}$ is also. It follows that $\mathbb{E}_{\mathbb{Q}}\left(e^{-\rho t_0} C_{t_0} \mid \mathcal{F}_t\right) = e^{-\rho t} C_t$, but substituting

$$C_{t_0} = \int_0^{t_0} e^{\rho(t_0-u)} D_u du + V_{t_0} = \int_0^{t_0} e^{\rho(t_0-u)} D_u du + C$$

into this relation, we obtain (5.35). It also follows that C_t is the value of a self-financing portfolio holding X_t units of stock and $B_t + \int_0^t e^{\rho(t_0-u)} D_u du$ in the bond maturing at time t_0.

To illustrate the changes in Theorem 5.2 and Corollary 5.2 for this situation, suppose that $D_t = k(S_t, t)$ for some appropriate function k. Then, if $V_t = p(S_t, t)$, $X_t = g(S_t, t)$ and $B_t = h(S_t, t)$ as before, we will have $p(x, t) = xg(x, t) + h(x, t)e^{-\rho(t_0 - t)}$ again and exactly as in the proof of Theorem 5.2 we require

$$\begin{aligned} dp(S_t, t) &= \left.\frac{\partial p}{\partial x}\right|_{(S_t, t)} dS_t + \left[\frac{\partial p}{\partial t} + \frac{1}{2}\sigma^2 x^2 \frac{\partial^2 p}{\partial x^2}\right]_{(S_t, t)} dt \\ &= g(S_t, t)\, dS_t + \rho h(S_t, t)\, e^{-\rho(t_0 - t)} dt - k(S_t, t)\, dt. \end{aligned}$$

Equating coefficients of dS_t and dt we see that g and h define a replicating portfolio if and only if (5.24) holds unchanged and (5.25) is replaced by

$$\frac{1}{2}\sigma^2 x^2 \frac{\partial g}{\partial x} + x\frac{\partial g}{\partial t} + e^{-\rho(t_0 - t)}\frac{\partial h}{\partial t} = -k.$$

In place of the Black–Scholes equation (5.28), the function p, defining the value of the replicating portfolio, will satisfy

$$\frac{1}{2}\sigma^2 x^2 \frac{\partial^2 p}{\partial x^2} + \rho x \frac{\partial p}{\partial x} + \frac{\partial p}{\partial t} - \rho p = -k. \tag{5.37}$$

5.3.3 General terminal-value claims

It was pointed out earlier that the method of pricing the European call option may be applied to find the price of any claim in the Black–Scholes model. The price at time 0 of a claim paying C at time t_0 is $\mathbb{E}_{\mathbb{Q}}\left(e^{-\rho t_0} C\right)$, where of course taking expectations with the martingale probability $\mathbb{Q}$ gives the same value as taking expectations with the original probabilities with the assumption that $\mu = \rho$; the price at time t will be $\mathbb{E}_{\mathbb{Q}}\left[e^{-\rho(t_0 - t)} C \mid \mathcal{F}_t\right]$. Here C may be any $\mathcal{F}_{t_0}$-random variable with $\mathbb{E}C^2 < \infty$; in a later section we consider examples when C may depend on the whole path of the stock price $\{S_t,\ 0 \leqslant t \leqslant t_0\}$ in the time interval $[0, t_0]$, but we consider here the simpler case when $C = f(S_{t_0})$ is a terminal-value claim which depends only on the value of the stock at the expiry time t_0. We illustrate the procedure with one example.

Example 5.2 *A power of the stock price.* Consider the case when the claim pays off the amount $C = \left(S_{t_0}\right)^\gamma$ for some real γ. Recall that $S_t = S_0 e^{\sigma W_t + (\mu - \sigma^2/2)t}$, so that the time-0 price is then

$$\begin{aligned} \mathbb{E}_{\mathbb{Q}}\left[e^{-\rho t_0}\left(S_{t_0}\right)^\gamma\right] &= e^{-\rho t_0}\,\mathbb{E}_{\mathbb{Q}}\left[(S_0)^\gamma\, e^{\gamma\sigma W_{t_0} + \gamma(\mu - \sigma^2/2)t_0}\right] \\ &= e^{-\rho t_0}\,(S_0)^\gamma\,\mathbb{E}\left[e^{\gamma\sigma W_{t_0} + \gamma(\rho - \sigma^2/2)t_0}\right] \\ &= (S_0)^\gamma\, e^{-(1-\gamma)(\rho + \gamma\sigma^2/2)t_0}, \end{aligned}$$

with the last equality following because $\mathbb{E}\left[e^{\theta W_{t_0}}\right] = e^{\theta^2 t_0/2}$, since W_{t_0} has the normal distribution with mean zero and variance t_0. To compute the time-t price the calculation is essentially the same to give

$$\mathbb{E}_{\mathbb{Q}}\left[e^{-\rho(t_0-t)}\left(S_{t_0}\right)^{\gamma} \mid \mathcal{F}_t\right] = (S_t)^{\gamma}\, e^{-(1-\gamma)(\rho+\gamma\sigma^2/2)(t_0-t)},$$

which should be compared with the situation in Example 5.1 to see that for this claim the hedging portfolio always maintains a fixed proportion, γ, of its value in stock. $\square$

For the case of a general terminal-value claim when $C = f(S_{t_0})$ for some function $f : (0,\infty) \to \mathbb{R}$, the time-0 price of the claim may be represented as

$$\begin{aligned} e^{-\rho t_0}\mathbb{E}_{\mathbb{Q}}\left[f(S_{t_0})\right] &= e^{-\rho t_0}\mathbb{E}_{\mathbb{Q}}\left[f\left(S_0 e^{\sigma W_{t_0}+(\mu-\sigma^2/2)t_0}\right)\right] \\ &= e^{-\rho t_0}\mathbb{E}\left[f\left(S_0 e^{\sigma Z\sqrt{t_0}+(\rho-\sigma^2/2)t_0}\right)\right], \end{aligned}$$

where Z is a random variable with the $N(0,1)$ distribution. To avoid technicalities we will assume that f is continuous and (twice) differentiable as required. The time-t price, conditional on $S_t = x$, is

$$\begin{aligned} p(x,t) &= e^{-\rho(t_0-t)}\mathbb{E}_{\mathbb{Q}}\left[f\left(S_{t_0}\right) \mid \mathcal{F}_t\right] \\ &= e^{-\rho(t_0-t)}\mathbb{E}\left[f\left(x e^{\sigma Z\sqrt{t_0-t}+(\rho-\sigma^2/2)(t_0-t)}\right)\right]. \end{aligned} \tag{5.38}$$

We may verify that (5.38) is indeed the correct price of the claim by checking that p satisfies (5.28) in the same way that we did for the Black–Scholes formula; that will show that $p(x,t)$ given by (5.38) is the price of a self-financing portfolio and setting $t = t_0$ it may be seen that $p(x,t_0) = f(x)$ so that the value of the portfolio at time t_0 coincides with the payoff of the claim. To carry this through, first compute the Delta of the claim, that is the holding in stock in the replicating portfolio, which is determined from

$$\begin{aligned} \frac{\partial p}{\partial x} &= e^{-\rho(t_0-t)}\mathbb{E}\left[e^{\sigma Z\sqrt{t_0-t}+(\rho-\sigma^2/2)(t_0-t)} f'\left(x e^{\sigma Z\sqrt{t_0-t}+(\rho-\sigma^2/2)(t_0-t)}\right)\right] \\ &= \mathbb{E}\left[f'\left(x e^{\sigma Z\sqrt{t_0-t}+(\rho+\sigma^2/2)(t_0-t)}\right)\right], \end{aligned} \tag{5.39}$$

with the second relation following because (A.15) implies that since the random variable Z has the standard $N(0,1)$-distribution, we have

$$\mathbb{E}\left[e^{\theta Z} g(Z)\right] = e^{\theta^2/2}\mathbb{E}\left[g(Z+\theta)\right]. \tag{5.40}$$

It is immediate from (5.39) that the holding in stock is non-negative when $f' \geqslant 0$, or non-positive when $f' \leqslant 0$; that is, the replicating portfolio is long in the stock when f is a non-decreasing function while it is short in the stock when f is a non-increasing function. Differentiating again with respect to x gives

$$\frac{\partial^2 p}{\partial x^2} = \mathbb{E}\left[e^{\sigma Z\sqrt{t_0-t}+(\rho+\sigma^2/2)(t_0-t)} f''\left(x e^{\sigma Z\sqrt{t_0-t}+(\rho+\sigma^2/2)(t_0-t)}\right)\right], \tag{5.41}$$

showing that when f is convex ($f'' \geqslant 0$) then the Gamma of the claim is non-negative, so the Delta is non-decreasing in the stock price x; when f is concave ($f'' \leqslant 0$) then the Gamma is non-positive, so the Delta is non-increasing in the stock price. That is, p is convex in x when f is convex, while p is concave when f is concave; this observation may also be obtained directly from the representation

$$p(x,t) = e^{-\rho(t_0-t)} \mathbb{E}\left[f\left(xe^{\sigma Z\sqrt{t_0-t}+(\rho-\sigma^2/2)(t_0-t)}\right)\right],$$

in (5.38) without necessarily assuming that f is differentiable. This shows that if the claim is a convex (respectively, concave) function of the final stock price, when the stock price goes up the holding in stock in the hedging portfolio increases (respectively, decreases).

To simplify the notation, write $\tilde{Z} = \sigma Z\sqrt{t_0 - t} + \left(\rho + \sigma^2/2\right)(t_0 - t)$ and then (5.39) and (5.41) may be expressed as

$$\frac{\partial p}{\partial x} = \mathbb{E}\left[f'\left(xe^{\tilde{Z}}\right)\right] \quad \text{and} \quad \frac{\partial^2 p}{\partial x^2} = \mathbb{E}\left[e^{\tilde{Z}} f''\left(xe^{\tilde{Z}}\right)\right]. \tag{5.42}$$

Differentiating (5.38) with respect to t we see that $\dfrac{\partial p}{\partial t}$ equals

$$\rho p - xe^{-\rho(t_0-t)} \mathbb{E}\left[\left(\frac{\sigma Z}{2\sqrt{t_0-t}} + \left(\rho - \frac{\sigma^2}{2}\right)\right) e^{\tilde{Z}-\sigma^2(t_0-t)} f'\left(xe^{\tilde{Z}-\sigma^2(t_0-t)}\right)\right]$$

and using (5.40) again, this gives

$$\begin{aligned} \frac{\partial p}{\partial t} &= \rho p - x\,\mathbb{E}\left[\left(\frac{\sigma Z}{2\sqrt{t_0-t}} + \rho\right) f'\left(xe^{\tilde{Z}}\right)\right] \\ &= \rho p - \frac{\sigma x}{2\sqrt{t_0-t}}\,\mathbb{E}\left[Zf'\left(xe^{\tilde{Z}}\right)\right] - \rho x\frac{\partial p}{\partial x}. \end{aligned} \tag{5.43}$$

Now, for the random variable Z with the standard normal distribution, use the identity

$$\mathbb{E}\left[Zg(Z)\right] = \mathbb{E}\left[g'(Z)\right], \tag{5.44}$$

which is a special case of (A.14) on page 199, to see that

$$\begin{aligned} \mathbb{E}\left[Zf'\left(xe^{\tilde{Z}}\right)\right] &= \mathbb{E}\left[Zf'\left(xe^{\sigma Z\sqrt{t_0-t}+(\rho+\sigma^2/2)(t_0-t)}\right)\right] \\ &= \sigma x\sqrt{t_0-t}\,\mathbb{E}\left[e^{\tilde{Z}} f''\left(xe^{\tilde{Z}}\right)\right] = \sigma x\sqrt{t_0-t}\,\frac{\partial^2 p}{\partial x^2}, \end{aligned}$$

from (5.42). Substitute this last expression into (5.43) to complete the verification that p satisfies (5.28).

Note that

$$\frac{\partial p}{\partial t} = \rho\left(p - x\frac{\partial p}{\partial x}\right) - \frac{1}{2}\sigma^2 x^2 \frac{\partial^2 p}{\partial x^2},$$

where $p - x\partial p/\partial x$ represents the value of the holding in bonds in the replicating portfolio. We may conclude from this that the price p is a non-increasing function of time t (the Theta is $\leqslant 0$) when the replicating portfolio is short in bonds and the Gamma of the claim is non-negative; that is, when

$$p - x\frac{\partial p}{\partial x} \leqslant 0 \quad \text{and} \quad \frac{\partial^2 p}{\partial x^2} \geqslant 0.$$

Suppose that f is a convex function and set $f(0+) = \lim_{\epsilon \downarrow 0} f(\epsilon)$; note that because f is convex this limit will always exist. As has been shown, when f is twice differentiable in this case the Gamma is necessarily non-negative. For the portfolio to be short in bonds, when f is convex and differentiable at all $x > 0$, it is sufficient for any of the following three equivalent conditions to hold:

(a) $f(0+) \leqslant 0$;
(b) $\lambda f(x) \leqslant f(\lambda x)$, for $\lambda \geqslant 1$ and $x > 0$; and
(c) $f(x) \leqslant x f'(x)$, for $x > 0$.

To see the equivalence of (a), (b) and (c) note that when (a) holds then by the convexity of f, for $\lambda \geqslant 1$ and $0 < \epsilon < x$,

$$f(x) \leqslant \left(\frac{x - \epsilon}{\lambda x - \epsilon}\right) f(\lambda x) + \left(\frac{x(\lambda - 1)}{\lambda x - \epsilon}\right) f(\epsilon);$$

then letting $\epsilon \downarrow 0$ gives

$$f(x) \leqslant \lambda^{-1} f(\lambda x) + \left(1 - \lambda^{-1}\right) f(0+) \leqslant \lambda^{-1} f(\lambda x),$$

which is (b). When (b) holds then for $\lambda > 1$ and $x > 0$

$$\frac{f(\lambda x) - f(x)}{(\lambda - 1)x} \geqslant \frac{f(x)}{x},$$

and letting $\lambda \downarrow 1$ gives (c). When (c) holds then letting $x \downarrow 0$ gives (a). When the convex function f is not differentiable, then in (c) replace $f'(x)$ by the right-hand derivative $f'(x+)$ where

$$f'(x+) = \lim_{y \downarrow x} \frac{f(y) - f(x)}{y - x}$$

always exists since f is convex; with this change, the equivalence of the three conditions (a), (b) and (c) still holds.

To see that any of these conditions imply that the portfolio is short in bonds, note that (from (5.39))

$$\left[p - x\frac{\partial p}{\partial x}\right]_{(S_t, t)} = e^{-\rho(t_0 - t)} E_{\mathbb{Q}}\left[f\left(S_{t_0}\right) - S_{t_0} f'\left(S_{t_0}\right) \mid \mathcal{F}_t\right] \leqslant 0, \tag{5.45}$$

from (c). In fact, when f is convex and differentiable it may be seen that the three conditions above are also necessary for the holding in bonds always to be negative because, if it is, then letting $t \uparrow t_0$ in (5.45) the right-hand expression converges (by the Martingale Convergence Theorem) to

$$f\left(S_{t_0}\right) - S_{t_0} f'\left(S_{t_0}\right) \leqslant 0,$$

which holds for all values of S_{t_0} and so case (c) is true.

In the same way it may be seen that the price p is a non-decreasing function of time, t, when the replicating portfolio is long in bonds and the Gamma of the claim is non-positive; that is, when

$$p - x\frac{\partial p}{\partial x} \geqslant 0 \quad \text{and} \quad \frac{\partial^2 p}{\partial x^2} \leqslant 0.$$

When f is concave then the Gamma is non-positive and for the portfolio to be long in bonds it is sufficient for any of the following equivalent conditions to hold:

(a′) $f(0+) \geqslant 0$;
(b′) $\lambda f(x) \geqslant f(\lambda x)$, for $\lambda \geqslant 1$ and $x > 0$; and
(c′) $f(x) \geqslant x f'(x)$, for $x > 0$.

The arguments go through immediately by replacing f by $-f$ in the above.

We will now consider the dependence of the price in (5.38) on the volatility, σ. Differentiating with respect to σ, we see that $\dfrac{\partial p}{\partial \sigma}$ equals

$$\mathbb{E}\left[x\left(Z\sqrt{t_0 - t} - \sigma\left(t_0 - t\right)\right) e^{\sigma Z\sqrt{t_0-t} - \sigma^2 (t_0 - t)/2} f'\left(e^{\sigma Z\sqrt{t_0-t} + (\rho - \sigma^2/2)(t_0 - t)}\right)\right],$$

but using (5.40) it follows that

$$\begin{aligned}\frac{\partial p}{\partial \sigma} &= x\sqrt{t_0 - t}\, \mathbb{E}\left[Z f'\left(x e^{\sigma Z\sqrt{t_0-t} + (\rho+\sigma^2/2)(t_0-t)}\right)\right] \\ &= \sigma x^2 \left(t_0 - t\right) e^{(\rho+\sigma^2/2)(t_0-t)}\, \mathbb{E}\left[e^{\sigma Z\sqrt{t_0-t}} f''\left(x e^{\sigma Z\sqrt{t_0-t} + (\rho+\sigma^2/2)(t_0-t)}\right)\right],\end{aligned}$$

with the second relation following from the identity (5.44). It follows immediately that when f is convex (so that $f'' \geqslant 0$) the price of the claim is non-decreasing as a function of the volatility, while when f is concave the price is non-increasing in σ.

5.3.4 Specific terminal-value claims

The most important terminal-value claim, other than the European call already considered, is the **European put option** with strike price c and expiry time t_0. This is the contract that entitles (but does not require) the holder to sell one unit of stock at the fixed strike price at the fixed expiry time t_0; on the other hand, an **American put option** entitles the holder to sell one unit of stock at the fixed strike price *at or* *before* the fixed expiry time and it is not a terminal-value claim. The European put pays $\left(c - S_{t_0}\right)_+$ at the expiry time t_0; its price at time t, P_t, is related to the stock price, S_t, and the price of a European call at the same strike, C_t, by

$$S_t + P_t = C_t + ce^{-\rho(t_0 - t)}. \tag{5.46}$$

This fact is known as **put-call parity**. This follows because for any real number x, $x = x_+ - x_-$, so that at time t_0,

$$\begin{aligned} S_{t_0} - c &= \left(S_{t_0} - c\right)_+ - \left(S_{t_0} - c\right)_- \\ &= \left(S_{t_0} - c\right)_+ - \left(c - S_{t_0}\right)_+ = C_{t_0} - P_{t_0}; \end{aligned} \tag{5.47}$$

then, using the martingale property,

$$\begin{aligned} S_t - ce^{-\rho(t_0-t)} &= \mathbb{E}_{\mathbb{Q}}\left(e^{-\rho(t_0-t)}\left(S_{t_0} - c\right) \mid \mathcal{F}_t\right) \\ &= \mathbb{E}_{\mathbb{Q}}\left(e^{-\rho(t_0-t)}\left(\left(S_{t_0} - c\right)_+ - \left(c - S_{t_0}\right)_+\right) \mid \mathcal{F}_t\right) = C_t - P_t. \end{aligned}$$

It should be noted that the put-call parity relation (5.46) holds outside the framework of the Black–Scholes model and will hold in any model in which there is no arbitrage. The argument is that the two sides of (5.47) represent two portfolios which have the same value at time t_0; on the left-hand side is a portfolio holding one unit of stock and short c in the bank while on the right-hand side is a portfolio holding one call and short one put. Since these two portfolios have the same value at time t_0 then they must have the same value at any time $t \leqslant t_0$, otherwise there would be an arbitrage opportunity, hence (5.46) holds. Within the Black–Scholes model the price of the European put may be obtained from the price of the call using put-call parity or it may be calculated directly (see Exercise 5.1).

We mention some other specific terminal-value options which are widely traded and which are among the simplest examples of what are known as 'exotic' options, which is the usual description applied to derivative contracts other than the 'plain vanilla' call and put options already considered. The (European) **digital** call is the contract which pays 1 at time t_0 if the stock price is above some pre-determined level c at that time. The reader should be warned that the naming of non-standard options varies between authors; for example, digital calls are sometimes known as **cash-or-nothing** calls or **binary** calls. The payoff of this contract at time t_0 is $C = I_{(S_{t_0} > c)}$ with its price at time t, $0 \leqslant t \leqslant t_0$, being

$$\begin{aligned} &\mathbb{E}_{\mathbb{Q}}\left[e^{-\rho(t_0-t)} I_{(S_{t_0} > c)} \,\middle|\, \mathcal{F}_t\right] \\ &\qquad = e^{-\rho(t_0-t)} \Phi\left(\frac{\ln\left(S_t/c\right) + \left(\rho - \sigma^2/2\right)\left(t_0 - t\right)}{\sigma\sqrt{t_0 - t}}\right). \end{aligned} \tag{5.48}$$

The **digital put** (or **binary put**) pays 1 if the stock price is at or below the pre-determined level c at time t_0 so that the payoff $C = I_{(S_{t_0} \leqslant c)}$ and the price is

$$\mathbb{E}_{\mathbb{Q}}\left[e^{-\rho(t_0-t)} I_{(S_{t_0} \leqslant c)} \,\middle|\, \mathcal{F}_t\right] = e^{-\rho(t_0-t)} \Phi\left(\frac{\ln\left(c/S_t\right) - \left(\rho - \sigma^2/2\right)\left(t_0 - t\right)}{\sigma\sqrt{t_0 - t}}\right),$$

at time t.

The **gap call** is an option which, for two pre-determined fixed levels c_1, c_2, pays the amount $S_{t_0} - c_2$ when the stock price is above c_1 at the expiry time t_0, otherwise it pays zero. The payoff is illustrated in Figure 5.3 and it may be represented as

$$C = \left(S_{t_0} - c_2\right) I_{\left(S_{t_0} > c_1\right)} = \left(S_{t_0} - c_1\right)_+ + (c_1 - c_2)\, I_{\left(S_{t_0} > c_1\right)},$$

which shows that holding the gap call is equivalent to holding the European call option with strike price c_1 together with $c_1 - c_2$ digital calls at strike c_1 (where the holding in digital calls is long or short according as $c_1 > c_2$ or $c_1 < c_2$). A special

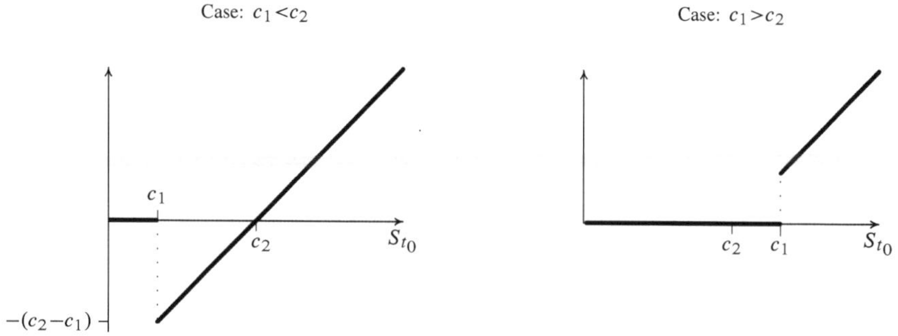

Figure 5.3: Payoff profile for the gap call

case is the **asset-or-nothing** option which pays $C = S_{t_0} I_{\left(S_{t_0} > c\right)}$, which is a gap call with $c_1 = c$ and $c_2 = 0$.

The **gap put** pays

$$C = \left(c_2 - S_{t_0}\right) I_{\left(S_{t_0} \leqslant c_1\right)} = \left(c_1 - S_{t_0}\right)_+ + (c_2 - c_1)\, I_{\left(S_{t_0} \leqslant c_1\right)},$$

and is equivalent similarly to holding a European put together with digital puts.

A **contingent-premium** call option allows the purchaser to pay some fixed proportion α, $0 \leqslant \alpha \leqslant 1$, of the initial price of a call option at the time of purchase (say, $t = 0$). The payment of the remaining premium is delayed until the expiry time t_0 and it is contingent upon the option ending in the money, that is on $S_{t_0} > c$; if the stock price ends at or below the strike price c, that is $S_{t_0} \leqslant c$, no further premium is paid. The terminal premium, d, must then satisfy

$$\mathbb{E}_{\mathbb{Q}}\left[d e^{-\rho t_0} I_{\left(S_{t_0} > c\right)}\right] = (1 - \alpha)\, \mathbb{E}_{\mathbb{Q}}\left[e^{-\rho t_0}\left(S_{t_0} - c\right)_+\right],$$

giving

$$d = (1 - \alpha)\, \frac{\mathbb{E}_{\mathbb{Q}}\left[e^{-\rho t_0}\left(S_{t_0} - c\right)_+\right]}{\mathbb{E}_{\mathbb{Q}}\left[e^{-\rho t_0} I_{\left(S_{t_0} > c\right)}\right]}.$$

This shows that d is $1-\alpha$ times the time-0 price of the standard European call option divided by the price of the digital call, with the respective expressions for these prices given in (5.3) and (5.48) (with $t = 0$). The special case when $\alpha = 0$ corresponds to the situation where the option requires no initial payment. As may be seen from

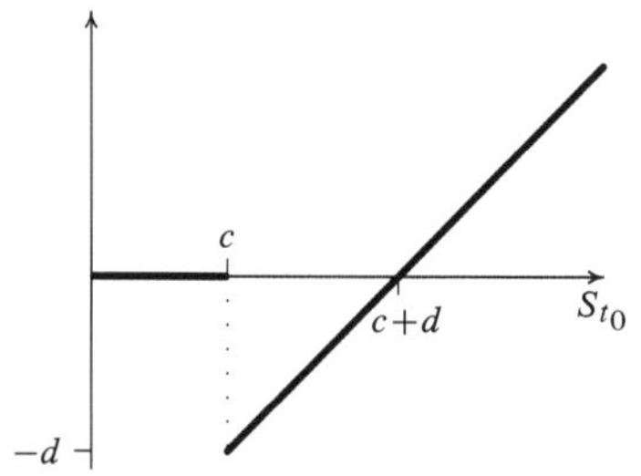

Figure 5.4: Net payoff for the contingent-premium call

Figure 5.4, the net payoff of the contingent-premium call, after the payment of the deferred premium if it is due, is the same as the gap call with the values $c_1 = c$ and $c_2 = c + d$; the net payoff is negative for S_{t_0} in the range $c < S_{t_0} < c + d$.

5.3.5 Utility maximization

As in the case of the binomial model, the methods of the previous section may be used to consider the problem of an investor with initial wealth w_0 at time 0, say, who wishes to maximize his expected utility of final wealth, C, at time t_0, for his given utility function $v(\cdot)$. The final wealth is achieved by trading in the stock and the bond (bank account) over the time interval $[0, t_0]$. This problem is equivalent to determining that contingent claim C which may be purchased at time 0 at a cost of w_0 so as to maximize $\mathbb{E}\, v(C)$; this is because the initial price of any claim is determined through considering how it is replicated exactly through continuous trading. As we have seen, this initial price is the same as $\mathbb{E}_{\mathbb{Q}}\left(e^{-\rho t_0} C\right)$ where $\mathbb{Q}$ is the equivalent martingale probability.

The problem may be formulated mathematically as that of choosing a random variable C (in general depending on $S_t, 0 \leqslant t \leqslant t_0$) so as to

$$\text{maximize} \quad \mathbb{E}\, v(C) \qquad \text{subject to} \quad \mathbb{E}_{\mathbb{Q}}\left(e^{-\rho t_0} C\right) = w_0. \tag{5.49}$$

Again, the constrained optimization problem may be tackled by considering the maximization of the Lagrangian,

$$\begin{aligned}\mathcal{L} &= \mathbb{E}\,v(C) + \lambda\left[w_0 - \mathbb{E}_{\mathbb{Q}}\left(e^{-\rho t_0}C\right)\right] \\ &= \mathbb{E}\,v(C) + \lambda\left[w_0 - e^{-\rho t_0}\,\mathbb{E}\left(\frac{d\mathbb{Q}}{d\mathbb{P}}C\right)\right] \\ &= \mathbb{E}\left[v(C) - \lambda e^{-\rho t_0}\frac{d\mathbb{Q}}{d\mathbb{P}}C\right] + \lambda w_0,\end{aligned} \tag{5.50}$$

for suitable constant Lagrange multiplier λ. Recall that the Radon–Nikodym derivative $d\mathbb{Q}/d\mathbb{P}$ is given by

$$\frac{d\mathbb{Q}}{d\mathbb{P}} = \exp\left(\frac{(\rho-\mu)}{\sigma}W_{t_0} - \frac{(\rho-\mu)^2}{2\sigma^2}t_0\right).$$

Since $S_t = S_0\exp\left(\sigma W_t + \left(\mu - \sigma^2/2\right)t\right)$, $d\mathbb{Q}/d\mathbb{P}$ may be expressed in terms of the stock price S_{t_0} as

$$\frac{d\mathbb{Q}}{d\mathbb{P}} = \left[e^{-(\rho+\mu-\sigma^2)t_0/2}\left(\frac{S_{t_0}}{S_0}\right)\right]^{\left(\frac{\rho-\mu}{\sigma^2}\right)}. \tag{5.51}$$

We may maximize the Lagrangian by maximizing the expression within the expectation in (5.50). When v is concave, this is achieved by setting the derivative with respect to C of the expression equal to zero to obtain

$$v'(C) - \lambda e^{-\rho t_0}\frac{d\mathbb{Q}}{d\mathbb{P}} = 0. \tag{5.52}$$

In principle, this equation may be solved for C in terms of λ, with the appropriate choice of λ obtained by substituting C back into the constraint in (5.49). The procedure is illustrated in the following two examples.

Example 5.3 *Logarithmic utility.* Consider the case when the utility function is $v(x) = \ln(x)$, then $v'(x) = 1/x$, and the equation (5.52) becomes

$$\frac{1}{C} = \lambda e^{-\rho t_0}\frac{d\mathbb{Q}}{d\mathbb{P}} \quad \text{or} \quad C = e^{\rho t_0}\Big/\left(\lambda\frac{d\mathbb{Q}}{d\mathbb{P}}\right).$$

Substituting into the constraint

$$\mathbb{E}_{\mathbb{Q}}\left(e^{-\rho t_0}C\right) = \mathbb{E}\left(\frac{d\mathbb{Q}}{d\mathbb{P}}e^{-\rho t_0}C\right) = w_0,$$

we find that $1/\lambda = w_0$ or $\lambda = 1/w_0$. It follows from (5.51) that the optimal choice of C is

$$C = w_0 e^{\rho t_0}\left[e^{-(\rho+\mu-\sigma^2)t_0/2}\left(\frac{S_{t_0}}{S_0}\right)\right]^{\left(\frac{\mu-\rho}{\sigma^2}\right)}.$$

Comparing this expression with Example 5.1 and Example 5.2 it may be seen that for an investor with logarithmic utility the optimal strategy is always to maintain a fixed proportion $\gamma = (\mu - \rho)/\sigma^2$ of his wealth in stock. Note that he maintains positive amounts in both stock and the bond only when $\rho < \mu < \rho + \sigma^2$. The case of logarithmic utility is also the situation where the investor maximizes the expected growth rate of his wealth over the time interval $[0, t_0]$, that is, $\mathbb{E}\left[\ln\left(C/w_0\right)\right]/t_0$. □

Example 5.4 *Exponential utility.* Now consider the situation where the utility function is $v(x) = \left(1 - e^{-ax}\right)/a$, for some $a > 0$, so that $v'(x) = e^{-ax}$ and (5.52) becomes the equation

$$e^{-aC} = \lambda e^{-\rho t_0} \frac{d\mathbb{Q}}{d\mathbb{P}} \quad \text{which gives} \quad C = \frac{1}{a}\left[\rho t_0 - \ln\lambda - \ln\left(\frac{d\mathbb{Q}}{d\mathbb{P}}\right)\right].$$

In order to obtain λ when we substitute into the constraint in (5.49) we must compute

$$\begin{aligned} \mathbb{E}_{\mathbb{Q}}\left[\ln\left(\frac{d\mathbb{Q}}{d\mathbb{P}}\right)\right] &= \mathbb{E}_{\mathbb{Q}}\left[\left(\frac{\rho-\mu}{\sigma}\right)W_{t_0} - \frac{1}{2}\left(\frac{\rho-\mu}{\sigma}\right)^2 t_0\right] \\ &= \left(\frac{\rho-\mu}{\sigma}\right)\mathbb{E}_{\mathbb{Q}}\left[W_{t_0}\right] - \frac{1}{2}\left(\frac{\rho-\mu}{\sigma}\right)^2 t_0 = \frac{1}{2}\left(\frac{\rho-\mu}{\sigma}\right)^2 t_0, \end{aligned}$$

the last equality coming from the observation in Section 5.1 that under the probability $\mathbb{Q}$ the stochastic process $\{W_t\}$ is a standard Brownian motion with drift $\left((\rho - \mu)/\sigma\right)$ so that

$$\mathbb{E}_{\mathbb{Q}}\left[W_{t_0}\right] = \mathbb{E}\left[W_{t_0} + \left(\frac{\rho-\mu}{\sigma}\right)t_0\right] = \left(\frac{\rho-\mu}{\sigma}\right)t_0,$$

since $\mathbb{E}\,W_{t_0} = 0$; alternatively, $\mathbb{E}_{\mathbb{Q}}\left[W_{t_0}\right]$ may be derived directly by observing that

$$\mathbb{E}_{\mathbb{Q}}\left[W_{t_0}\right] = \mathbb{E}\left[\frac{d\mathbb{Q}}{d\mathbb{P}} W_{t_0}\right] = \mathbb{E}\left[W_{t_0} e^{\left(\frac{\rho-\mu}{\sigma}\right)W_{t_0} - \frac{1}{2}\left(\frac{\rho-\mu}{\sigma}\right)^2 t_0}\right],$$

and then using the fact that $\mathbb{E}\left[W_{t_0} e^{\theta W_{t_0}}\right] = \theta t_0 e^{\theta^2 t_0/2}$. Substituting into the constraint $w_0 = \mathbb{E}_{\mathbb{Q}}\left[e^{-\rho t_0} C\right]$ gives

$$\begin{aligned} w_0 e^{\rho t_0} &= \frac{1}{a}\left(\rho t_0 - \ln\lambda - \mathbb{E}_{\mathbb{Q}}\left(\ln\left(\frac{d\mathbb{Q}}{d\mathbb{P}}\right)\right)\right) \\ &= \frac{1}{a}\left(\rho t_0 - \ln\lambda - \frac{1}{2}\left(\frac{\rho-\mu}{\sigma}\right)^2 t_0\right), \end{aligned}$$

which shows that the optimal choice of C is

$$C = w_0 e^{\rho t_0} + \frac{1}{a}\left[\frac{1}{2}\left(\frac{\rho-\mu}{\sigma}\right)^2 t_0 - \ln\left(\frac{d\mathbb{Q}}{d\mathbb{P}}\right)\right]$$

and from (5.51) this gives

$$
\begin{aligned}
C &= w_0 e^{\rho t_0} + \left(\frac{\rho-\mu}{a\sigma^2}\right)\left[\frac{1}{2}(\rho-\mu)\, t_0 + \frac{1}{2}\left(\rho+\mu-\sigma^2\right) t_0 - \ln\left(\frac{S_{t_0}}{S_0}\right)\right] \\
&= w_0 e^{\rho t_0} + \left(\frac{\rho-\mu}{a\sigma^2}\right)\left[\left(\rho-\frac{\sigma^2}{2}\right) t_0 - \ln\left(\frac{S_{t_0}}{S_0}\right)\right].
\end{aligned}
$$

We may compute $E_{\mathbb{Q}}\left[e^{-\rho(t_0-t)} C \mid \mathcal{F}_t\right]$, the value of the replicating portfolio at time t, as

$$
w_0 e^{\rho t} + e^{-\rho(t_0-t)}\left(\frac{\rho-\mu}{a\sigma^2}\right)\left[\left(\rho-\frac{\sigma^2}{2}\right) t_0 - E_{\mathbb{Q}}\left[\ln\left(\frac{S_{t_0}}{S_0}\right) \,\Big|\, \mathcal{F}_t\right]\right]. \tag{5.53}
$$

But $\ln\left(S_{t_0}/S_0\right) = \ln\left(S_{t_0}/S_t\right) + \ln\left(S_t/S_0\right)$, with $\ln\left(S_{t_0}/S_t\right)$ being independent of $\mathcal{F}_t$ and $\ln\left(S_t/S_0\right)$ known given $\mathcal{F}_t$, whence

$$
E_{\mathbb{Q}}\left[\ln\left(\frac{S_{t_0}}{S_0}\right) \,\Big|\, \mathcal{F}_t\right] = \ln\left(\frac{S_t}{S_0}\right) + E_{\mathbb{Q}}\left[\ln\left(\frac{S_{t_0}}{S_t}\right)\right],
$$

with

$$
\begin{aligned}
E_{\mathbb{Q}}\left[\ln\left(\frac{S_{t_0}}{S_t}\right)\right] &= E_{\mathbb{Q}}\left[\sigma\left(W_{t_0}-W_t\right) + \left(\mu-\frac{\sigma^2}{2}\right)(t_0-t)\right] \\
&= \left(\rho-\frac{\sigma^2}{2}\right)(t_0-t),
\end{aligned}
$$

showing that the expression in (5.53) is

$$
w_0 e^{\rho t} + e^{-\rho(t_0-t)}\left(\frac{\rho-\mu}{a\sigma^2}\right)\left[\left(\rho-\frac{\sigma^2}{2}\right) t - \ln\left(\frac{S_t}{S_0}\right)\right].
$$

The value of this portfolio at time t, given $S_t = x$, is

$$
p(x,t) = w_0 e^{\rho t} + e^{-\rho(t_0-t)}\left(\frac{\rho-\mu}{a\sigma^2}\right)\left[\left(\rho-\frac{\sigma^2}{2}\right) t - \ln\left(\frac{x}{S_0}\right)\right],
$$

and so

$$
\frac{\partial p}{\partial x} = e^{-\rho(t_0-t)}\left(\frac{\mu-\rho}{a\sigma^2}\right)\frac{1}{x}
$$

showing, from the discussion in Section 5.3.1, that the value of the holding in stock within the portfolio at time t is

$$
x\frac{\partial p}{\partial x} = e^{-\rho(t_0-t)}\left(\frac{\mu-\rho}{a\sigma^2}\right)
$$

which is constant in 'real' terms, that is when it is discounted to time 0 by the discount factor $e^{-\rho t}$ to allow for the change in the value of the bank account due to the interest rate. □

The arguments of this section extend to the case when the investor gains utility from consumption throughout the period $(0, t_0)$ in addition to the utility derived from his final wealth C. Now suppose that when at time t he consumes wealth (or takes a dividend) at the rate D_t/unit time, where D_t is known given $\mathcal{F}_t$, then he gains utility at rate $v_1(D_t)$; suppose that he gains utility $v_2(C)$) from his final wealth C, where $v_1(\cdot)$ and $v_2(\cdot)$ are given functions. The problem becomes

$$\begin{aligned} &\text{maximize} \quad \mathbb{E}\left[\int_0^{t_0} v_1(D_t)dt + v_2(C)\right] \\ &\text{subject to} \quad \mathbb{E}_{\mathbb{Q}}\left[e^{-\rho t_0}\left(\int_0^{t_0} e^{\rho(t_0-t)} D_t dt + C\right)\right] = w_0, \end{aligned} \tag{5.54}$$

where w_0 is again the investor's initial wealth. Adopting the same approach as before, we consider the Lagrangian

$$\begin{aligned} \mathcal{L} &= \mathbb{E}\left[\int_0^{t_0} v_1(D_t)dt + v_2(C)\right] + \lambda\left(w_0 - \mathbb{E}_{\mathbb{Q}}\left[\int_0^{t_0} e^{-\rho t} D_t dt + e^{-\rho t_0} C\right]\right) \\ &= \int_0^{t_0} \mathbb{E}\left[v_1(D_t) - \lambda \mathbb{E}\left(\frac{d\mathbb{Q}}{d\mathbb{P}} \,\Big|\, \mathcal{F}_t\right) e^{-\rho t} D_t\right] dt \\ &\qquad\qquad + \mathbb{E}\left[v_2(C) - \lambda \frac{d\mathbb{Q}}{d\mathbb{P}} e^{-\rho t_0} C\right] + \lambda w_0. \end{aligned}$$

Here, we are using the facts that

$$\mathbb{E}_{\mathbb{Q}}(C) = \mathbb{E}\left[\frac{d\mathbb{Q}}{d\mathbb{P}} C\right] \text{ and } \mathbb{E}_{\mathbb{Q}}(D_t) = \mathbb{E}\left(\frac{d\mathbb{Q}}{d\mathbb{P}} D_t\right) = \mathbb{E}\left[\mathbb{E}\left(\frac{d\mathbb{Q}}{d\mathbb{P}} \,\Big|\, \mathcal{F}_t\right) D_t\right],$$

the second relation holding because D_t is known given $\mathcal{F}_t$ and so may be taken outside the conditional expectation. Now maximize inside the expectations in each of the terms in $\mathcal{L}$, in the first case with respect to D_t and in the second with respect to C, to give

$$v_1'(D_t) = \lambda e^{-\rho t} \mathbb{E}\left(\frac{d\mathbb{Q}}{d\mathbb{P}} \,\Big|\, \mathcal{F}_t\right) \quad \text{and} \quad v_2'(C) = \lambda e^{-\rho t_0} \frac{d\mathbb{Q}}{d\mathbb{P}}. \tag{5.55}$$

Example 5.5 *The logarithmic case.* Suppose that $v_1(x) = a \ln(x)$ and $v_2(x) = b \ln(x)$, where a and b are non-negative constants, then (5.55) gives

$$D_t = a e^{\rho t} \Big/ \left(\lambda \mathbb{E}\left(\frac{d\mathbb{Q}}{d\mathbb{P}} \,\Big|\, \mathcal{F}_t\right)\right) \quad \text{and} \quad C = b e^{\rho t_0} \Big/ \left(\lambda \frac{d\mathbb{Q}}{d\mathbb{P}}\right).$$

Substitute into the constraint in (5.54) to see that

$$\begin{aligned} w_0 &= \mathbb{E}_{\mathbb{Q}}\left[e^{-\rho t_0}\left(\int_0^{t_0} e^{\rho(t_0-t)} D_t dt + C\right)\right] \\ &= \int_0^{t_0} e^{-\rho t}\,\mathbb{E}\left(D_t \mathbb{E}\left(\frac{d\mathbb{Q}}{d\mathbb{P}} \,\Big|\, \mathcal{F}_t\right)\right) dt + e^{-\rho t_0}\,\mathbb{E}\left(\frac{d\mathbb{Q}}{d\mathbb{P}} C\right) \\ &= \int_0^{t_0} e^{-\rho t}\left(\frac{a e^{\rho t}}{\lambda}\right) dt + e^{-\rho t_0}\left(\frac{b e^{\rho t_0}}{\lambda}\right) = \frac{a t_0 + b}{\lambda}, \end{aligned}$$

which gives $\lambda = (a t_0 + b)/w_0$. Then since

$$\begin{aligned} \mathbb{E}\left(\frac{d\mathbb{Q}}{d\mathbb{P}} \,\Big|\, \mathcal{F}_t\right) &= \exp\left(\frac{(\rho-\mu)}{\sigma} W_t - \frac{(\rho-\mu)^2}{2\sigma^2} t\right) \\ &= \left[e^{-(\rho+\mu-\sigma^2)t/2}\left(\frac{S_t}{S_0}\right)\right]^{\left(\frac{\rho-\mu}{\sigma^2}\right)}, \end{aligned} \tag{5.56}$$

expressions for C and D_t may be obtained in terms of the stock price by using the relation (5.51). □

Finally, we observe that the approach of this section may be used in some cases to study the inverse problem: what form of utility function would induce a utility-maximizing investor to purchase a particular terminal-value contingent claim such as, for example, a European call option. That is, for a terminal-value claim $f(S_{t_0})$ what utility function $v(\cdot)$ would ensure that the maximizing C in (5.50) is proportional to $f(S_{t_0})$, so that $C = \alpha f(S_{t_0})$, for some constant α, which would show that the optimal investment strategy for the investor with that utility is to buy a fixed number, α, of these contracts each paying $f(S_{t_0})$ at time t_0.

For a given function $f(\cdot)$, put $C = \alpha f(S_{t_0})$ in (5.52) and use (5.51) to obtain

$$v'\left(\alpha f\left(S_{t_0}\right)\right) = \lambda e^{-\rho t_0}\left[e^{-(\rho+\mu-\sigma^2)t_0/2}\left(\frac{S_{t_0}}{S_0}\right)\right]^{\left(\frac{\rho-\mu}{\sigma^2}\right)},$$

and putting $y = S_{t_0}$ and $\gamma = (\rho-\mu)/\sigma^2$ we see that

$$v'\left(\alpha f(y)\right) \propto y^{\gamma}.$$

Now set $z = \alpha f(y)$ or $y = f^{-1}(z/\alpha)$, assuming that the inverse exists for an appropriate range of values of z, then

$$v'(z) \propto \left[f^{-1}(z/\alpha)\right]^{\gamma} \quad \text{or} \quad v(z) \propto \int_0^z \left[f^{-1}(u/\alpha)\right]^{\gamma} du.$$

For example, consider the case of the European call option so that $f(y) = (y-c)_+$. If $z > 0$ then $y > c$ and $y = c + z/\alpha$ with $v(z) \propto (z + \alpha c)^{\gamma+1}$. Note that for (5.52) to yield a maximum of (5.50) we would require $\gamma \leqslant 0$ and for $v(\cdot)$ to be a non-decreasing utility function we need $\gamma + 1 \geqslant 0$ which together impose the restrictions

$\sigma^2 + \rho \geqslant \mu \geqslant \rho$ on the parameters μ, ρ and σ^2; for such a range of parameters any utility function of the form $v(x) = (x + \beta)^{\gamma+1}$ for $x \geqslant 0$ (and $v(x) = -\infty$ for $x < 0$) will result in the investor purchasing call options.

5.3.6 American claims

The treatment of American claims in Section 2.2.6 for the binomial model carries through with some changes to the Black–Scholes case. For a payoff function $f(\cdot)$ the holder of the corresponding American claim may choose a stopping time $T \leqslant t_0$ at which time he receives $f(S_T)$. Here, a stopping time T is a random time for which the event $(T \leqslant t) \in \mathcal{F}_t$ for all t, $0 \leqslant t \leqslant t_0$; as we have seen, a stopping time has the same intuitive interpretation as in discrete time, in that the decision to stop at a time $T = t$ depends only on the evolution of the stock-price process up to and including time t. When the holder of the claim chooses a stopping time $T \leqslant t_0$ at which to exercise the claim he receives the amount $f(S_T)$ at time T which is equivalent to receiving the amount $C_T = e^{\rho(t_0 - T)} f(S_T)$ at time t_0. In this section the results require conditions on f that we will not spell out at each stage, they would be true for example when f is a continuous function and $\sup_{0 \leqslant t \leqslant t_0} E_{\mathbb{Q}} |f(S_t)| < \infty$.

Suppose that the claim has not been exercised prior to the time t, then for a stopping time T, with $t \leqslant T \leqslant t_0$, the value to the holder of the claim at time t if he adopts the strategy T will be

$$V_t^T = E_{\mathbb{Q}}\left[e^{-\rho(t_0 - t)} C_T \mid \mathcal{F}_t\right] = E_{\mathbb{Q}}\left[e^{-\rho(T-t)} f(S_T) \mid \mathcal{F}_t\right]. \tag{5.57}$$

We will see that we obtain the overall value of the claim at time t (if it has not been exercised already) when we maximize V_t^T appropriately over stopping times T, with $t \leqslant T \leqslant t_0$. We take the essential supremum of V_t^T (see page 196 for an introduction to the notion of the essential supremum of a collection of random variables)

$$V_t = \operatorname*{ess\,sup}_{t \leqslant T \leqslant t_0} V_t^T = \operatorname*{ess\,sup}_{t \leqslant T \leqslant t_0} E_{\mathbb{Q}}\left[e^{-\rho(T-t)} f(S_T) \mid \mathcal{F}_t\right]. \tag{5.58}$$

We note that if we take stopping times T_i with $t \leqslant T_i \leqslant t_0$, $i = 1, 2$, and define

$$\overline{T} = T_1 I_{\left(V_t^{T_1} \geq V_t^{T_2}\right)} + T_2 I_{\left(V_t^{T_1} < V_t^{T_2}\right)}$$

then $\overline{T}$ is again a stopping time satisfying $t \leqslant \overline{T} \leqslant t_0$. Furthermore, let

$$A = \left(V_t^{T_1} \geq V_t^{T_2}\right) \quad \text{and} \quad A^c = \left(V_t^{T_1} < V_t^{T_2}\right)$$

be the events on which $\overline{T} = T_1$ and $\overline{T} = T_2$, respectively; since A and A^c are events in $\mathcal{F}_t$, we may argue that

$$\begin{aligned} V_t^{\overline{T}} &= E_{\mathbb{Q}}\left[e^{-\rho(\overline{T}-t)} f(S_{\overline{T}}) \mid \mathcal{F}_t\right] \\ &= E_{\mathbb{Q}}\left[e^{-\rho(T_1 - t)} f(S_{T_1}) I_A + e^{-\rho(T_2 - t)} f(S_{T_2}) I_{A^c} \mid \mathcal{F}_t\right] \\ &= V_t^{T_1} I_{\left(V_t^{T_1} \geq V_t^{T_2}\right)} + V_t^{T_2} I_{\left(V_t^{T_1} < V_t^{T_2}\right)} = \max\left(V_t^{T_1}, V_t^{T_2}\right). \end{aligned}$$

This shows that the collection of random variables $\left\{V_t^T;\ t \leqslant T \leqslant t_0\right\}$ is closed under the operation of taking the maximum of two of the elements, that is it is directed upwards. It follows that there exists a sequence of stopping time $\{T_k\}$, with $t \leqslant T_k \leqslant t_0$ for each k, with

$$V_t^{T_k} \uparrow V_t = \operatorname*{ess\,sup}_{t \leqslant T \leqslant t_0} V_t^T, \quad \text{as} \quad k \uparrow \infty.$$

A consequence of this is that for $u < t \leqslant t_0$,

$$\begin{aligned} \mathbb{E}_{\mathbb{Q}}\left[V_t \mid \mathcal{F}_u\right] &= \mathbb{E}_{\mathbb{Q}}\Big[\operatorname*{ess\,sup}_{t \leqslant T \leqslant t_0} V_t^T \mid \mathcal{F}_u\Big] \\ &= \mathbb{E}_{\mathbb{Q}}\Big[\lim_{k \uparrow \infty} V_t^{T_k} \mid \mathcal{F}_u\Big] = \lim_{k \to \infty} \mathbb{E}_{\mathbb{Q}}\left[V_t^{T_k} \mid \mathcal{F}_u\right] \end{aligned}$$

where the interchange of the conditional expectation and the limit is justified by monotone convergence, and using (5.57) together with the tower property of conditional expectations, we see that this

$$\begin{aligned} &= \lim_{k \to \infty} \mathbb{E}_{\mathbb{Q}}\left[e^{-\rho(T_k - t)} f\left(S_{T_k}\right) \mid \mathcal{F}_u\right] \\ &\leqslant \operatorname*{ess\,sup}_{t \leqslant T \leqslant t_0} \mathbb{E}_{\mathbb{Q}}\left[e^{-\rho(T-t)} f(S_T) \mid \mathcal{F}_u\right] \\ &\leqslant e^{-\rho(u-t)} \operatorname*{ess\,sup}_{u \leqslant T \leqslant t_0} \mathbb{E}_{\mathbb{Q}}\left[e^{-\rho(T-u)} f(S_T) \mid \mathcal{F}_u\right] = e^{-\rho(u-t)} V_u, \end{aligned}$$

with the second inequality holding because the set of stopping times T, satisfying $T \geqslant u$, is larger than the set for which $T \geqslant t$, since $u < t$. This demonstrates that the process $\left\{e^{-\rho t} V_t, \mathcal{F}_t, 0 \leqslant t \leqslant t_0\right\}$ is a supermartingale under the probability $\mathbb{Q}$; furthermore by considering $T \equiv t$ we see that $V_t \geqslant f(S_t)$, so that $e^{-\rho t} V_t$ is a supermartingale under $\mathbb{Q}$ dominating the discounted payoffs $e^{-\rho t} f(S_t)$.

In fact $e^{-\rho t} V_t$ is the smallest supermartingale (under $\mathbb{Q}$) dominating $e^{-\rho t} f(S_t)$; to see this, suppose $\{W_t, \mathcal{F}_t, 0 \leqslant t \leqslant t_0\}$ is a supermartingale under $\mathbb{Q}$ which satisfies $W_t \geqslant e^{-\rho t} f(S_t)$ for each t, then for a stopping time $T \geqslant t$, using the Optional Sampling Theorem for supermartingales we have

$$W_t \geqslant \mathbb{E}_{\mathbb{Q}}\left[W_T \mid \mathcal{F}_t\right] \geqslant \mathbb{E}_{\mathbb{Q}}\left[e^{-\rho T} f(S_T) \mid \mathcal{F}_t\right]$$

so that

$$W_t \geqslant e^{-\rho t} \operatorname*{ess\,sup}_{t \leqslant T \leqslant t_0} \mathbb{E}_{\mathbb{Q}}\left[e^{-\rho(T-t)} f(S_T) \mid \mathcal{F}_t\right] = e^{-\rho t} V_t.$$

As mentioned in Section 2.2.6, $\left\{e^{-\rho t} V_t\right\}$ is known as the Snell envelope of the process $\left\{e^{-\rho t} f(S_t)\right\}$.

Now, for each t define the stopping time

$$T_t^* = \inf\{u \geqslant t : V_u = f(S_u)\};$$

we have $V_{t_0} = f\left(S_{t_0}\right)$ so necessarily $t \leqslant T_t^* \leqslant t_0$. It may be shown that

$$V_t = \mathbb{E}_{\mathbb{Q}}\left[e^{-\rho(T_t^*-t)} f\left(S_{T_t^*}\right) \mid \mathcal{F}_t\right], \tag{5.59}$$

although the argument requires more technicalities than the corresponding derivation in the discrete framework of the binomial model so it will be omitted here.

We may argue now that V_t is the price of the American claim at time t if it has not been exercised prior to that instant. It is clear from (5.58) that V_t is a lower bound for the price, while if the price is greater that V_t an arbitrage may be formed by selling the American claim and buying a claim paying $e^{\rho(t_0-T_t^*)} f\left(S_{T_t^*}\right)$ at time t_0 for which the price at time t is V_t from (5.59).

When $f(\cdot)$ is a convex function satisfying one (and hence all) of the three equivalent conditions (a), (b) and (c) of page 133 then the argument in Section 2.2.6 carries through similarly to show that $\left\{e^{-\rho t} f(S_t)\right\}$ is a submartingale under $\mathbb{Q}$. For $u < t < t_0$, use the conditional form of Jensen's inequality and then the martingale property of $\left\{e^{-\rho t} S_t\right\}$ under $\mathbb{Q}$ to see that

$$\begin{aligned} \mathbb{E}_{\mathbb{Q}}\left[e^{-\rho t} f(S_t) \mid \mathcal{F}_u\right] &\geqslant e^{-\rho t} f\left(\mathbb{E}_{\mathbb{Q}}\left[S_t \mid \mathcal{F}_u\right]\right) \\ &= e^{-\rho t} f\left(e^{\rho(t-u)} S_u\right) \geqslant e^{-\rho u} f(S_u), \end{aligned}$$

with the final inequality coming from condition (b) (taking $\lambda = e^{\rho(t-u)} \geqslant 1$). Now use the Optional Sampling Theorem for submartingales, for any stopping time T, with $t \leqslant T \leqslant t_0$,

$$e^{-\rho t} f(S_t) \leqslant \mathbb{E}_{\mathbb{Q}}\left[e^{-\rho T} f\left(S_T\right) \mid \mathcal{F}_t\right] \leqslant \mathbb{E}_{\mathbb{Q}}\left[e^{-\rho t_0} f\left(S_{t_0}\right) \mid \mathcal{F}_t\right];$$

this shows that

$$e^{-\rho t} V_t = \operatorname*{ess\,sup}_{t \leqslant T \leqslant t_0} \mathbb{E}_{\mathbb{Q}}\left[e^{-\rho T} f\left(S_T\right) \mid \mathcal{F}_t\right] = \mathbb{E}_{\mathbb{Q}}\left[e^{-\rho t_0} f\left(S_{t_0}\right) \mid \mathcal{F}_t\right]$$

with the stopping time $T \equiv t_0$ attaining the essential supremum. This shows that in this case it is always optimal to wait until the expiry time to terminate the American claim. The remarks about the American call at the end of Section 2.2.6 apply here also as does the observation that when the interest rate is zero and the function f is convex (without any further condition) then it is optimal to wait until expiry.

Note that the price of an American claim at time t is necessarily a non-increasing function of $t \leqslant t_0$; this follows because for $t_1 < t_2 < t_0$ the holder of the claim at time t_1 is maximizing over a larger set of possible exercise times T, $t_1 \leqslant T \leqslant t_0$, than at time t_2, when $t_2 \leqslant T \leqslant t_0$. When it is the case that it is optimal not to exercise an American claim before expiry then it follows that the common Theta of the American claim and the corresponding European claim must be non-positive; in the case when the payoff of the European claim is $f\left(S_{t_0}\right)$ and f is convex, this could be seen directly from the above remarks and the discussion on page 133.

5.4 Path-dependent claims

5.4.1 Forward-start and lookback options

We consider now some contracts where the payoff C may depend on the path of the stock price between purchase of the contract and the expiry time t_0 and not just on the value of the stock price at the expiry time. As we have observed previously, for any random variable C (with $\mathbb{E}\left(C^2\right) < \infty$) which depends on $\mathcal{F}_{t_0}$ and so is determined by the path $\{S_t, 0 \leqslant t \leqslant t_0\}$, the price of the corresponding contract at time t is $\mathbb{E}_{\mathbb{Q}}\left[e^{-\rho(t_0-t)}C \mid \mathcal{F}_t\right]$, where $\mathbb{Q}$ is the martingale probability as usual. Valuation of the contract reduces to the probabilistic problem of calculating this expectation.

The simplest non-trivial examples occur when the payoff depends on S_{t_0} and the stock price S_{t_1} at some other fixed time t_1, $0 < t_1 < t_0$. The **forward-start call** option is a European call where the strike price is set to be the level of the stock price at a pre-determined fixed time t_1, so the payoff $C = \left(S_{t_0} - S_{t_1}\right)_+$. At any time t in the range $t_1 \leqslant t \leqslant t_0$, the value of S_{t_1} is known and it is clear that holding this option at this time is equivalent to holding the standard European call with 'fixed' strike $c = S_{t_1}$. The price at time t, $0 \leqslant t \leqslant t_1$, is

$$
\begin{aligned}
&\mathbb{E}_{\mathbb{Q}}\left[e^{-\rho(t_0-t)}\left(S_{t_0} - S_{t_1}\right)_+ \,\Big|\, \mathcal{F}_t\right] \\
&\qquad = \mathbb{E}_{\mathbb{Q}}\left[e^{-\rho(t_1-t)}\mathbb{E}_{\mathbb{Q}}\left[e^{-\rho(t_0-t_1)}\left(S_{t_0} - S_{t_1}\right)_+ \,\Big|\, \mathcal{F}_{t_1}\right] \,\Big|\, \mathcal{F}_t\right],
\end{aligned}
\tag{5.60}
$$

by the tower property of conditional expectations. The inner expectation is the price at time t_1 of the ordinary at-the-money ($c = S_{t_1}$) call and so the price in (5.60) is

$$
\begin{aligned}
&\mathbb{E}_{\mathbb{Q}}\Bigg[e^{-\rho(t_1-t)}S_{t_1}\Bigg(\Phi\left(\frac{2\rho+\sigma^2}{2\sigma}\sqrt{t_0-t_1}\right) \\
&\qquad\qquad - e^{-\rho(t_0-t_1)}\Phi\left(\frac{2\rho-\sigma^2}{2\sigma}\sqrt{t_0-t_1}\right)\Bigg) \,\Bigg|\, \mathcal{F}_t\Bigg] \\
&\quad = S_t\left(\Phi\left(\frac{2\rho+\sigma^2}{2\sigma}\sqrt{t_0-t_1}\right) - e^{-\rho(t_0-t_1)}\Phi\left(\frac{2\rho-\sigma^2}{2\sigma}\sqrt{t_0-t_1}\right)\right),
\end{aligned}
$$

since $\mathbb{E}_{\mathbb{Q}}\left[e^{-\rho(t_1-t)}S_{t_1} \mid \mathcal{F}_t\right] = S_t$ because $\left\{e^{-\rho t}S_t\right\}$ is a martingale under $\mathbb{Q}$. It is immediate that for $t \leqslant t_1$, before the strike price is known, holding the forward-start call is equivalent to holding a replicating portfolio which consists of just the fixed amount

$$
\Phi\left(\frac{2\rho+\sigma^2}{2\sigma}\sqrt{t_0-t_1}\right) - e^{-\rho(t_0-t_1)}\Phi\left(\frac{2\rho-\sigma^2}{2\sigma}\sqrt{t_0-t_1}\right) \tag{5.61}
$$

of stock, and after time t_1 the replicating portfolio is that for the standard call with strike S_{t_1}. This accords with intuition as the value of the option at time t_1 is just the proportion (5.61) of the stock price S_{t_1} which is not known at time $t < t_1$; so to hedge the option prior to t_1 requires holding that fixed proportion of the stock.

The **forward-start put**, for which the strike price of the put is fixed as S_{t_1} at time t_1, may be dealt with in the same way.

A number of path-dependent claims have payoffs which are functions of the maximum (or minimum) level that the stock price achieves during the life of the contract, say the time interval $[0, t_0]$. The first that we will consider is the **fixed-strike lookback call** option which pays off the amount $C = \left(\sup_{0 \leqslant u \leqslant t_0} S_u - c\right)_+$ at time t_0, for a pre-determined strike price c, so that the holder is permitted to exercise the call option at expiry at the highest price attained. In order to determine the price of this option at time 0 we must calculate first an expression for $\mathbb{E}_{\mathbb{Q}}\left(c \vee \sup_{0 \leqslant u \leqslant t_0} S_u\right)$, which will follow from formulae derived in the previous chapter. Note that

$$\begin{aligned} \mathbb{E}_{\mathbb{Q}}\left(c \vee \sup_{0 \leqslant u \leqslant t_0} S_u\right) &= \mathbb{E}_{\mathbb{Q}}\left[c \vee \left(S_0 e^{\sup_{0 \leqslant u \leqslant t_0}(\sigma W_u + (\mu - \sigma^2/2)u)}\right)\right] \\ &= \mathbb{E}\left[c \vee \left(S_0 e^{\sigma M_{t_0}^{\nu}}\right)\right], \end{aligned} \tag{5.62}$$

where $M_{t_0}^{\nu} = \sup_{0 \leqslant u \leqslant t_0}(W_u + \nu u)$ and here $\nu = \left(2\rho - \sigma^2\right)/(2\sigma)$, since evaluating the expectation in (5.62) under $\mathbb{Q}$ corresponds to taking expectations under $\mathbb{P}$ with $\mu = \rho$. Necessarily $M_{t_0}^{\nu} \geqslant 0$, and by using Girsanov's Theorem we see that the expression in (5.62) equals

$$\begin{aligned} &(c \vee S_0)\mathbb{P}\left(\sigma M_{t_0}^{\nu} \leqslant \ln((c \vee S_0)/S_0)\right) \\ &\qquad + S_0 \mathbb{E}\left[e^{\sigma M_{t_0} + \nu W_{t_0} - \nu^2 t_0/2} I_{(\sigma M_{t_0} > \ln((c \vee S_0)/S_0))}\right], \end{aligned} \tag{5.63}$$

where $M_{t_0} = M_{t_0}^0 = \sup_{0 \leqslant u \leqslant t_0} W_u$ is the maximum of the standard Brownian motion with no drift. The probability and the expectation in (5.63) may be evaluated using equations (4.9), (4.10) and (4.28) of Chapter 4, and then the Black–Scholes formula may be used to show that

$$\begin{aligned} \mathbb{E}_{\mathbb{Q}}\left(c \vee \sup_{0 \leqslant u \leqslant t_0} S_u\right) &= \mathbb{E}_{\mathbb{Q}}\left(c \vee S_0 \vee S_{t_0}\right) + e^{\rho t_0} r_1(S_0, c \vee S_0, t_0) \\ &= c \vee S_0 + \mathbb{E}_{\mathbb{Q}}\left(S_{t_0} - (c \vee S_0)\right)_+ + e^{\rho t_0} r_1(S_0, c \vee S_0, t_0), \end{aligned} \tag{5.64}$$

where the function r_1 is defined by

$$r_1(x, c, t) = \begin{cases} \sigma^2 x\left[\Phi(d + a) - e^{-2ad}\Phi(d - a)\right]/(2\rho) & \text{when } \rho > 0, \\ \sigma x\sqrt{t}\,[\phi(d) + d\Phi(d)] & \text{when } \rho = 0, \end{cases} \tag{5.65}$$

with $d = d(x, c, t) = \left(\ln(x/c) + \sigma^2 t/2\right)/\left(\sigma\sqrt{t}\right)$, and $a = a(t) = \rho\sqrt{t}/\sigma$. We may now compute the time-0 price of the fixed-strike lookback call as

$$\begin{aligned} &e^{-\rho t_0}\mathbb{E}_{\mathbb{Q}}\left[\sup_{0 \leqslant u \leqslant t_0} S_u - c\right]_+ \\ &\quad = e^{-\rho t_0}\mathbb{E}_{\mathbb{Q}}\left[\left(c \vee \sup_{0 \leqslant u \leqslant t_0} S_u\right) - c\right] \\ &\quad = e^{-\rho t_0}\left[(S_0 - c)_+ + \mathbb{E}_{\mathbb{Q}}\left(S_{t_0} - (c \vee S_0)\right)_+\right] + r_1(S_0, c \vee S_0, t_0) \end{aligned}$$

showing that the price of the lookback call at fixed strike c is the price of the standard European call at strike price $c \vee S_0$, with the same expiry time t_0, together with an additional amount

$$e^{-\rho t_0}(S_0 - c)_+ + r_1(S_0, c \vee S_0, t_0);$$

note that since $a \geqslant 0$ the function $r_1 \geqslant 0$. To calculate the time-t price, when $0 \leqslant t \leqslant t_0$, recall that S_u/S_t is independent of $\mathcal{F}_t$ for $u > t$, so conditional on $S_t = x$ and $\sup_{0 \leqslant u \leqslant t} S_u = y \geqslant x$, we have

$$\begin{aligned} \mathbb{E}_{\mathbb{Q}}\left[\left(\sup_{0\leqslant u\leqslant t_0} S_u - c\right)_+ \middle| \mathcal{F}_t\right] &= \mathbb{E}_{\mathbb{Q}}\left[\left(\left(y \vee \left(x \sup_{t\leqslant u\leqslant t_0} \frac{S_u}{S_t}\right)\right) - c\right)_+\right] \\ &= \mathbb{E}_{\mathbb{Q}}\left[y \vee c \vee \left(x \sup_{t\leqslant u\leqslant t_0} \frac{S_u}{S_t}\right)\right] - c. \end{aligned}$$

Use the fact that $\sup_{t\leqslant u\leqslant t_0}(S_u/S_t)$ and $\sup_{0\leqslant u\leqslant t_0-t}(S_u/S_0)$ have the same distribution (under the probability $\mathbb{Q}$ as well as the probability $\mathbb{P}$) and using the above expressions we see that, at the instant t,

$$e^{-\rho(t_0-t)}\mathbb{E}_{\mathbb{Q}}\left[\left(\sup_{0\leqslant u\leqslant t_0} S_u - c\right)_+ \middle| \mathcal{F}_t\right],$$

which is the time-t price of the lookback call with fixed strike c and expiry time t_0, equals the time-t price of a standard European call with strike price $c \vee \sup_{0\leqslant u\leqslant t} S_u$ and the same expiry **plus** an additional amount equal to

$$e^{-\rho(t_0-t)}\left(\sup_{0\leqslant u\leqslant t} S_u - c\right)_+ + r_1\left(S_t, c \vee \sup_{0\leqslant u\leqslant t} S_u, t_0 - t\right),$$

where the function r_1 is again given by (5.65).

The **floating-strike lookback put** option pays $C = \left(\sup_{0\leqslant u\leqslant t_0} S_u - S_{t_0}\right)$ at the expiry time t_0, so that it sets the strike price of the put at the largest level that the stock price attains during the lifetime, $[0, t_0]$, of the option. From (5.64), by setting $c = 0$, we have that the time-0 price of the option is

$$\begin{aligned} e^{-\rho t_0}\mathbb{E}_{\mathbb{Q}}\left[\sup_{0\leqslant u\leqslant t_0} S_u - S_{t_0}\right] &= e^{-\rho t_0}\mathbb{E}_{\mathbb{Q}}\left((S_0 \vee S_{t_0}) - S_{t_0}\right) + r_1(S_0, S_0, t_0) \\ &= e^{-\rho t_0}\mathbb{E}_{\mathbb{Q}}\left(S_0 - S_{t_0}\right)_+ + r_1(S_0, S_0, t_0), \end{aligned}$$

showing that the price equals the price of a standard European put option at strike price $c = S_0$ together with the extra amount $r_1(S_0, S_0, t_0)$, which is determined from (5.65). To evaluate the time-t price, $0 \leqslant t \leqslant t_0$, argue as previously by splitting the supremum as

$$\sup_{0\leqslant u\leqslant t_0} S_u = \left(\sup_{0\leqslant u\leqslant t} S_u\right) \vee \left(S_t \sup_{t\leqslant u\leqslant t_0}\left(S_u/S_t\right)\right)$$

and then from (5.64) we see that

$$
\begin{aligned}
&e^{-\rho(t_0-t)}\mathbb{E}_{\mathbb{Q}}\left[\sup_{0\leqslant u\leqslant t_0} S_u - S_{t_0} \,\middle|\, \mathcal{F}_t\right] \\
&= e^{-\rho(t_0-t)}\mathbb{E}_{\mathbb{Q}}\left[\left(\left(\sup_{0\leqslant u\leqslant t} S_u\right)\vee S_{t_0}\right) - S_{t_0} \,\middle|\, \mathcal{F}_t\right] + r_1\left(S_t, \sup_{0\leqslant u\leqslant t} S_u, t_0 - t\right) \\
&= e^{-\rho(t_0-t)}\mathbb{E}_{\mathbb{Q}}\left[\left(\sup_{0\leqslant u\leqslant t} S_u - S_{t_0}\right)_+ \,\middle|\, \mathcal{F}_t\right] + r_1\left(S_t, \sup_{0\leqslant u\leqslant t} S_u, t_0 - t\right);
\end{aligned}
$$

this shows that, at the instant t, the price of the floating-strike lookback put equals that of a standard European put, with the same expiry time and with strike price $c = \sup_{0\leqslant u\leqslant t} S_u$, plus the additional amount

$$
r_1\left(S_t, \sup_{0\leqslant u\leqslant t} S_u, t_0 - t\right)
$$

which, as before, is determined by the formula (5.65).

Two options which mirror the cases just considered but which involve the lowest level achieved by the stock price during the life of the contract are the **fixed-strike lookback put** which pays $C = \left(c - \inf_{0\leqslant u\leqslant t_0} S_u\right)_+$ and the **floating-strike lookback call** which pays $C = \left(S_{t_0} - \inf_{0\leqslant u\leqslant t_0} S_u\right)$. By observing that the random variable $\sigma \inf_{0\leqslant u\leqslant t_0}(W_u + \nu u)$ has the same distribution as $-\sigma \sup_{0\leqslant u\leqslant t_0}(W_u - \nu u)$, since $\{-W_t, t \geqslant 0\}$ is again a standard Brownian motion under $\mathbb{P}$, we may compute as in (5.62),

$$
\begin{aligned}
\mathbb{E}_{\mathbb{Q}}\left(c \wedge \inf_{0\leqslant u\leqslant t_0} S_u\right) &= \mathbb{E}_{\mathbb{Q}}\left[c \wedge \left(S_0 e^{\inf_{0\leqslant u\leqslant t_0}(\sigma W_u + (\mu - \sigma^2/2)u)}\right)\right] \\
&= \mathbb{E}\left[c \wedge \left(S_0 e^{-\sigma M_{t_0}^{-\nu}}\right)\right],
\end{aligned}
\tag{5.66}
$$

where again $\nu = \left(2\rho - \sigma^2\right)/(2\sigma)$. The expression in (5.66) is

$$
\begin{aligned}
&(c \wedge S_0)\mathbb{P}\left(\sigma M_{t_0}^{-\nu} \leq \ln\left(S_0/(c\wedge S_0)\right)\right) \\
&\qquad + S_0 \mathbb{E}\left[e^{-\sigma M_{t_0} - \nu W_{t_0} - \nu^2 t_0/2} I_{(\sigma M_{t_0} > \ln(S_0/(c\wedge S_0)))}\right],
\end{aligned}
$$

which may be evaluated in the same way as (5.63) to give

$$
\begin{aligned}
\mathbb{E}_{\mathbb{Q}}\left(c \wedge \inf_{0\leqslant u\leqslant t_0} S_u\right) &= \mathbb{E}_{\mathbb{Q}}\left(c \wedge S_0 \wedge S_{t_0}\right) - e^{\rho t_0} r_2\left(S_0, c\wedge S_0, t_0\right) \\
&= c \wedge S_0 - \mathbb{E}_{\mathbb{Q}}\left((c\wedge S_0) - S_{t_0}\right)_+ - e^{\rho t_0} r_2\left(S_0, c\wedge S_0, t_0\right),
\end{aligned}
\tag{5.67}
$$

where the function r_2 is defined by

$$
r_2(x,c,t) = \begin{cases} \dfrac{\sigma^2 x}{2\rho}\left[e^{-2ad}\Phi(-d+a) - \Phi(-d-a)\right] & \text{when } \rho > 0, \\[2ex] \sigma x\sqrt{t}\left[\phi(d) - d\Phi(-d)\right] & \text{when } \rho = 0, \end{cases}
\tag{5.68}
$$

with $d = d(x,c,t) = \left(\ln(x/c) + \sigma^2 t/2\right) / \left(\sigma\sqrt{t}\right)$ and $a = a(t) = \rho\sqrt{t}/\sigma$, as before. Note that since $a \geqslant 0$ the function $r_2 \geqslant 0$. Now, by observing that

$$\left(c - \inf_{0 \leqslant u \leqslant t_0} S_u\right)_+ = c - \left(c \wedge \inf_{0 \leqslant u \leqslant t_0} S_u\right),$$

we see from (5.67) that the time-0 price of the fixed-strike lookback put is

$$e^{-\rho t_0} \mathbb{E}_{\mathbb{Q}} \left[c - \inf_{0 \leqslant u \leqslant t_0} S_u\right]_+ = e^{-\rho t_0} \left[(c - S_0)_+ + \mathbb{E}_{\mathbb{Q}} \left((c \wedge S_0) - S_{t_0}\right)_+\right] + r_2\left(S_0, c \wedge S_0, t_0\right)$$

showing that the price of the lookback put at fixed strike c is the price of the standard European put at strike price $c \wedge S_0$, with the same expiry time t_0, together with an additional amount

$$e^{-\rho t_0} (c - S_0)_+ + r_2\left(S_0, c \wedge S_0, t_0\right).$$

For the time-t price of the lookback put with fixed strike c and expiry time t_0, the calculation is essentially the same as for the fixed-strike lookback call replacing maxima by minima and now using (5.67), and we may see that it equals the time-t price of a standard European put with strike price $c \wedge \inf_{0 \leqslant u \leqslant t} S_u$ and the same expiry **plus** an additional amount equal to

$$e^{-\rho(t_0 - t)} \left(c - \inf_{0 \leqslant u \leqslant t} S_u\right)_+ + r_2\left(S_t, c \wedge \inf_{0 \leqslant u \leqslant t} S_u, t_0 - t\right),$$

where the function r_2 is again given by (5.68).

For the floating-strike lookback call the time-0 price may be calculated immediately from (5.67) (by setting $c = \infty$) as

$$e^{-\rho t_0} \mathbb{E}_{\mathbb{Q}} \left[S_{t_0} - \inf_{0 \leqslant u \leqslant t_0} S_u\right] = e^{-\rho t_0} \mathbb{E}_{\mathbb{Q}} \left(S_{t_0} - S_0\right)_+ + r_2(S_0, S_0, t_0),$$

showing that the price equals the price of a standard European call option at strike price $c = S_0$ together with the extra amount $r_2(S_0, S_0, t_0)$, where the function r_2 is defined in (5.68). With a similar argument to the case of the put, the time-t price of the floating-strike lookback call equals that of a standard European call, with the same expiry time and with strike price $c = \inf_{0 \leqslant u \leqslant t} S_u$, plus the additional amount $r_2\left(S_t, \inf_{0 \leqslant u \leqslant t} S_u, t_0 - t\right)$.

5.4.2 Barrier options

In order to reduce the initial cost of certain options some contracts contain a provision that the contract is cancelled if the share price hits some prescribed level, known as a barrier, during the life of the option; alternatively, there may be a provision that the contract pays out only if the price hits some prescribed level during the life of the option. Such contingent claims are known as **barrier** options.

For example, if b is the barrier let $\tau_b = \inf\{t \geqslant 0 : S_t \geqslant b\}$ when $b \geqslant S_0$ and $\tau_b = \inf\{t \geqslant 0 : S_t \leqslant b\}$ when $b \leqslant S_0$ (where in both cases the infimum of the empty set is taken to be $+\infty$), so that τ_b denotes the first time, if ever, that the stock price crosses the level b. Of course in the Black–Scholes world where S_t is continuous in t then τ_b may be defined as $\tau_b = \inf\{t \geqslant 0 : S_t = b\}$. When a claim that pays C at time t_0 is modified to pay $CI_{(\tau_b \leqslant t_0)}$ then it is known usually as an **up-and-in claim** if $b > S_0$ or a **down-and-in claim** if $b < S_0$, although alternative terminology is a **knock-in claim**; an up-and-in or down-and-in claim only pays out if the barrier is hit in the lifetime, $[0, t_0]$, of the contract. Similarly, the claim paying $CI_{(\tau_b > t_0)}$ would be an **up-and-out claim** if $b > S_0$ and a **down-and-out claim** if $b < S_0$, so a payout only occurs if the barrier is **not** reached in the lifetime of the contract; these are also known as **knock-out claims**.

To evaluate the prices of up-and-in terminal-value claims it is necessary to consider the joint probability $\mathbb{Q}(S_t > x, \tau_b \leqslant t)$ under the martingale probability $\mathbb{Q}$. In this case $b > S_0$, and then for $x > b$, since $S_t > x$ implies that $\tau_b \leqslant t$, it follows that

$$\mathbb{Q}(S_t > x, \tau_b \leqslant t) = \mathbb{Q}(S_t > x), \tag{5.69}$$

while for $x \leqslant b$, we have

$$\begin{aligned}
\mathbb{Q}(S_t > x, \tau_b \leqslant t) &= \mathbb{Q}\left(S_t > x, \sup_{0 \leqslant u \leqslant t} S_u \geqslant b \right) \\
&= \mathbb{P}\left(S_0 e^{\sigma W_t + (\rho - \sigma^2/2)t} > x, S_0 \sup_{0 \leqslant u \leqslant t} e^{\sigma W_u + (\rho - \sigma^2/2)u} \geqslant b \right) \\
&= \mathbb{P}\left(W_t^\nu > \ln(x/S_0)/\sigma, M_t^\nu \geqslant \ln(b/S_0)/\sigma \right),
\end{aligned}$$

where $\nu = \left(2\rho - \sigma^2\right)/(2\sigma)$. Now for $a \geqslant y$, $a > 0$,

$$\begin{aligned}
&\mathbb{P}\left(W_t^\nu > y, M_t^\nu \geqslant a\right) \\
&\quad = 1 - \mathbb{P}\left(W_t^\nu \leqslant y\right) - \mathbb{P}\left(M_t^\nu < a\right) + \mathbb{P}\left(W_t^\nu \leqslant y, M_t^\nu < a\right) \qquad (5.70) \\
&\quad = \mathbb{P}\left(W_t^\nu > a\right) + e^{2a\nu}\left[\mathbb{P}\left(W_t^\nu > y - 2a\right) - \mathbb{P}\left(W_t^\nu > -a\right)\right],
\end{aligned}$$

with the last equality following from (4.25) of Chapter 4. Now, set $a = \ln(b/S_0)/\sigma$ and $y = \ln(x/S_0)/\sigma$, to obtain

$$\begin{aligned}
&\mathbb{Q}(S_t > x, \tau_b \leqslant t) \\
&\quad = \mathbb{Q}(S_t > b) + (1/\kappa)^{\nu/\sigma}\left[\mathbb{Q}(S_t > \kappa x) - \mathbb{Q}(S_t > \kappa b)\right],
\end{aligned} \tag{5.71}$$

for $x \leqslant b$ where $\kappa = (S_0/b)^2$. Differentiating the expressions (5.69) and (5.71) with respect to x, we have

$$\mathbb{Q}(S_t \in dx, \tau_b \leqslant t) = \begin{cases} (1/\kappa)^{\nu/\sigma}\,\mathbb{Q}(S_t/\kappa \in dx) & \text{when } x \leqslant b, \\ \mathbb{Q}(S_t \in dx) & \text{when } x > b. \end{cases}$$

When the payoff of the claim depends just on the terminal value, so that we have $C = f(S_{t_0})$, from (5.69) and (5.71) when $b > S_0$ we may calculate immediately that

$$\begin{aligned} &\mathbb{E}_{\mathbb{Q}}\left[f(S_{t_0})\, I_{(\tau_b \leqslant t_0)}\right] \\ &\quad = \mathbb{E}_{\mathbb{Q}}\left[f(S_{t_0})\, I_{(S_{t_0} > b)}\right] + (1/\kappa)^{\nu/\sigma}\, \mathbb{E}_{\mathbb{Q}}\left[f(S_{t_0}/\kappa)\, I_{(S_{t_0} \leqslant \kappa b)}\right]. \end{aligned} \tag{5.72}$$

The time-0 price of the up-and-in claim is $e^{-\rho t_0}\mathbb{E}_{\mathbb{Q}}\left[f(S_{t_0})\, I_{(\tau_b \leqslant t_0)}\right]$, and we see from the identity (5.72) that this price is the same as that of the terminal-value claim with payoff $g(S_{t_0})$, where

$$g(x) = f(x) I_{(x>b)} + (1/\kappa)^{\nu/\sigma} f(x/\kappa) I_{(x \leqslant \kappa b)}. \tag{5.73}$$

Remark It is important to note that the price of the up-and-in claim is only the same as this terminal-value claim with payoff determined by (5.73) instantaneously at time $t = 0$, since as the time t evolves the value of κ changes; the analogous statement at time t may be expressed in terms of $\kappa_t = S_t/b$ (assuming the barrier has not been hit by time t). □

Example 5.6 *The up-and-in digital call.* The up-and-in digital call pays the amount $I_{(\tau_b \leqslant t_0,\, S_{t_0} > c)}$ at the expiry time t_0 so that the payoff is 1 if the barrier b, $b > S_0$, is hit before expiry and the stock price ends above the strike price c, otherwise it is zero. Here we may take $b \geqslant c$, otherwise the payoff is the same as the ordinary digital call with strike price c since if $c > b$, $S_{t_0} > c$ will imply that $\tau_b \leqslant t_0$. The price at time 0 is then the same as that of the terminal-value claim paying

$$I_{(S_{t_0}>b)} + (1/\kappa)^{\nu/\sigma} I_{(\kappa b \geqslant S_{t_0} > \kappa c)} = I_{(S_{t_0}>b)} + (1/\kappa)^{\nu/\sigma}\left[I_{(S_{t_0}>\kappa c)} - I_{(S_{t_0}>\kappa b)}\right],$$

or, in other words, the price is the same as that of the digital call with strike price b plus $(1/\kappa)^{\nu/\sigma}$ digital calls at strike price κc less that of $(1/\kappa)^{\nu/\sigma}$ digital calls at strike price κb; the prices of these digital calls may be obtained from (5.48). □

Example 5.7 *The up-and-in European call.* This is the up-and-in version of the standard European call at strike price c, say, with payoff $(S_{t_0} - c)_+ I_{(\tau_b \leqslant t_0)}$ (where we need consider only the case $c \leqslant b$ for the same reasons as in the previous example). Using (5.73) its price at time 0 is the same as that of the terminal-value claim with payoff

$$(S_{t_0} - c)_+ I_{(S_{t_0}>b)} + \left(\frac{1}{\kappa}\right)^{\nu/\sigma}\left(\frac{S_{t_0}}{\kappa} - c\right)_+ I_{(S_{t_0} \leqslant \kappa b)}$$

which equals

$$\begin{aligned} &(S_{t_0} - b)_+ + (b-c) I_{(S_{t_0}>b)} \\ &\qquad + (1/\kappa)^{1+\nu/\sigma}\left[(S_{t_0} - \kappa c)_+ - (S_{t_0} - \kappa b)_+ - \kappa(b-c) I_{(S_{t_0}>\kappa b)}\right]; \end{aligned}$$

this shows how the price of the up-and-in call may be expressed in terms of the prices of standard European calls at strikes b, κc and κb and the prices of standard digital calls of strikes b and κb. Alternatively, the price may be expressed in terms of the prices of a standard European call and gap calls in an obvious way. □

For up-and-out claims where $C = f(S_{t_0})$, we may make similar calculations since it follows immediately from (5.72) that

$$\begin{aligned} &\mathbb{E}_{\mathbb{Q}}\left[f(S_{t_0})\, I_{(\tau_b > t_0)}\right] \\ &\quad = \mathbb{E}_{\mathbb{Q}}\left[f(S_{t_0})\right] - \mathbb{E}_{\mathbb{Q}}\left[f(S_{t_0})\, I_{(\tau_b \leqslant t_0)}\right] \\ &\quad = \mathbb{E}_{\mathbb{Q}}\left[f(S_{t_0})\, I_{(S_{t_0} \leqslant b)}\right] - (1/\kappa)^{\nu/\sigma}\, \mathbb{E}_{\mathbb{Q}}\left[f(S_{t_0}/\kappa)\, I_{(S_{t_0} \leqslant \kappa b)}\right], \end{aligned}$$

and so the time-0 price of the up-and-out claim is the same as that of the terminal-value claim with payoff $g(S_{t_0})$, where

$$g(x) = f(x) I_{(x \leqslant b)} - (1/\kappa)^{\nu/\sigma} f(x/\kappa) I_{(x \leqslant \kappa b)}. \tag{5.74}$$

The warning in the remark on page 152 that the two claims only have the same values instantaneously at time 0 applies in this case also.

For down-and-in claims, when $b < S_0$, for $x \leqslant b$ we have that

$$\mathbb{Q}(S_t \leqslant x,\ \tau_b \leqslant t) = \mathbb{Q}(S_t \leqslant x), \tag{5.75}$$

while for $x > b$, with $a = \ln(b/S_0)/\sigma$ and $y = \ln(x/S_0)/\sigma$ as before (so that $a \leqslant 0$),

$$\begin{aligned} \mathbb{Q}(S_t \leqslant x,\ \tau_b \leqslant t) &= \mathbb{Q}\left(S_t \leqslant x,\ \inf_{0 \leqslant u \leqslant t} S_u \leqslant b \right) \\ &= \mathbb{P}\left(S_0 e^{\sigma W_t + (\rho - \sigma^2/2)t} \leqslant x,\ S_0 \inf_{0 \leqslant u \leqslant t} e^{\sigma W_u + (\rho - \sigma^2/2)u} \leqslant b \right) \end{aligned}$$

so that we obtain

$$\begin{aligned} \mathbb{Q}(S_t \leqslant x,\ \tau_b \leqslant t) &= \mathbb{P}\left(W_t^\nu \leqslant y,\ \inf_{0 \leqslant u \leqslant t} W_u^\nu \leqslant a \right) \\ &= \mathbb{P}\left(W_t^{-\nu} \geqslant -y,\ M_t^{-\nu} \geqslant -a\right), \end{aligned}$$

where the last equality comes from the fact that $\{-W_u,\ 0 \leqslant u \leqslant t\}$ is again a standard Brownian motion under $\mathbb{P}$. Since $-a \geqslant -y$ and $-a \geqslant 0$, use (5.70) and the relation that $\mathbb{P}(W_t^{-\nu} \geqslant -y) = \mathbb{P}(W_t^\nu \leqslant y)$ to express this last probability as

$$\mathbb{P}\left(W_t^\nu \leqslant a\right) + e^{2a\nu}\left[\mathbb{P}\left(W_t^\nu \leqslant y - 2a\right) - \mathbb{P}\left(W_t^\nu \leqslant -a\right)\right],$$

from which we conclude that, for $x > b$,

$$\begin{aligned} &\mathbb{Q}(S_t \leqslant x,\ \tau_b \leqslant t) \\ &\quad = \mathbb{Q}(S_t \leqslant b) + (1/\kappa)^{\nu/\sigma}\left[\mathbb{Q}(S_t \leqslant \kappa x) - \mathbb{Q}(S_t \leqslant \kappa b)\right], \end{aligned} \tag{5.76}$$

where $\kappa = (S_0/b)^2$, as previously. For a terminal-value claim with $C = f\left(S_{t_0}\right)$, for $b < S_0$ we may derive the analogue of (5.72) from (5.75) and (5.76),

$$\begin{aligned} &\mathbb{E}_{\mathbb{Q}}\left[f\left(S_{t_0}\right) I_{(\tau_b \leqslant t_0)}\right] \\ &\quad = \mathbb{E}_{\mathbb{Q}}\left[f\left(S_{t_0}\right) I_{\left(S_{t_0} \leqslant b\right)}\right] + (1/\kappa)^{\nu/\sigma}\, \mathbb{E}_{\mathbb{Q}}\left[f\left(S_{t_0}/\kappa\right) I_{\left(S_{t_0} > \kappa b\right)}\right], \end{aligned} \tag{5.77}$$

and conclude that the time-0 price of the down-and-in claim, which has the payoff $f\left(S_{t_0}\right) I_{(\tau_b \leqslant t_0)}$, is the same as that of the terminal-value claim with payoff $g\left(S_{t_0}\right)$, where

$$g(x) = f(x) I_{(x \leqslant b)} + (1/\kappa)^{\nu/\sigma} f(x/\kappa) I_{(x > \kappa b)}. \tag{5.78}$$

Down-and-out claims may be priced in the same way from the observation that

$$\begin{aligned} &\mathbb{E}_{\mathbb{Q}}\left[f\left(S_{t_0}\right) I_{(\tau_b > t_0)}\right] \\ &\quad = \mathbb{E}_{\mathbb{Q}}\left[f\left(S_{t_0}\right) I_{\left(S_{t_0} > b\right)}\right] - (1/\kappa)^{\nu/\sigma}\, \mathbb{E}_{\mathbb{Q}}\left[f\left(S_{t_0}/\kappa\right) I_{\left(S_{t_0} > \kappa b\right)}\right], \end{aligned} \tag{5.79}$$

which implies immediately that the time-0 price of the down-and-out claim with payoff $f\left(S_{t_0}\right) I_{(\tau_b \leqslant t_0)}$ is the same as that of the terminal-value claim with payoff $g\left(S_{t_0}\right)$, where

$$g(x) = f(x) I_{(x > b)} - (1/\kappa)^{\nu/\sigma} f(x/\kappa) I_{(x > \kappa b)}. \tag{5.80}$$

Again, recall the remark on page 152 in connection with (5.78) and (5.80).

The formulae given in this section may be used also to provide the prices of the respective barrier options at any time t, $0 \leqslant t \leqslant t_0$, although in each case the price will depend on whether the barrier b has been reached or not by time t; that is, whether $\tau_b \leqslant t$ or $\tau_b > t$. For example, in the case of knock-in claims, if $\tau_b \leqslant t$ so that the barrier has already been hit then the time-t price is just that of the ordinary claim that pays $C = f\left(S_{t_0}\right)$ at time t_0, while if $\tau_b > t$ the time-t price may be obtained from the time-0 price by replacing S_0 by S_t, t_0 by $t_0 - t$ and taking $\kappa = (S_t/b)^2$.

Note that in general at the instant τ_b (when $\tau_b \leqslant t_0$) at which the barrier is first hit, while the price of these claims will be continuous, there will be a change in the mathematical form of the price; this may lead in turn to a discontinuity in the replicating portfolio at this instant which may in practice give rise to problems of rebalancing the portfolio. We illustrate by considering the down-and-in European call.

Example 5.8 *The down-and-in European call.* Suppose that the strike price of the call is c, with $c > b$, then if $\tau_b \leqslant t$ the price at time t is just given by the usual Black–Scholes formula discussed earlier in this chapter, that is $p(S_t, t)$ where

$$p(x,t) = x\Phi\left(d_1(x,t)\right) - ce^{-\rho(t_0 - t)}\Phi\left(d_2(x,t)\right),$$

with $d_1(x,t)$ and $d_2(x,t)$ defined in (5.4) and (5.5). When $\tau_b > t$, by taking $f(x) = (x-c)_+$ in (5.78) it may be seen that, instantaneously at time t, the time-t price is the same as that of the terminal-value claim paying

$$(b/S_t)^{1+2\rho/\sigma^2}\left(S_{t_0} - c\,(S_t/b)^2\right)_+;$$

by calculating the price of the European call with strike price $c\,(S_t/b)^2$ it follows that the price of this down-and-in call may be expressed as $q\,(S_t,t)$ where

$$q(x,t) = (b/x)^{2\nu/\sigma}\, p\left(b^2/x, t\right).$$

It follows that for all t, $0 \leqslant t \leqslant t_0$, the time-$t$ price of the down-and-in call is given by

$$r\,(S_t,t) = p(S_t,t)\, I_{(\tau_b \leqslant t)} + q\,(S_t,t)\, I_{(\tau_b > t)}.$$

The holding in stock in the replicating portfolio for this option when $\tau_b \leqslant t$ is given by

$$\left.\frac{\partial r}{\partial x}\right|_{(S_t,t)} = \left.\frac{\partial p}{\partial x}\right|_{(S_t,t)} = \Phi\left(d_1\,(S_t,t)\right) > 0, \tag{5.81}$$

while, when $\tau_b > t$, it is

$$\left.\frac{\partial r}{\partial x}\right|_{(S_t,t)} = \left.\frac{\partial q}{\partial x}\right|_{(S_t,t)}$$

where

$$\begin{aligned}\frac{\partial q}{\partial x} &= -\left(\frac{2\nu}{\sigma x}\right) q - \left(\frac{b}{x}\right)^{2+2\nu/\sigma} \Phi\left(d_1\left(b^2/x,t\right)\right) \\ &= -\left(\frac{2\rho}{\sigma^2 x}\right) q - \frac{c}{x}\left(\frac{b}{x}\right)^{2\nu/\sigma} e^{-\rho(t_0-t)}\,\Phi\left(d_2\left(b^2/x,t\right)\right) < 0,\end{aligned} \tag{5.82}$$

since $q \geqslant 0$ and $\rho \geqslant 0$. This shows that while the price of the option changes

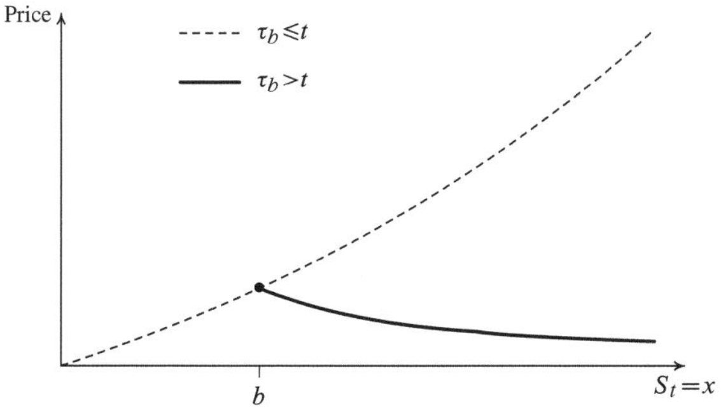

Figure 5.5: Time-t price of a down-and-in call

continuously, there is a discontinuity in the composition of the replicating portfolio at the time τ_b, when $0 \leqslant \tau_b \leqslant t_0$. This is because, as $t \uparrow \tau_b$, S_t tends to b from above, it follows from (5.82) that the portfolio is short in the stock with the holding being negative, while at the instant τ_b the price of the option becomes that of the ordinary European call and the replicating portfolio is long in the stock with the positive holding determined from (5.81). Figure 5.5 illustrates the time-t price which is decreasing (and strictly convex since we may calculate that $\partial^2 q/\partial x^2 > 0$) in $S_t = x$ when $\tau_b > t$ and increasing and strictly convex in x when $\tau_b \leqslant t$. □

5.5 Dividend-paying assets

Up to now we have considered the case when the underlying asset pays no dividends during the life of an option or claim. When this restriction is relaxed there is an effect on the pricing of claims because the evolution of the stock price is changed. Typically there will be predetermined fixed times $0 < t_1 < t_2 < \cdots < t_r < t_0$, with a dividend $\theta_i > 0$ per unit of stock being paid at time t_i, $i = 1, \ldots, r$; we will deal below with the case of dividends which are paid continuously. Here the amount θ_i may in general be random as, for example, it may depend on the past evolution of the stock, but its level will usually be fixed at some time prior to t_i making its value predictable as the time of the dividend payment approaches. The usual situation where the $\{\theta_i\}$ are predetermined fixed amounts does not fit very naturally with the modelling of the stock price in the Black–Scholes model as there is a possibility that the stock price may become negative; the case where dividends are proportional to the stock price just before the instants t_i is more easily treated (see below). Furthermore, for the purpose of illustration we will restrict to the case of just one dividend payment since the extension to multiple dividends is straightforward.

First consider the effect on the stock price of the dividend payment at time t_1. For each unit of stock held at time t_1 an amount θ_1 is paid so that this payment will have been figured into the stock price prior to t_1; consequently, there will be a discontinuity in the stock price when the stock goes 'ex-dividend' at time t_1. Let $S_{t_1-} = \lim_{t \uparrow t_1} S_t$ be the price of the stock just before the instant t_1, then the price at t_1 will be $S_{t_1} = S_{t_1-} - \theta_1$. To model the process $\{S_t, 0 \leqslant t \leqslant t_0\}$, we might represent $S_t = S_0 e^{\sigma W_t + (\mu - \sigma^2/2)t}$, for $t < t_1$, where $\{W_t, t \geqslant 0\}$ is a standard Brownian motion, and

$$S_t = S_{t_1} e^{\sigma(W_t - W_{t_1}) + (\mu - \sigma^2/2)(t - t_1)}$$

for $t > t_1$. There is a problem here in that S_{t_1} would be negative in the situation when $\theta_1 > S_{t_1-}$, which could happen when θ_1 is a constant, but that difficulty does not arise in the proportional case when $\theta_1 = \theta S_{t_1-}$, for some $0 < \theta < 1$.

Now consider the behaviour of the price of some terminal-value claim, such as the standard European call option, across the point of discontinuity of the stock price; we observe that in the Black–Scholes world the claim price will be continuous at the dividend time. Remember that the price of the claim is the same as the value of the hedging portfolio that replicates the claim and that there will be conservation of value of the portfolio as the stock goes ex-dividend since the change in the stock price is balanced by the payment of the dividend; value in the holding in stock in the portfolio is transferred to 'cash in hand' in the form of the dividend. Of course, at the instant of the discontinuity, the composition of the hedging portfolio is discontinuous as we saw could be the case for barrier options at the instant at which the barrier is reached; this may cause practical difficulties in the real world in implementing the rebalancing of the portfolio to accommodate the changed stock price, but it does not cause any difficulties in the mathematical model.

We can see again that when $\mathcal{Q}$ is the martingale probability that transforms the standard Brownian motion $\{W_t,\ 0 \leqslant t \leqslant t_0\}$ with no drift into the Brownian motion with linear drift $(\rho - \mu)/\sigma$ then, as before, the time-t price of the claim paying the amount C at time t_0 is $V_t = \mathbb{E}_{\mathcal{Q}}\left[e^{-\rho(t_0-t)}C \mid \mathcal{F}_t\right]$. Note that now, under $\mathcal{Q}$, the process $\left\{e^{-\rho t}S_t,\ 0 \leqslant t < t_1\right\}$ is a martingale, as is the process $\left\{e^{-\rho t}S_t,\ t_1 \leqslant t \leqslant t_0\right\}$; because of the discontinuity at t_1, in general $\left\{e^{-\rho t}S_t,\ 0 \leqslant t \leqslant t_0\right\}$ will not be a martingale. When θ_1 is a constant then the process $\left\{e^{-\rho t}\overline{S}_t,\ 0 \leqslant t \leqslant t_0\right\}$ is a martingale under $\mathcal{Q}$, where

$$\overline{S}_t = \begin{cases} S_t & \text{for } t < t_1, \\ S_t + e^{\rho(t-t_1)}\theta_1 & \text{for } t_1 \leqslant t \leqslant t_0. \end{cases}$$

For the interval $t_1 \leqslant t \leqslant t_0$, the price, C_t, of a claim will be the same as the case with no dividends, while the price for $t < t_1$ may be obtained by evaluating the expression $\mathbb{E}_{\mathcal{Q}}\left[e^{-\rho(t_1-t)}C_{t_1} \mid \mathcal{F}_t\right]$. For the standard European call, when θ_1 is a constant there is no closed-form expression for the option price for $t < t_1$, while in the proportional case when $\theta_1 = \theta\left(S_{t_1-}\right)$, with $0 < \theta < 1$, the price for $t < t_1$ is just that for the usual situation with S_t replaced by $(1-\theta)\,S_t$. This is because, in the proportional case, S_{t_0} may be represented as

$$S_{t_0} = (1-\theta)\,S_t e^{\sigma\left(W_{t_0}-W_t\right)+\left(\mu-\sigma^2/2\right)(t_0-t)}, \quad \text{for } t < t_1,$$

so the result may be seen immediately by considering

$$\mathbb{E}_{\mathcal{Q}}\left[e^{-\rho(t_0-t)}\left(S_{t_0}-c\right)_+ \mid \mathcal{F}_t\right]$$

and comparing it with the usual calculation.

We turn to the case where the underlying asset pays a dividend continuously in time and which is proportional to the stock price, so that at t the rate of dividend payment is θS_t per unit time where θ is constant. The stochastic differential equation governing the stock price then becomes

$$dS_t = \mu S_t dt - \theta S_t dt + \sigma S_t dW_t.$$

In this case of continuous dividend yield, the martingale probability required is such that under $\mathcal{Q}$, the process $\{e^{-(\rho-\theta)t} S_t,\ 0 \leqslant t \leqslant t_0\}$ is a martingale and the time-t price of a claim paying C at time t_0 is $\mathbb{E}_{\mathcal{Q}}\left[e^{-\rho(t_0-t)}C \mid \mathcal{F}_t\right]$. To see this, when C_t is the value of a portfolio holding X_t units of stock and B_t bonds, the self-financing condition becomes

$$dC_t = X_t dS_t + \left(\rho B_t e^{-\rho(t_0-t)} + \theta X_t S_t\right) dt,$$

because the amount $X_t S_t$ held in stock produces the amount $\theta X_t S_t dt$ in cash in the interval $[t, t+dt]$; using this condition and the relation $C_t = X_t S_t + B_t e^{-\rho(t_0-t)}$ gives

$$d\left(e^{-\rho t} C_t\right) = e^{-\rho t} dC_t - \rho e^{-\rho t} C_t dt = e^{-\theta t} X_t \left[d\left(e^{-(\rho-\theta)t} S_t\right)\right].$$

It follows that $\{e^{-\rho t} C_t,\ 0 \leqslant t \leqslant t_0\}$ is a martingale under the probability $\mathcal{Q}$ that makes $\{e^{-(\rho-\theta)t} S_t,\ 0 \leqslant t \leqslant t_0\}$ a martingale, from which it is immediate that

$$C_t = \mathbb{E}_{\mathcal{Q}}\left[e^{-\rho(t_0-t)} C_{t_0} \mid \mathcal{F}_t\right] = \mathbb{E}_{\mathcal{Q}}\left[e^{-\rho(t_0-t)} C \mid \mathcal{F}_t\right],$$

when $C_{t_0} \equiv C$.

With a continuous dividend yield, to ensure that a portfolio with value

$$C_t = f(S_t, t) = g(S_t, t)\, S_t + h(S_t, t)\, e^{-\rho(t_0-t)}$$

is self-financing, then we argue as before that

$$\begin{aligned} df(S_t, t) &= \left.\frac{\partial f}{\partial x}\right|_{(S_t, t)} dS_t + \left[\frac{\partial f}{\partial t} + \frac{1}{2}\sigma^2 x^2 \frac{\partial^2 f}{\partial x^2}\right]_{(S_t, t)} dt \\ &= g(S_t, t)\, dS_t + \left(\rho h(S_t, t)\, e^{-\rho(t_0-t)} + \theta S_t g(S_t, t)\right) dt, \end{aligned}$$

and equating coefficients of dS_t and dt, we get (5.24) as before and

$$\rho h e^{-\rho(t_0-t)} + \theta x g = \frac{\partial f}{\partial t} + \frac{1}{2}\sigma^2 x^2 \frac{\partial^2 f}{\partial x^2} \tag{5.83}$$

which, by a calculation similar to the previous one, yields

$$\frac{1}{2}\sigma^2 x^2 \frac{\partial g}{\partial x} + x \frac{\partial g}{\partial t} + e^{-\rho(t_0-t)} \frac{\partial h}{\partial t} = \theta x g, \tag{5.84}$$

in place of (5.25). The conditions (5.24) and (5.84) are necessary and sufficient for the portfolio to be self-financing and the Black–Scholes equation (5.28) for contingent claims must be replaced by

$$\frac{1}{2}\sigma^2 x^2 \frac{\partial^2 f}{\partial x^2} + (\rho - \theta)\, x \frac{\partial f}{\partial x} + \frac{\partial f}{\partial t} - \rho f = 0. \tag{5.85}$$

In this situation the martingale probability $\mathcal{Q}$ is such that, under $\mathcal{Q}$, the stock-price process $\{S_t,\ 0 \leqslant t \leqslant t_0\}$ has the same distribution as the process

$$\left\{ S_0 e^{\sigma W_t + (\rho - \theta - \sigma^2/2)t},\ 0 \leqslant t \leqslant t_0 \right\},$$

where $\{W_t,\ 0 \leqslant t \leqslant t_0\}$ is a standard Brownian motion under $\mathcal{Q}$; so calculations under $\mathcal{Q}$ are equivalent to setting $\mu = \rho - \theta$ under the original probability. For any terminal-value claim, paying $f\left(S_{t_0}\right)$ at time t_0, with a continuously paid dividend yield at rate θ on the stock, the time-t price is $e^{-\rho(t_0-t)} E_{\mathcal{Q}}\left[f\left(S_{t_0}\right) \mid \mathcal{F}_t\right]$, which is the same as $e^{-\theta(t_0-t)}$ times the price of the same claim when no dividend is paid and the interest rate ρ is replaced by $\rho - \theta$. Notice that if $p = p(x,t)$ satisfies (5.28) with ρ replaced by $\rho - \theta$, that is

$$\frac{1}{2}\sigma^2 x^2 \frac{\partial^2 p}{\partial x^2} + (\rho - \theta)\, x \frac{\partial p}{\partial x} + \frac{\partial p}{\partial t} - (\rho - \theta)\, p = 0,$$

then it is trivial to verify that $q = q(x,t) = e^{-\theta(t_0-t)} p(x,t)$ satisfies (5.85).

In particular, the time-t price of the European call at strike price c when the stock has a continuous dividend yield is then

$$S_t e^{-\theta(t_0-t)} \Phi\left(d_1\left(S_t,t\right)\right) - c e^{-\rho(t_0-t)} \Phi\left(d_2\left(S_t,t\right)\right),$$

where here

$$d_1(x,t) = \frac{\ln(x/c) + (\rho - \theta + \sigma^2/2)(t_0 - t)}{\sigma\sqrt{t_0 - t}},$$
$$d_2(x,t) = d_1(x,t) - \sigma\sqrt{t_0 - t},$$

in place of the expressions in (5.4) and (5.5).

Finally in this context, we observe that we may combine the discussion given in Section 5.3.2 for a dividend-paying claim with the situation of a dividend-paying stock. It is easy to see that for a claim paying $D_t = k(S_t,t)$/unit time at time t for a stock with continuous dividend yield, the value, f, of a replicating portfolio satisfies

$$\frac{1}{2}\sigma^2 x^2 \frac{\partial^2 f}{\partial x^2} + (\rho - \theta)\, x \frac{\partial f}{\partial x} + \frac{\partial f}{\partial t} - \rho f = -k,$$

in place of (5.85) (or the Black–Scholes equation).

5.6 Exercises

Exercise 5.1 Show *directly* that the price at time t in the Black–Scholes model of a European put option at strike price c and with expiry time t_0 is $q(S_t,t)$ where S_t is the stock price and

$$q(x,t) = -x\Phi\left(-d_1(x,t)\right) + ce^{-\rho(t_0-t)}\Phi\left(-d_2(x,t)\right),$$

with d_1 and d_2 given in (5.4) and (5.5).

Using the Black–Scholes formula for the price of a call option, verify that the put-call parity relation (5.46) on page 135 is satisfied in this case.

Show that the price q is a strictly decreasing, strictly convex function of the stock price. Show that it is strictly increasing and strictly convex as a function of the strike price c and strictly decreasing in the interest rate ρ. Is the price of the put always a decreasing function of the time t?

Exercise 5.2 Calculate the time-0 prices in the Black–Scholes model of the claims that pay the following amounts at time t_0:

$$\text{(i)} \quad \int_0^{t_0} S_t\, dt; \qquad \text{(ii)} \quad \left(\ln S_{t_0}\right)^2.$$

Exercise 5.3 Consider a claim in the Black–Scholes model which pays the holder at the rate $k(S_t) = \theta \ln(S_t)$ per unit time up until the expiry time t_0, where θ is constant. Calculate the price of this claim and the holding in stock in the hedging portfolio at time t. Suppose that $p(S_t, t)$ represents this price; verify that the function $p(x,t)$ satisfies the equation (5.37), that is

$$\frac{1}{2}\sigma^2 x^2 \frac{\partial^2 p}{\partial x^2} + \rho x \frac{\partial p}{\partial x} + \frac{\partial p}{\partial t} - \rho p = -k.$$

Exercise 5.4 For a terminal-value claim in the Black–Scholes model paying $f\left(S_{t_0}\right)$ at time t_0, let $p(\rho,\sigma)$ be the time-0 price considered as a function of the interest rate ρ and the volatility σ. Suppose that ρ is a random variable with the $N\left(\rho_0, \tau^2\right)$-distribution while the volatility σ and the initial stock price S_0 are fixed. Show that

$$\mathbb{E}\, p(\rho,\sigma) = p\left(\rho_0 - \tau^2 t_0/2, \sqrt{\sigma^2 + \tau^2 t_0}\right),$$

where the expectation is over the values taken on by ρ.

Note that here the randomness in the interest rate about its mean value ρ_0 has the same effect as decreasing the interest rate from its mean by the amount $\tau^2 t_0/2$, while at the same time increasing the volatility from σ to $\sqrt{\sigma^2 + \tau^2 t_0}$.

Exercise 5.5 An investor in the Black–Scholes model, with initial wealth w_0 at time 0, wishes to maximize the expected utility of his final wealth at time t_0 when his utility function is $v(x) = px^{1/p}$, where $p > 1$ is a constant. Determine the claim that he should purchase to achieve the maximum expected utility and show that holding this claim is equivalent to holding a portfolio with a certain fixed 'proportion' of wealth in stock, where this holding is positive if $\mu > \rho$ and negative if $\mu < \rho$.

Exercise 5.6 Calculate the time-0 price of the following claims that pay the following amounts at time t_0:

$$\text{(i)}\quad (1/t_0)\int_0^{t_0} (S_t - S_0)_+\, dt; \qquad \text{(ii)}\quad (1/t_0)\int_0^{t_0} \left(S_{t_0} - S_t\right)_+ dt.$$

The claim (i) is the **average-expiry call** while (ii) is the **average-forward-start call**.

You may find it helpful first to establish the following identities for the integral $I = \int_0^a e^{bx}\Phi\left(c\sqrt{x}\right) dx$ with $a > 0$:

when $b \neq 0$ and $c^2 > 2b$,

$$I = \frac{1}{2b}\left[2e^{ab}\Phi\left(c\sqrt{a}\right) - 1 - \frac{c}{\sqrt{c^2 - 2b}}\left(2\Phi\left(\sqrt{a\left(c^2 - 2b\right)}\right) - 1\right)\right];$$

when $b \neq 0$ and $c^2 = 2b$,

$$I = \frac{1}{c^2}\left[2e^{ac^2/2}\Phi\left(c\sqrt{a}\right) - 1 - c\sqrt{2a/\pi}\right];$$

when $b = 0$, $c \neq 0$,

$$I = \frac{1}{c^2}\left[\left(ac^2 - 1\right)\Phi\left(c\sqrt{a}\right) + c\sqrt{a}\phi\left(c\sqrt{a}\right) + \frac{1}{2}\right];$$

finally, when $b = c = 0$, $I = a/2$.

Exercise 5.7 For the up-and-in digital call with strike price c, expiry time t_0 and barrier $b > S_0 \vee c$, suppose that the barrier is hit for the first time at the time τ_b where $\tau_b < t_0$. Prove that at this instant a positive amount of stock must be sold to rebalance the replicating portfolio.

You may wish to use the fact that

$$a\Phi(x + a) + \phi(x + a) < a\Phi(a) + \phi(a), \text{ for } x > 0 \text{ and } -\infty < a < \infty. \tag{5.86}$$

Exercise 5.8 An **American digital barrier call** pays 1 at the instant the barrier $b > S_0$ is first reached by the stock price if this instant occurs at, or before, the expiry time t_0; otherwise it pays nothing. Determine the price at time 0 of this option in the Black–Scholes model.
[Strictly, this option is not an American option in that the time it is exercised is not under the control of the holder since it is a predetermined, but random, time.]

Exercise 5.9 An **up-down-and-in claim** pays $f\left(S_{t_0}\right)$ at time t_0 if and only if the barrier $b_1 > S_0$ is reached during the lifetime of the claim $[0, t_0]$ and, subsequently, the barrier $b_2 < b_1$ is also attained before the expiry time t_0. Such a claim might be of interest to an individual who expected the stock price first to rise and then to fall before expiry. Show that the price at time 0 of this claim is the same as that of a standard terminal-value claim paying $g\left(S_{t_0}\right)$ at time t_0, where

$$g(x) = (\kappa_1/\kappa_2)^{\nu/\sigma} f(\kappa_1 x/\kappa_2) I_{(x \geqslant \kappa_2 b_2/\kappa_1)} + (1/\kappa_1)^{\nu/\sigma} f(x/\kappa_1) I_{(x \leqslant \kappa_1 b_2)},$$

with $\kappa_i = (S_0/b_i)^2$, $i = 1, 2$, and $\nu = \left(2\rho - \sigma^2\right) / (2\sigma)$.

Exercise 5.10 Suppose that the stock has a continuous dividend yield of rate θ. For a general terminal-value claim paying $f\left(S_{t_0}\right)$ at time t_0, let $q(\theta, \sigma)$ denote the time-0 price considered as a function of the dividend yield and of the volatility. Suppose that θ is a random variable and has the $N\left(\theta_0, \tau^2\right)$-distribution while σ and the other parameters determining the price are fixed. Show that

$$\mathbb{E} q(\theta, \sigma) = q\left(\theta_0 - \tau^2 t_0/2, \sqrt{\sigma^2 + \tau^2 t_0}\right),$$

where the expectation is over the values taken on by θ.

Exercise 5.11 For the standard Black–Scholes model, consider a path-dependent claim which pays $f\left(\left(\prod_{i=1}^{n} S_{t_i}\right)^{1/n}\right)$ at time t_0, where $0 \leqslant t_1 \leqslant \cdots \leqslant t_n \leqslant t_0$ are fixed times; that is, the payoff is a function of the geometric average of the stock prices at designated times during the lifetime, $[0, t_0]$, of the contract. Show that the price of this claim at time 0 is the same as that for a claim with payoff $f\left(S_{t_0}\right)$ at time t_0 when the stock has volatility $\overline{\sigma}$ and pays a continuous dividend θ, where

$$\overline{\sigma}^2 = \frac{\sigma^2}{n^2 t_0} \sum_{i=1}^{n} (2n - 2i + 1) t_i \text{ and } \theta = \left(\rho - \overline{\sigma}^2/2\right) - \left(\rho - \sigma^2/2\right) \sum_{i=1}^{n} t_i / (n t_0).$$

In particular, when $t_i = t_0 - (n - i)\delta/n$, for $0 \leqslant \delta \leqslant t_0$, and n is large, then

$$\overline{\sigma} \approx \sigma \sqrt{1 - 2\delta/(3t_0)} \quad \text{and} \quad \theta \approx \delta\left(\rho + \sigma^2/6\right) / (2t_0).$$

Exercise 5.12 In the context of the Black–Scholes formula for a European call option with d_2 as defined in (5.5), show from first principles (without using the results in Section 5.3.1), that for $u < t < t_0$,

$$\mathbb{E}_{\mathbb{Q}}\left[\Phi\left(d_2\left(S_t, t\right)\right) \,\middle|\, \mathcal{F}_u\right] = \Phi\left(d_2\left(S_u, u\right)\right)$$

so that the holding in bonds in the replicating portfolio, $B_t = -c\Phi\left(d_2\left(S_t, t\right)\right)$, is a martingale under $\mathbb{Q}$. You may wish to use the fact (see (A.17)) that for a random variable X with the $N\left(\nu, \tau^2\right)$-distibution,

$$\mathbb{E}\,\Phi(X) = \Phi\left(\nu/\sqrt{1 + \tau^2}\right).$$

Use a similar argument to show that the discounted value of the holding in stock, $e^{-\rho t} S_t \Phi\left(d_1\left(S_1, t\right)\right)$, is also a martingale under $\mathbb{Q}$.

Exercise 5.13 Suppose that $p = p(x,t)$ is a function satisfying the Black–Scholes equation (5.28) for $t < t_0$, with the boundary condition $p(x, t_0) \equiv f(x)$. By using Itô's Lemma and the arguments that gave (5.34), verify directly that $\left\{e^{-\rho t} p(S_t, t)\right\}$ is a martingale under the martingale probability $\mathbb{Q}$ and that $p(S_t, t)$ may be represented as

$$p(S_t, t) = E_{\mathbb{Q}}\left[e^{-\rho(t_0 - t)} f\left(S_{t_0}\right) \,\middle|\, \mathcal{F}_t\right].$$

Also, use a direct argument to show that the discounted value of the holding in stock, $\left\{e^{-\rho t} g(S_t, t)\, S_t\right\}$ is a martingale under $\mathbb{Q}$, where $g = \partial p / \partial x$.

Chapter 6

INTEREST-RATE MODELS

6.1 Introduction

When formulating a mathematical model for interest rates, the principal difference from previous sections is that, unlike a stock price, for which the changes may be represented by a one-parameter stochastic process, interest rates are naturally represented by a two-parameter process. One of the parameters is the time that the loan matures while the other is 'real' time, so that the process being modelled is a random surface. In interest-rate markets the tradeable assets are bonds and we will assume that we are dealing with a random surface $\{P_{s,t} : 0 \leqslant s \leqslant t\}$, where $P_{s,t}$ represents the price at time $s \geqslant 0$ of a bond paying one unit at time $t \geqslant s$. These are known as **zero-coupon bonds** in that they just pay out the fixed amount 1 at the maturity time t and make no payments ('coupons') at other times during their lifetime. We will not allow the possibility of default for these bonds.

We will assume sufficient regularity so that $P_{s,t}$ may be represented as

$$P_{s,t} = \exp\left(-\int_s^t F_{s,u} du\right),$$

where $F_{s,t}$ denotes the instantaneous forward interest rate for the date t at time s, $0 \leqslant s \leqslant t$, giving $F_{s,t} = -\partial \ln P_{s,t}/\partial t$; that is, $F_{s,t}$ is the rate of interest paid to borrow (or lend) between times t and $t + dt$, for (infinitesimally) small dt, when the contract is entered at time $s < t$. The **short** rate (or **spot** rate) at time s is $R_s = F_{s,s}$. It is a consequence of the representation that $P_{s,t} \to 1$ as $s \to t$, so that bond prices tend to their face values at maturity.

The dependence of interest rates on the maturity time is often referred to as the **term structure of interest rates** and the rates are usually expressed in terms of the **yield**

$$Y_{s,t} = -\frac{1}{t-s} \ln P_{s,t} = \frac{1}{t-s} \int_s^t F_{s,u} du,$$

which is the equivalent continuously compounded fixed interest rate for a loan entered at time s for the time interval $(s, t]$.

We will assume that for each time s the yields $Y_{s,t}$ (or equivalently, the bond prices $P_{s,t}$) are available for each $t \geqslant s$. In practice however, at any time s what may be observed in the market is usually only a fixed number, n, of yields of the form $Y_{s,t_1}, \ldots, Y_{s,t_n}$ (or the bond prices $P_{s,t_1}, \ldots, P_{s,t_n}$) for fixed times $t_1, \ldots, t_n$ with $s \leqslant t_1 < \cdots < t_n$. This leads to questions about how to obtain the whole **yield curve** $\{Y_{s,s+d}, d \geqslant 0\}$ at time s by interpolation or approximation, but we will not concern ourselves with such issues here.

In the Black–Scholes model the bank account was used as a 'numeraire'; that is a particular asset whose value provides a unit of measurement for the other assets, with the reciprocal of its value being the discount factor. It is relative to this unit of measurement that the stock prices being martingales characterized the martingale probability which enabled the prices of contingent claims to be evaluated. The natural numeraire in the interest-rate market is the **money-market account**; the value of this account at time s is the amount obtained by investing one unit at time 0 and then rolling it over continuously at the short rate between times 0 and s so that it is worth $\exp\left(\int_0^s R_u du\right)$ at time s. We will denote the discounted bond prices as

$$Z_{s,t} = P_{s,t} \exp\left(-\int_0^s R_u du\right).$$

We will assume that the information available to an investor in this market at time s is $\mathcal{F}_s = \sigma\left(P_{u,v} : 0 \leqslant u \leqslant s, u \leqslant v < \infty\right)$, so that at time s an investor knows the prices of bonds of all maturities at all times up to time s; from the representation of $P_{s,t}$ in terms of the forward rates $F_{s,t}$, this is the same as assuming that $\mathcal{F}_s = \sigma\left(F_{u,v} : 0 \leqslant u \leqslant s,\ u \leqslant v < \infty\right)$, so that the information available is equivalent to observing all the instantaneous forward rates up to time s. It is customary in discussing interest rates to concentrate on investigating the evolution of the rates under the martingale probability; that is, the probability under which, for each $t \geqslant 0$, the process $\{Z_{s,t}, \mathcal{F}_s, 0 \leqslant s \leqslant t\}$ is a martingale. The following result provides the principal characterization of this probability.

Theorem 6.1 *The discounted bond-price process $\{Z_{s,t}, \mathcal{F}_s, 0 \leqslant s \leqslant t\}$ is a martingale, for each $t \geqslant 0$, if and only if the bond prices $\{P_{s,t},\ 0 \leqslant s \leqslant t\}$ may be represented in terms of the short-rate process $\{R_s,\ s \geqslant 0\}$ as*

$$P_{s,t} = \mathbb{E}\left[e^{-\int_s^t R_u du} \,\middle|\, \mathcal{F}_s\right], \quad \text{for all} \quad 0 \leqslant s \leqslant t. \tag{6.1}$$

Proof. Firstly, when (6.1) holds then, since $\int_0^s R_u du$ is known given $\mathcal{F}_s$,

$$Z_{s,t} = e^{-\int_0^s R_u du}\, \mathbb{E}\left[e^{-\int_s^t R_u du} \,\middle|\, \mathcal{F}_s\right] = \mathbb{E}\left[e^{-\int_0^t R_u du} \,\middle|\, \mathcal{F}_s\right] \tag{6.2}$$

and so for fixed t, $\{Z_{s,t}, \mathcal{F}_s, 0 \leqslant s \leqslant t\}$ is a martingale. Conversely, when the process $\{Z_{s,t}, \mathcal{F}_s, 0 \leqslant s \leqslant t\}$ is a martingale then, since $P_{t,t} = 1$,

$$Z_{s,t} = \mathbb{E}\left[Z_{t,t} \mid \mathcal{F}_s\right] = \mathbb{E}\left[e^{-\int_0^t R_u du} P_{t,t} \,\middle|\, \mathcal{F}_s\right] = \mathbb{E}\left[e^{-\int_0^t R_u du} \,\middle|\, \mathcal{F}_s\right],$$

and the result follows by reversing the step in (6.2). ∎

Much effort has been given in the literature to devising appropriate models for the short-rate process, $\{R_s,\ s \geqslant 0\}$, and for the filtration, $\{\mathcal{F}_s,\ s \geqslant 0\}$, which in general will contain more information than that generated just by the short-rate process. In the next section, we present a brief survey of some standard interest-rate models, starting with the case where the filtration is indeed generated by the short rate; in this situation, all bond prices $\{P_{u,t},\ u \leqslant s, u \leqslant t\}$ are determined by $\{R_u,\ u \leqslant s\}$. Before that, we introduce several of the more common contracts on interest rates.

A **caplet** at strike rate d, say, for the period from t to $t + \Delta$ may be regarded as a European option on the forward rate $F^{\Delta}_{s,t}$ where

$$F^{\Delta}_{s,t} = \frac{1}{\Delta} \int_t^{t+\Delta} F_{s,u} du = \frac{1}{\Delta} \ln\left(P_{s,t}/P_{s,t+\Delta}\right).$$

Note that the forward rate $F^{\Delta}_{s,t}$ is the continuously compounded interest rate for the period between the times t and $t + \Delta$ implied by the bond prices at time s, $s \leqslant t$.

The option is exercised at time t when $F^{\Delta}_{t,t} = Y_{t,t+\Delta} > d$, yielding a payoff at time $t + \Delta$ of

$$\begin{aligned}\left[\left(e^{\Delta F^{\Delta}_{t,t}} - 1\right) - \left(e^{\Delta d} - 1\right)\right]_+ &= \left[e^{\Delta F^{\Delta}_{t,t}} - e^{\Delta d}\right]_+ \\ &= \left[1/P_{t,t+\Delta} - e^{\Delta d}\right]_+ .\end{aligned} \tag{6.3}$$

An **interest-rate cap** consists normally of a string of such caplets for successive time periods; it is sufficient to consider just one period and the price of the cap follows by adding the prices of the individual caplets together. The payoff of the caplet at time $t + \Delta$, discounted back to time $s \leqslant t$, is

$$e^{-\left(\int_s^{t+\Delta} R_u du\right)} \left(e^{\Delta F^{\Delta}_{t,t}} - e^{\Delta d}\right)_+, \tag{6.4}$$

and its price at time $s \leqslant t$ would be determined by calculating the conditional expectation of the expression in (6.4),

$$\mathbb{E}\left[e^{-\left(\int_s^{t+\Delta} R_u du\right)} \left(e^{\Delta F^{\Delta}_{t,t}} - e^{\Delta d}\right)_+ \,\Big|\, \mathcal{F}_s\right], \tag{6.5}$$

where the expectation is with respect to the martingale probability of Theorem 6.1. This price is calculated in the context of a particular Gaussian model of interest rates in Section 6.3.2.

On the other hand, a **floorlet** for the period from t to $t + \Delta$ at strike d is the corresponding contract that pays off when the forward rate for the period at time t is below the strike level d; the payoff to the holder of the contract at time $t + \Delta$ would be

$$\left[\left(e^{\Delta d} - 1\right) - \left(e^{\Delta F^{\Delta}_{t,t}} - 1\right)\right]_+ = \left[e^{\Delta d} - e^{\Delta F^{\Delta}_{t,t}}\right]_+ \tag{6.6}$$

and, corresponding to (6.5), its price at time s would be

$$\mathbb{E}\left[e^{-\left(\int_s^{t+\Delta} R_u du\right)} \left(e^{\Delta d} - e^{\Delta F^{\Delta}_{t,t}}\right)_+ \,\Big|\, \mathcal{F}_s\right]. \tag{6.7}$$

An **interest-rate floor** consists of a string of floorlets for successive time periods and its price is determined by adding the prices of the individual floorlets together.

A **forward-rate agreement** is a contract between two parties to exchange the payments accrued from a fixed interest rate d and from a floating-rate set at time t over a given period from t to $t+\Delta$; the first party agrees to pay the second the amount $e^{\Delta d}-1$, while the second agrees to pay the first the amount $e^{\Delta F_{t,t}^{\Delta}}-1$, so that the net amount received by the first party is

$$e^{\Delta F_{t,t}^{\Delta}} - e^{\Delta d} = \left[e^{\Delta F_{t,t}^{\Delta}} - e^{\Delta d}\right]_+ - \left[e^{\Delta d} - e^{\Delta F_{t,t}^{\Delta}}\right]_+; \tag{6.8}$$

typically, this amount is paid at time t, rather than at time $t+\Delta$ which is the case for the caplet or floorlet, so that its price at time $s \leqslant t$ would be

$$\mathbb{E}\left[e^{-\left(\int_s^t R_u du\right)}\left(e^{\Delta F_{t,t}^{\Delta}} - e^{\Delta d}\right) \,\middle|\, \mathcal{F}_s\right]. \tag{6.9}$$

When we amend the forward-rate agreement so that the time of payment is $t+\Delta$ rather than t then (6.9) would be replaced by

$$\mathbb{E}\left[e^{-\left(\int_s^{t+\Delta} R_u du\right)}\left(e^{\Delta F_{t,t}^{\Delta}} - e^{\Delta d}\right) \,\middle|\, \mathcal{F}_s\right]. \tag{6.10}$$

We may see immediately from (6.5), (6.7), (6.8) and (6.10) that there is a put-call parity relationship between the price of the amended forward-rate agreement and the prices of the caplet and floorlet. An **interest-rate swap** consists of a string of such amended forward-rate agreements for successive time periods and its price is determined by adding together the prices of the agreements for the individual periods.

6.2 Survey of interest-rate models

6.2.1 One-factor models

The simplest models assume that there is only one source of randomness in interest rates and that may be captured by specifying the short-rate process $\{R_s, s \geqslant 0\}$. Typically the evolution of this process is described by a stochastic differential equation of the form

$$dR_s = \mu_s ds + \sigma_s dW_s, \tag{6.11}$$

where $\{W_s, s \geqslant 0\}$ is a standard Brownian motion and μ_s, σ_s are suitable processes adapted to $\mathcal{F}_s = \sigma(W_u, 0 \leqslant u \leqslant s) = \sigma(R_u, 0 \leqslant u \leqslant s)$; such a model is known as a **one-factor** model, because all of the randomness derives from the one Brownian motion driving the equation (6.11). The usual case considered is when μ_s and σ_s are just functions of the short rate R_s and the time s, so that $\mu_s = \mu(R_s, s)$ and $\sigma_s = \sigma(R_s, s)$.

Example 6.1 *Vasicek model.* The case when $\mu_s = \alpha(\beta - R_s)$ and $\sigma_s \equiv \sigma$, a constant, was first introduced by Vasicek, and (when $\alpha > 0$) it gives the situation where the short-rate process is an Ornstein–Uhlenbeck process, mean-reverting to the level β, and we know that, for $u > s$,

$$R_u = e^{-\alpha(u-s)} R_s + \left(1 - e^{-\alpha(u-s)}\right) \beta + \sigma \int_s^u e^{-\alpha(u-v)} d W_v.$$

It follows that for $t > s$, when $\alpha \neq 0$,

$$\int_s^t R_u du = \frac{1}{\alpha}\left[1 - e^{-\alpha(t-s)}\right](R_s - \beta) + \beta(t-s) + \frac{\sigma}{\alpha}\int_s^t \left(1 - e^{-\alpha(t-v)}\right) d W_v, \tag{6.12}$$

while in the case $\alpha = 0$,

$$\int_s^t R_u du = (t-s) R_s + \sigma \int_s^t (t-v) d W_v. \tag{6.13}$$

Note that, when $\alpha \neq 0$, the quantity $(\sigma/\alpha)\int_s^t \left(1 - e^{-\alpha(t-v)}\right) d W_v$ is independent of the sigma field $\mathcal{F}_s = \sigma\left(R_u,\ u \leqslant s\right\}$ and has the normal distribution with mean 0 and variance

$$\frac{\sigma^2}{\alpha^2}\int_s^t \left(1 - e^{-\alpha(t-v)}\right)^2 dv = \frac{\sigma^2}{\alpha^2}\left[t - s - \frac{2}{\alpha}\left(1 - e^{-\alpha(t-s)}\right) + \frac{1}{2\alpha}\left(1 - e^{-2\alpha(t-s)}\right)\right]; \tag{6.14}$$

in the case when $\alpha = 0$, the quantity $\sigma \int_s^t (t-v) d W_v$ is similarly independent of $\mathcal{F}_s$ and has the normal distribution with mean 0 and variance

$$\sigma^2 \int_s^t (t-v)^2 dv = \sigma^2 (t-s)^3 / 3. \tag{6.15}$$

We may now deduce from (6.1), when we set $b_{s,t} = \left(1 - e^{-\alpha(t-s)}\right)/\alpha$ in the case when $\alpha \neq 0$, that the bond price is given by

$$\begin{aligned} P_{s,t} &= \mathbb{E}\left[e^{-\int_s^t R_u du} \,\middle|\, \mathcal{F}_s\right] \\ &= e^{-b_{s,t}(R_s - \beta) - \beta(t-s)}\, \mathbb{E}\left[e^{-(\sigma/\alpha)\int_s^t \left(1 - e^{-\alpha(t-v)}\right) d W_v}\right]; \end{aligned}$$

use the expression for the moment-generating function of a random variable with the normal distribution (in (A.16) on page 199) to see that the bond price takes the form

$$P_{s,t} = \exp\left[a_{s,t} - b_{s,t} R_s\right], \tag{6.16}$$

where

$$a_{s,t} = \frac{(b_{s,t} - t + s)\left(\alpha^2\beta - \sigma^2/2\right)}{\alpha^2} - \frac{\sigma^2 b_{s,t}^2}{4\alpha}.$$

In the case when $\alpha = 0$, the bond price again takes the form (6.16) but now with $b_{s,t} = t - s$ and $a_{s,t} = \sigma^2 (t-s)^3/6$.

For this example, the calculation of the form of the exact expression for the bond prices is possible because of the Markovian nature of the short-rate process and the relatively tractable form of the distributions, since the process is Gaussian.

An alternative derivation of (6.16) is that essentially given originally by Vasicek, who postulated that $P_{s,t} = f(R_s, s, t)$ for some appropriate function $f = f(r, s, t)$. Then, using Itô's Lemma on $Z_{s,t} = e^{-\int_0^s R_u du} P_{s,t}$ we obtain

$$\begin{aligned} dZ_{s,t} &= e^{-\int_0^s R_u du}\left[\frac{1}{2}\sigma^2\frac{\partial^2 f}{\partial r^2}ds + \frac{\partial f}{\partial r}dR_s + \frac{\partial f}{\partial s}ds - R_s f ds\right] \\ &= e^{-\int_0^s R_u du}\left[\left(\frac{1}{2}\sigma^2\frac{\partial^2 f}{\partial r^2} + \alpha(\beta - R_s)\frac{\partial f}{\partial r} + \frac{\partial f}{\partial s} - R_s f\right)ds + \sigma\frac{\partial f}{\partial r}dW_s\right]; \end{aligned}$$

for $Z_{s,t}$ to be a martingale we need the coefficient of ds in this expression to vanish, that is we require $f = f(r, s, t)$ to satisfy the equation

$$\frac{1}{2}\sigma^2\frac{\partial^2 f}{\partial r^2} + \alpha(\beta - r)\frac{\partial f}{\partial r} + \frac{\partial f}{\partial s} - rf = 0. \tag{6.17}$$

Fix t, and look for a solution of (6.17) of the form $f = e^{a-br}$, where $a = a(s)$ and $b = b(s)$, then substituting into (6.17) and dividing through by f we obtain

$$\frac{1}{2}\sigma^2 b^2 - \alpha(\beta - r)b + a' - rb' - r = 0.$$

This equation must hold for all values of r, so we may equate the coefficient of r to zero to give the two equations

$$\alpha b - b' - 1 = 0 \quad \text{and} \quad \frac{1}{2}\sigma^2 b^2 - \alpha\beta b + a' = 0;$$

recall that $P_{t,t} = 1$ so we must have $b(t) = 0$ and from the first equation we obtain $b(s) = b_{s,t}$ as given above and substituting this expression into the second equation yields $a(s) = a_{s,t}$. □

There are a number of other one-factor models, involving different choices of the drift μ_s and volatility σ_s, which may be handled in a similar fashion to produce bond prices as a function of the short-rate process $\{R_s,\ s \geqslant 0\}$; some of these are set out in the exercises in Section 6.4. In general, the equation (6.11) will lead to the equation

$$\frac{1}{2}\sigma_s^2\frac{\partial^2 f}{\partial r^2} + \mu_s\frac{\partial f}{\partial r} + \frac{\partial f}{\partial s} - rf = 0, \tag{6.18}$$

in place of (6.17), which will be satisfied by the function f determining the bond prices $P_{s,t} = f(R_s, s, t)$. The drift and volatility are usually chosen to ensure some particular desirable feature of the short-rate process (such as maintaining positivity of the process $\{R_s : s \geqslant 0\}$, which is the case for the Cox, Ingersoll and Ross model introduced in Exercise 6.2). Note that the equation (6.18) is the analogue of the Black–Scholes equation (5.28) in this context.

One of the principal drawbacks of a one-factor model such as the Vasicek model is that it is not in general possible to calibrate it so that it fits the presently observed term structure; that is, if the present time is $s = 0$, and the rates $\{F_{0,t},\ t \geqslant 0\}$ (or equivalently $\{P_{0,t},\ t \geqslant 0\}$) are observed, then because of the functional form of the bond prices as given in (6.16) which depend on just three parameters α, β and σ, it is not possible to choose values of those parameters so that the observed bond prices are fitted exactly by the model. It is possible to adapt the Vasicek model to allow fitting of the initial term structure by adding a time-dependent function in the drift term of the driving stochastic differential equation (6.11) as follows.

Example 6.2 *Extended Vasicek model* (also known as the *Hull and White model*). The short rate is assumed to satisfy

$$dR_s = \alpha\,(\theta_s - R_s)\,ds + \sigma dW_s,$$

where θ_s is deterministic but time dependent and we will assume that $\alpha \neq 0$, so that the process is mean-reverting around the moving target θ_s. Proceeding as before we have that, for $u > s$,

$$R_u = e^{-\alpha(u-s)}R_s + \int_s^u \alpha e^{-\alpha(u-v)}\theta_v dv + \sigma \int_s^u e^{-\alpha(u-v)} dW_v,$$

and conditional on $\mathcal{F}_s = \sigma\,(R_v,\ v \leqslant s)$, $\int_s^t R_u du$ has the normal distribution with mean

$$b_{s,t}R_s + \int_s^t \left(1 - e^{-\alpha(t-u)}\right)\theta_u du$$

and variance given again by (6.14), where $b_{s,t} = \left(1 - e^{-\alpha(t-s)}\right)/\alpha$. It follows that $P_{s,t}$ again takes the form (6.16) but now with

$$a_{s,t} = -\int_s^t \left(1 - e^{-\alpha(t-v)}\right)\theta_v dv - \frac{\sigma^2}{2\alpha}\left(\frac{b_{s,t} - t + s}{\alpha} + \frac{b_{s,t}^2}{2}\right). \tag{6.19}$$

Assume that the initial term structure $\{F_{0,t},\ t \geqslant 0\}$ is observed, or equivalently that the bond prices $\{P_{0,t},\ t \geqslant 0\}$ are observed, then we may solve for the function θ_t, since we must have

$$a_{0,t} - b_{0,t}R_0 = -\int_0^t F_{0,u}du;$$

differentiating with respect to t, gives

$$\alpha e^{-\alpha t}\int_0^t e^{\alpha u}\theta_u du - \frac{\sigma^2}{2\alpha^2}\left(1 - e^{-\alpha t}\right)^2 - e^{-\alpha t}R_0 = F_{0,t}.$$

Multiply through by $e^{\alpha t}/\alpha$ and assume that we may differentiate again to obtain

$$\theta_t = \frac{1}{\alpha} e^{-\alpha t} \frac{d\left(e^{\alpha t} F_{0,t}\right)}{dt} + \frac{\sigma^2}{\alpha^2} e^{-\alpha t} \sinh(\alpha t);$$

substituting this function into (6.19) enables us to express the function $a_{s,t}$ in terms of the initial bond prices (and forward rates) as

$$a_{s,t} = \ln\left(\frac{P_{0,t}}{P_{0,s}}\right) + b_{s,t} F_{0,s} - \frac{\sigma^2}{2\alpha} e^{-\alpha s} \sinh(\alpha s) b_{s,t}^2.$$

While this model may then be calibrated to the initial term structure the major drawback to the implementation of this and all finite-factor models is that as time evolves they must be constantly re-calibrated to fit the changing term structure. □

Any one-factor model leading to bond prices of the form (6.16), with the terms $a_{s,t} = a(t-s)$ and $b_{s,t} = b(t-s)$ for some functions a and b, is known as an **affine model**; the terminology is because it follows that the yield

$$Y_{s,t} = -\ln\left(P_{s,t}\right)/(t-s) = \left[b(t-s)R_s - a(t-s)\right]/(t-s)$$

is then an affine function (linear plus a constant) of the short rate R_s. In Exercise 6.4 it may be seen that affine models are essentially those for which the drift, μ_s, and the squared volatility, σ_s^2, of the short rate are affine functions in R_s.

6.2.2 Forward-rate and market models

We conclude this brief survey of interest-rate models with a mention of the two models which have generated the most interest in recent years.

The first centres on specifying the evolution of the forward rates $\{F_{s,t},\ 0 \leqslant s \leqslant t\}$ and inferring bond prices and the prices of derivative contracts from these. The **Heath, Jarrow and Morton model** assumes that for each fixed t, the rate $F_{s,t}$ satisfies

$$F_{s,t} = F_{0,t} + \int_0^s \alpha_{u,t} du + \sum_{i=1}^k \int_0^s \sigma_{i,u,t} dW_u^i, \quad 0 \leqslant s \leqslant t, \tag{6.20}$$

where $W^1, \ldots, W^k$ are standard Brownian motions, not necessarily independent; here the information at time s is $\mathcal{F}_s = \sigma\left(W_u^i,\ u \leqslant s,\ i = 1, \ldots, k\right)$ while for $s \geqslant 0$, $\{\alpha_{s,t},\ t \geqslant s\}$ and $\{\sigma_{i,s,t}, t \geqslant s\}$ are appropriate processes adapted to $\mathcal{F}_s$. The movement of the whole forward-rate curve through time is generated by a finite number of Brownian motions.

Under certain conditions, the uniqueness of the equivalent martingale measure for the HJM model may be established, in contrast to studies of many other models which have proceeded along the lines of giving conditions to ensure that the original probability distribution gives rise to discounted bond-price processes which are martingales and then pricing contingent claims using this distribution. In the generality of (6.20) one cannot expect to make exact computations; in the special case when the $\alpha_{u,t}$ and $\sigma_{i,u,t}$ are deterministic functions the joint distributions of the $\{F_{s,t}\}$ become Gaussian; it would follow that the bond prices then have log-normal distributions. In the remaining sections of this chapter, we will look in detail at a model which generalizes this particular situation; the generalization allows possibly infinitely many Brownian motions driving the evolution of the instantaneous forward rates and we will consider it principally for its mathematical tractability.

One of the principal criticisms of the one-factor models of the previous section, or of the HJM model, is that the specification of the dynamics of the movements of interest rates is in terms of quantities for which the values are not directly observable in the market; in the case of the one-factor models the movement of the spot rate, R_s, is specified while in the case of the HJM model it is the movement of the instantaneous forward rates, $F_{s,t}$. Neither of these quantities are directly observable but their values must be approximated from other observable data.

As mentioned in Section 6.1 the data available from the market at time s are bond prices $P_{s,t_1}, \ldots, P_{s,t_n}$ where n is a fixed number and the times $t_1 < \cdots < t_n$ are fixed, with $s \leqslant t_1$. Now set

$$L_s^i = \frac{1}{t_{i+1} - t_i}\left(\frac{P_{s,t_i}}{P_{s,t_{i+1}}} - 1\right) \quad \text{for} \quad s \leqslant t_i;$$

the quantity L_s^i is referred to as a London Interbank Offer Rate (LIBOR) at time s for the period (t_i, t_{i+1}). It is an effective forward interest rate (on an annualized basis, *not* continuously compounded) for a loan for the period (t_i, t_{i+1}) contracted at time s.

The observed quantities $L_s^1, \ldots, L_s^n$ at time s are taken to be the basic building blocks for constructing the **LIBOR market model**, for which it is assumed that each L_s^i satisfies a stochastic differential equation of the form

$$dL_s^i = L_s^i\left(\alpha_s^i + \sum_{j=1}^{k} \sigma_s^{i,j}\, dW_s^j\right) \quad \text{for} \quad s \leqslant t_i, \tag{6.21}$$

where $W^1, \ldots, W^k$ are standard Brownian motions, not necessarily independent, and α_s^i and $\sigma_s^{i,j}$ are deterministic functions of s for each i and j, $1 \leqslant i \leqslant n$, $1 \leqslant j \leqslant k$; this last assumption and the form of (6.21) ensure that the distributions of the L_s^i are log-normal which make for tractable calculations of derivative prices in the model. In particular, it leads to Black–Scholes-like formulae for the prices of contracts such as calls and puts on interest rates (as with Black's formula given in (6.35) on page 178).

6.3 Gaussian random-field model

6.3.1 Introduction

The **Gaussian random-field model** assumes that the collection of random variables $\{F_{s,t},\ 0 \leqslant s \leqslant t\}$ is Gaussian, in that any finite sub-collection has a multivariate normal distribution. A random field for the present context is just a stochastic process $\{X_{s,t}\}$ indexed by two parameters s and t, and for interest rates we are only interested in the situation where the indices lie in the region $0 \leqslant s \leqslant t$. The Gaussian assumption on the forward rates means that the bond prices have log-normal distributions. The most important example of a Gaussian random field is the following.

Example 6.3 *Brownian sheet.* The most basic random field of the type which we will consider is one generalization of the standard Brownian motion to a two-parameter process and is known as the **Brownian sheet**. This stochastic process $\{W_{s,t},\ s \geqslant 0,\ t \geqslant 0\}$ is such that all finite-dimensional distributions are Gaussian, so that for any choices of $(s_1, t_1), \ldots, (s_n, t_n)$, the joint distribution of the random variables $W_{s_1,t_1}, \ldots, W_{s_n,t_n}$ is normal and the means and covariances are given by

$$\mathbb{E}\, W_{s,t} = 0 \qquad \text{and} \qquad \mathbb{C}ov\left(W_{s_1,t_1}, W_{s_2,t_2}\right) = (s_1 \wedge s_2)(t_1 \wedge t_2)\,.$$

It may be shown that this process can be taken to have continuous sample paths; a sample path of the process in this case is a surface indexed by s and t. When one considers a section through this surface for either fixed s giving the one-parameter process $\{W_{s,t},\ t \geqslant 0\}$, or for fixed t giving the one-parameter process $\{W_{s,t},\ s \geqslant 0\}$, one obtains a Brownian motion, with variance parameter s or t, respectively. □

Recall that multivariate Gaussian (normal) distributions are specified completely by their means and covariances (see Section A.3). It will be assumed throughout that the forward rates are of the form

$$F_{s,t} = \mu_{s,t} + X_{s,t}, \qquad 0 \leqslant s \leqslant t,$$

where $X_{s,t}$ is a centred (mean-zero) continuous Gaussian random field with covariance structure specified by

$$\mathbb{C}ov\left(F_{s_1,t_1}, F_{s_2,t_2}\right) = \mathbb{C}ov\left(X_{s_1,t_1}, X_{s_2,t_2}\right) = \Gamma(s_1, t_1, s_2, t_2),$$

for $0 \leqslant s_i \leqslant t_i$, $i = 1, 2$. As we will see below, when we are interested in the probability measure under which the discounted bond prices are martingales then necessarily the covariance takes the form

$$\Gamma(s_1, t_1, s_2, t_2) = c(s_1 \wedge s_2, t_1, t_2), \quad 0 \leqslant s_i \leqslant t_i, \quad i = 1, 2, \qquad (6.22)$$

for some appropriate function c; it is necessary that $c(s_1 \wedge s_2, t_1, t_2)$ is symmetric in t_1 and t_2 and is non-negative definite in (s_1, t_1) and (s_2, t_2). We assume that the drift function $\mu_{s,t}$ is deterministic and continuous in $0 \leqslant s \leqslant t$. Conditions on the covariance function c sufficient to ensure that $\{X_{s,t}\}$ may be taken to be continuous may be given but we will not consider them here; the examples we will deal with will be just be deterministic time changes of the Brownian sheet considered above but the model covers a wider class of processes.

The situation when the covariance function is specified as a function of $s_1 \wedge s_2$ ensures that the Gaussian random field $X_{s,t}$ has independent increments in the s-direction, in the sense that for any $0 \leqslant s \leqslant s' \leqslant t$, the random variable $X_{s',t} - X_{s,t}$ is independent of the σ-field

$$\mathcal{F}_s = \sigma\left(X_{u,v} : u \leqslant v,\ u \leqslant s\right) = \sigma\left(F_{u,v} : u \leqslant v,\ u \leqslant s\right); \tag{6.23}$$

this follows since for $u \leqslant s$, $u \leqslant v$,

$$\begin{aligned} \mathcal{C}ov\left(X_{s',t} - X_{s,t}, X_{u,v}\right) &= c(s' \wedge u, t, v) - c(s \wedge u, t, v) \\ &= c(u, t, v) - c(u, t, v) = 0, \end{aligned}$$

which implies the independence by the Gaussian assumption. We will refer to (6.22) as the **independent-increments property**. We will assume that the σ-field specified in (6.23) represents the information available in this model at time s; this means that at the instant s the whole term structure for all times $u \leqslant s$ may be observed. The principal result for analyzing this model is the following.

Theorem 6.2 *For the Gaussian random-field model, the following three statements are equivalent:*

(a) *for each $t \geqslant 0$, the discounted bond-price process $\{Z_{s,t}, \mathcal{F}_s, 0 \leqslant s \leqslant t\}$ is a martingale;*

(b) *the covariance* $\Gamma(s_1, t_1, s_2, t_2)$ *takes the form* (6.22) *with*

$$\mu_{s,t} = \mu_{0,t} + \int_0^t \left[c(s \wedge v, v, t) - c(0, v, t)\right] dv, \text{ for all } 0 \leqslant s \leqslant t; \tag{6.24}$$

(c) $$P_{s,t} = \mathbb{E}\left[e^{-\int_s^t R_u du} \,\middle|\, \mathcal{F}_s\right], \quad \textit{for all} \quad 0 \leqslant s \leqslant t. \tag{6.25}$$

Proof. The equivalence of (a) and (c) is the content of Theorem 6.1. To show that (a) implies (b), first observe that when (a) holds, then the covariance Γ is necessarily of the form (6.22). To see this, for $0 \leqslant s_2 \leqslant s_1 \leqslant t$, notice that

$$\begin{aligned} \frac{Z_{s_1,t}}{Z_{s_2,t}} &= \exp\left[-\int_{s_1}^t F_{s_1,u} du - \int_0^{s_1} R_u du + \int_{s_2}^t F_{s_2,u} du + \int_0^{s_2} R_u du\right] \\ &= \exp\left[-\int_{s_1}^t \left(F_{s_1,u} - F_{s_2,u}\right) du - \int_{s_2}^{s_1} \left(F_{u,u} - F_{s_2,u}\right) du\right]. \end{aligned}$$

The martingale property is $\mathbb{E}\left[Z_{s_1,t} \mid \mathcal{F}_{s_2}\right] = Z_{s_2,t}$ or

$$\mathbb{E}\left[e^{-C} \mid \mathcal{F}_{s_2}\right] \equiv 1, \tag{6.26}$$

where $C = A + B$, with

$$A = \int_{s_1}^{t} \left(F_{s_1,u} - F_{s_2,u}\right) du \quad \text{and} \quad B = \int_{s_2}^{s_1} \left(F_{u,u} - F_{s_2,u}\right) du;$$

the equivalence in (6.26) is to mean that the left-hand side is 1 with probability 1. But the Gaussian property means that the left-hand side of (6.26) is

$$\exp\left[-\mathbb{E}\left(C \mid \mathcal{F}_{s_2}\right) + \mathbb{V}ar\left(C \mid \mathcal{F}_{s_2}\right)/2\right].$$

The quantity $\mathbb{V}ar\left(C \mid \mathcal{F}_{s_2}\right)$ is constant with probability 1, whence $\mathbb{E}\left(C \mid \mathcal{F}_{s_2}\right)$ is constant with probability 1, so that $\mathbb{E}\left(C \mid \mathcal{F}_{s_2}\right) \equiv \mathbb{E}(C)$. This implies that the Gaussian random variable C is independent of the σ-field $\mathcal{F}_{s_2}$, whence for any $s \leqslant s_2$, $s \leqslant v$, we have $\mathbb{C}ov\ (C, F_{s,v}) = 0$ which gives

$$\int_{s_1}^{t} \left(\Gamma(s_1, u, s, v) - \Gamma(s_2, u, s, v)\right) du + \int_{s_2}^{s_1} \left(\Gamma(u, u, s, v) - \Gamma(s_2, u, s, v)\right) du \equiv 0.$$

Differentiate with respect to t to see that

$$\Gamma(s_1, t, s, v) = \Gamma(s_2, t, s, v) \quad \text{holds for all} \quad s \leqslant s_2 < s_1 \leqslant t;\ s \leqslant v,$$

which implies (6.22).

To prove now that (b) holds, since A and B are (jointly) normally distributed, we have

$$\mathbb{E}\left[\exp\{-(A + B)\}\right] = \exp\{\mathbb{V}ar(A + B)/2 - \mathbb{E}A - \mathbb{E}B\};$$

because (6.22) holds A and B are independent of $\mathcal{F}_{s_2}$, which shows (using (6.26)) that the process $\{Z_{s,t}, \mathcal{F}_s,\ 0 \leqslant s \leqslant t\}$ is a martingale if and only if, for all choices of $0 \leqslant s_2 \leqslant s_1 \leqslant t$,

$$\mathbb{V}ar(A + B) = 2(\mathbb{E}A + \mathbb{E}B). \tag{6.27}$$

First, note that

$$\mathbb{E}A = \int_{s_1}^{t} \left(\mu_{s_1,u} - \mu_{s_2,u}\right) du \quad \text{and} \quad \mathbb{E}B = \int_{s_2}^{s_1} \left(\mu_{u,u} - \mu_{s_2,u}\right) du. \tag{6.28}$$

Next, calculate that

$$\begin{aligned}
\mathbb{V}ar(A) &= \mathbb{V}ar\left(\int_{s_1}^{t} \left(F_{s_1,u} - F_{s_2,u}\right) du\right) \\
&= \int_{s_1}^{t} \int_{s_1}^{t} \mathbb{C}ov\left(F_{s_1,u} - F_{s_2,u}, F_{s_1,v} - F_{s_2,v}\right) du dv \\
&= \int_{s_1}^{t} \int_{s_1}^{t} \left(c(s_1, u, v) - c(s_2, u, v)\right) du dv.
\end{aligned}$$

Introduce the notation

$$g(r,s,t) = \int\limits_{u=s}^{t} \int\limits_{v=s}^{u} c(r,u,v)dudv \quad \text{and} \quad h(s,t) = \int\limits_{u=s}^{t} \int\limits_{v=s}^{u} c(v,u,v)dudv,$$

then, by the symmetry of c in its last two components, the calculation above shows that

$$\mathbb{V}ar(A) = 2\left[g(s_1,s_1,t) - g(s_2,s_1,t)\right]. \tag{6.29}$$

Similar calculations give

$$\begin{aligned} \mathbb{V}ar(B) &= 2\left[h(s_2,s_1) - g(s_2,s_2,s_1)\right], \quad \text{and} \\ \mathbb{C}ov\,(A,B) &= h(s_2,t) - h(s_2,s_1) - h(s_1,t) \\ &\quad - g(s_2,s_2,t) + g(s_2,s_2,s_1) + g(s_2,s_1,t). \end{aligned} \tag{6.30}$$

Substituting from (6.28)–(6.30) into (6.27) shows that $\{Z_{s,t}, \mathcal{F}_s,\ 0 \leqslant s \leqslant t\}$ is a martingale if and only if, for all $0 \leqslant s_2 \leqslant s_1 \leqslant t$,

$$\begin{aligned} \int_{s_1}^{t} \left(\mu_{s_1,u} - \mu_{s_2,u}\right) du + \int_{s_2}^{s_1} \left(\mu_{u,u} - \mu_{s_2,u}\right) du \\ = g(s_1,s_1,t) - g(s_2,s_2,t) - h(s_1,t) + h(s_2,t), \end{aligned}$$

which is equivalent to the statement that the expression

$$\int_{s}^{t} \mu_{s,u} du + \int_{0}^{s} \mu_{u,u} du - g(s,s,t) + h(s,t)$$

does not depend on s, $0 \leqslant s \leqslant t$. Setting $s = 0$, observe that this, in turn, is equivalent to

$$\begin{aligned} \int_{s}^{t} \mu_{s,u} du + \int_{0}^{s} \mu_{u,u} du \\ = \left[g(s,s,t) - g(0,0,t)\right] - \left[h(s,t) - h(0,t)\right] + \int_{0}^{t} \mu_{0,u} du. \end{aligned} \tag{6.31}$$

Differentiating (6.31) with respect to t gives (6.24), which completes the argument that (a) implies (b).

Conversely, suppose that (b) holds, setting $s = t = u$ in (6.24) and integrating gives

$$\int_{0}^{s} \mu_{u,u} du = \int_{0}^{s} \mu_{0,u} du + h(0,s) - g(0,0,s), \tag{6.32}$$

while integrating the relation in (6.24) gives

$$\begin{aligned} \int_{s}^{t} \mu_{s,u} du = \int_{s}^{t} \mu_{0,u} du + h(0,t) - h(0,s) - h(s,t) \\ + g(s,s,t) - g(0,0,t) + g(0,0,s). \end{aligned} \tag{6.33}$$

Adding (6.32) and (6.33) gives (6.31) which, in turn, implies (6.27) which completes the proof of the equivalence of (a) and (b). □

Before we illustrate the use of Theorem 6.2 for pricing interest-rate derivatives we consider an example of an appropriate form for the covariance function $c(s, t_1, t_2)$.

Example 6.4 Suppose that $\sigma, \tau : [0, \infty) \to [0, \infty)$ are continuous and monotone increasing and monotone non-increasing functions respectively, and set

$$c(r, s, t) = \sigma(r)\tau(s \vee t). \tag{6.34}$$

Then $X_{s,t}$ may be represented as $X_{s,t} = W_{\sigma(s),\tau(t)}$, where W is the standard Brownian sheet described in Example 6.3. That is, the forward rates $F_{s,t}$ are a deterministic time change of the Brownian sheet plus the drift surface $\mu_{s,t}$. The assumption that τ is decreasing corresponds to assuming that the volatility of $F_{s,t}, 0 \leqslant s \leqslant t$, decreases as the maturity time t increases. The case where τ increases might also be considered with the corresponding interpretation in terms of the volatility. Note that, for fixed s,

$$\mathcal{C}orr\,(X_{s,t_1}, X_{s,t_2}) = \sqrt{\frac{\tau(t_1 \vee t_2)}{\tau(t_1 \wedge t_2)}}.$$

Thus, for s and t_1 fixed, as t_2 increases from $t_2 = t_1$, the correlation between F_{s,t_1} and F_{s,t_2} decays from 1. In particular, if $\tau(t) = e^{-\alpha t}$, $\alpha > 0$, then $\mathrm{Corr}(F_{s,t_1}, F_{s,t_2}) = e^{-\alpha|t_1 - t_2|/2}$, giving the case where the correlation between forward rates of different maturities tails off exponentially as the distance in maturity time between them, $|t_1 - t_2|$, increases. □

6.3.2 Pricing a caplet on forward rates

As an example of the application of Theorem 6.2 we consider the problem of pricing a caplet at strike rate d, for the period from t to $t + \Delta$ as described in Section 6.1.

Theorem 6.3 *The price at time s, $0 \leqslant s \leqslant t$, of a caplet on the forward rate $F^{\Delta}_{t,t}$ at strike rate d for the period $[t, t + \Delta]$ is given by*

$$\begin{aligned} P_{s,t}\,\Phi &\left(\frac{\ln\left(P_{s,t}/P_{s,t+\Delta}\right) - \Delta d}{\sigma(s)} + \frac{\sigma(s)}{2}\right) \\ &- P_{s,t+\Delta}e^{\Delta d}\,\Phi\left(\frac{\ln\left(P_{s,t}/P_{s,t+\Delta}\right) - \Delta d}{\sigma(s)} - \frac{\sigma(s)}{2}\right), \end{aligned} \tag{6.35}$$

where

$$\begin{aligned} \sigma^2(s) &= \mathbb{V}ar\left(\Delta F^{\Delta}_{t,t} \mid \mathcal{F}_s\right) = \mathbb{V}ar\left(\ln\left(P_{t,t+\Delta}\right) \mid \mathcal{F}_s\right) \\ &= 2\int_t^{t+\Delta}\int_t^u \left[c(t,u,v) - c(s,u,v)\right] du dv. \end{aligned} \tag{6.36}$$

Proof. The price at time s of the caplet is given by the expression (6.5) on page 167, where the expectation is taken under the assumption that the three statements of Theorem 6.2 hold; that is the underlying probability is the martingale probability and the covariance has the independent-increments property. Suppose that we have two random variables (N_1, N_2) which are jointly normally distributed with $\mathbb{E} N_i = \mu_i$, $\mathbb{V}ar(N_i) = \sigma_i^2$ and $\mathbb{C}orr(N_1, N_2) = \rho$, so that $\mathbb{C}ov(N_1, N_2) = \rho\sigma_1\sigma_2$, we may calculate as follows,

$$\begin{aligned}\mathbb{E}\left[e^{-N_2}\left(e^{N_1}-e^{\gamma}\right)_+\right] &= \mathbb{E}\left[\mathbb{E}\left[\left(e^{N_1}-e^{\gamma}\right)_+ e^{-N_2} \mid N_1\right]\right] \\ &= \mathbb{E}\left[\left(e^{N_1}-e^{\gamma}\right)_+ \mathbb{E}\left[e^{-N_2} \mid N_1\right]\right] \\ &= \mathbb{E}\left[\left(e^{N_1}-e^{\gamma}\right)_+ e^{-\mathbb{E}(N_2|N_1)+\mathbb{V}ar(N_2|N_1)/2}\right].\end{aligned}$$

Now substitute the expressions

$$\mathbb{E}\left(N_2 \mid N_1\right) = \mu_2 + \rho\sigma_2\left(N_1-\mu_1\right)/\sigma_1, \quad \text{and} \quad \mathbb{V}ar\left(N_2 \mid N_1\right) = \sigma_2^2\left(1-\rho^2\right);$$

these relations follow because $N_2 - \rho\sigma_2 N_1/\sigma_1$ and N_1 are independent (since they are jointly normal and their covariance equals 0) and hence

$$\begin{aligned}\mathbb{E}\left(N_2 \mid N_1\right) - \rho\sigma_2 N_1/\sigma_1 &= \mathbb{E}\left(N_2 - \rho\sigma_2 N_1/\sigma_1 \mid N_1\right) \\ &= \mathbb{E}\left(N_2 - \rho\sigma_2 N_1/\sigma_1\right) = \mu_2 - \rho\sigma_2\mu_1/\sigma_1,\end{aligned}$$

and

$$\begin{aligned}\mathbb{V}ar\left(N_2 \mid N_1\right) &= \mathbb{V}ar\left(N_2 - \rho\sigma_2 N_1/\sigma_1 \mid N_1\right) \\ &= \mathbb{V}ar\left(N_2 - \rho\sigma_2 N_1/\sigma_1\right) = \sigma_2^2\left(1-\rho^2\right).\end{aligned}$$

Observe that

$$\mathbb{E}\left[e^{-\rho\sigma_2 N_1/\sigma_1}\left(e^{N_1}-e^{\gamma}\right)_+\right] = e^{\left(\rho^2\sigma_2^2/2\right)-\rho\sigma_2\mu_1/\sigma_1}\,\mathbb{E}\left[\left(e^{N_1-\rho\sigma_1\sigma_2}-e^{\gamma}\right)_+\right]$$

and use Lemma 5.1 to deduce that $\mathbb{E}\left[e^{-N_2}\left(e^{N_1}-e^{\gamma}\right)_+\right]$ equals

$$\begin{aligned}& e^{\mu_1-\mu_2+\mathbb{V}ar(N_1-N_2)/2}\,\Phi\left(\frac{\mu_1-\gamma+\sigma_1^2-\mathbb{C}ov\left(N_1,N_2\right)}{\sigma_1}\right) \\ &\qquad - e^{\gamma-\mu_2+\sigma_2^2/2}\,\Phi\left(\frac{\mu_1-\gamma-\mathbb{C}ov\left(N_1,N_2\right)}{\sigma_1}\right).\end{aligned} \tag{6.37}$$

Now take $N_1 = \Delta F_{t,t}^{\Delta} = \int_t^{t+\Delta} F_{t,u}du$, $N_2 = \int_s^{t+\Delta} R_u du = \int_s^{t+\Delta} F_{u,u}du$ and $\gamma = \Delta d$. Conditional on the information $\mathcal{F}_s$, the random variables (N_1, N_2) are jointly normal with means $\mu_i = \mathbb{E}\left(N_i \mid \mathcal{F}_s\right)$, and variances $\sigma_i^2 = \mathbb{V}ar\left(N_i \mid \mathcal{F}_s\right)$, $i = 1, 2$, and covariance $\mathbb{C}ov\,(N_1, N_2 \mid \mathcal{F}_s)$. Using the independent-increments property and Theorem 6.2, calculate that

$$\begin{aligned}\mu_1 &= \int_t^{t+\Delta} F_{s,u}du + \mathbb{E}\left[\int_t^{t+\Delta} (F_{t,u} - F_{s,u}) \,\middle|\, \mathcal{F}_s\right] \\ &= \int_t^{t+\Delta} F_{s,u}du + \mathbb{E}\left[\int_t^{t+\Delta} (F_{t,u} - F_{s,u})\right] \\ &= \int_t^{t+\Delta} F_{s,u}du + \int_t^{t+\Delta} (\mu_{t,u} - \mu_{s,u})\, du.\end{aligned}$$

From (6.24),

$$\begin{aligned}\mu_{t,u} - \mu_{s,u} &= \int_0^u (c(t \wedge v, u, v) - c(s \wedge v, u, v))\, dv \\ &= \int_s^u (c(t \wedge v, u, v) - c(s, u, v))\, dv,\end{aligned}$$

from which we deduce that

$$\begin{aligned}\mu_1 &= \int_t^{t+\Delta} F_{s,u}du + \int_{u=t}^{t+\Delta} \int_{v=s}^{u} (c(t \wedge v, u, v) - c(s, u, v))\, dvdu \\ &= \int_t^{t+\Delta} F_{s,u}du + \int_{u=t}^{t+\Delta} \int_{v=s}^{t} (c(v, u, v) - c(s, u, v))\, dvdu + \sigma^2(s)/2,\end{aligned}$$

where $\sigma^2(s)$ is defined in (6.36). Similar calculations yield

$$\mu_2 = \int_s^{t+\Delta} F_{s,u}du + \int_{u=s}^{t+\Delta} \int_{v=s}^{u} (c(v, u, v) - c(s, u, v))\, dvdu; \quad \sigma_1^2 = \sigma^2(s);$$

$$\sigma_2^2 = 2\int_{u=s}^{t+\Delta} \int_{v=s}^{u} (c(v, u, v) - c(s, u, v))\, dvdu;$$

and

$$\mathbb{C}ov\,(N_1, N_2 \mid \mathcal{F}_s) = \int_{u=t}^{t+\Delta} \int_{v=s}^{t} (c(v, u, v) - c(s, u, v))\, dvdu + \sigma^2(s).$$

Using these relations, obtain

$$\mu_1 - \mu_2 + \mathbb{V}ar(N_1 - N_2 \mid \mathcal{F}_s)/2 = -\int_s^t F_{s,u}du = \ln(P_{s,t}); \tag{6.38}$$

$$-\mu_2 + \sigma_2^2/2 = -\int_s^{t+\Delta} F_{s,u}du = \ln(P_{s,t+\Delta}); \tag{6.39}$$

and

$$\begin{aligned}\mu_1 + \sigma_1^2 - \mathbb{C}ov\,(N_1, N_2 \mid \mathcal{F}_s) &= \int_t^{t+\Delta} F_{s,u}du + \sigma^2(s)/2 \\ &= \ln(P_{s,t}/P_{s,t+\Delta}) + \sigma^2(s)/2.\end{aligned} \tag{6.40}$$

Substituting from (6.38)–(6.40) into (6.37) gives (6.35). ∎

Remarks

1. The expression in (6.35) is known as **Black's formula**. It may be rearranged as

$$e^{-\left(\int_s^{t+\Delta} F_{s,u}du\right)}\left[e^{\Delta F_{s,t}^{\Delta}}\Phi\left(\frac{\Delta\left(F_{s,t}^{\Delta}-d\right)}{\sigma(s)}+\frac{\sigma(s)}{2}\right)\right.$$
$$\left.-\,e^{\Delta d}\,\Phi\left(\frac{\Delta\left(F_{s,t}^{\Delta}-d\right)}{\sigma(s)}-\frac{\sigma(s)}{2}\right)\right].$$

The formula shows that the caplet is equivalent to holding the portfolio which at time s is long

$$\Phi\left(\frac{\ln\left(P_{s,t}/P_{s,t+\Delta}\right)-\Delta d}{\sigma(s)}+\frac{\sigma(s)}{2}\right)$$

bonds of maturity date t and short

$$e^{\Delta d}\,\Phi\left(\frac{\ln\left(P_{s,t}/P_{s,t+\Delta}\right)-\Delta d}{\sigma(s)}-\frac{\sigma(s)}{2}\right)$$

bonds of maturity date $t+\Delta$.

2. Notice that the formula for the cap price is not quite as tidy if the payoff from the option is earned at time t rather than at $t+\Delta$. The discount factor is changed so that the time-s price is

$$\mathbb{E}\left[e^{-\left(\int_s^t R_u du\right)}\left(e^{\Delta F_{t,t}^{\Delta}}-e^{\Delta d}\right)_+\,\middle|\,\mathcal{F}_s\right],$$

with a calculation similar to the above showing that (6.35) should be replaced by

$$e^{-\left(\int_s^t F_{s,u}du\right)}\left[e^{\Delta F_{s,t}^{\Delta}+\sigma^2(s)}\Phi\left(\frac{\Delta\left(F_{s,t}^{\Delta}-d\right)}{\sigma(s)}+\frac{3\sigma(s)}{2}\right)\right.$$
$$\left.-\,e^{\Delta d}\,\Phi\left(\frac{\Delta\left(F_{s,t}^{\Delta}-d\right)}{\sigma(s)}+\frac{\sigma(s)}{2}\right)\right],$$

as the time-s price of the option.

3. It is interesting to compare the price for the above cap with that for a European call option on the bond of maturity date $t+\Delta$ with expiry time t and strike price k. The time-s price of such an option is

$$\mathbb{E}\left[e^{-\left(\int_s^t R_u du\right)}\left(P_{t,t+\Delta}-k\right)_+\,\middle|\,\mathcal{F}_s\right] \tag{6.41}$$

when the equivalent statements in Theorem 6.2 hold. The caplet may be thought of as being equivalent to a European call option with expiry time t on a 'bond' of price $P_{s,t}/P_{s,t+\Delta}$ at time s. A similar calculation to that already given shows that (6.41) equals

$$P_{s,t+\Delta}\Phi\left(\frac{\ln\left(P_{s,t+\Delta}/P_{s,t}\right)-\ln(k)}{\sigma(s)}+\frac{\sigma(s)}{2}\right) \qquad (6.42)$$
$$-\,kP_{s,t}\Phi\left(\frac{\ln\left(P_{s,t+\Delta}/P_{s,t}\right)-\ln(k)}{\sigma(s)}-\frac{\sigma(s)}{2}\right)$$

where $\sigma(s)$ is given in (6.36) (see Exercise 6.6). In contrast with the caplet, this option is equivalent to a portfolio long in bonds of maturity date $t+\Delta$ and short in bonds of maturity date t. The fact that this formula is so similar to that for the caplet is not surprising since the distributions of both bond prices and their ratios are log-normal.

4. Black's formula and the expression in Remark 3 demonstrate the robustness of the Gaussian model presented here. Being similar in form to the Black–Scholes formula, both contain only one quantity requiring estimation, that is $\sigma(s)$, which is proportional to the standard deviation of the forward rate for $[t, t+\Delta]$. □

6.3.3 Markov properties

We will consider what covariance structures are possible for the Gaussian-random-field model under certain structural properties (such as appropriate formulations of the Markov property and stationarity). It will be seen that such properties limit greatly the form of the function $\Gamma(\cdot,\cdot,\cdot,\cdot)$, specifying the covariance, which appears in the independent-increments property in Theorem 6.2.

We investigate first an appropriate formulation of the Markov property in the context of the random field of instantaneous forward interest rates. For random quantities A, B and C (or equivalently for σ-fields) write $A \perp B \mid C$ for the statement that A and B are conditionally independent given C. Note that if A, B and C are random variables having a joint normal distribution then $A \perp B \mid C$ if and only if

$$\mathbb{C}ov\,(A,B)\mathbb{V}ar(C) = \mathbb{C}ov\,(A,C)\mathbb{C}ov\,(B,C);$$

this is because

$$\mathbb{C}ov\,(A,B \mid C) = \mathbb{C}ov\,(A,B) - \mathbb{C}ov\,(A,C)\,\mathbb{C}ov\,(B,C)\,/\mathbb{V}ar\,(C)\,.$$

The usual specification of the Markov property for a stochastic process is to stipulate that the past and the future of the process are conditionally independent given the present. Here we assume that the past at time s is represented by the σ-field $\mathcal{F}_s$, defined above; the future is represented by the σ-field $\mathcal{H}_s = \sigma\{F_{u,v},\ s \leqslant u \leqslant v\}$ and the present by $\mathcal{G}_s = \sigma\{F_{s,v},\ v \geqslant s\}$. The most natural formulation of the Markov property would be that

$$\mathcal{F}_s \perp \mathcal{H}_s \mid \mathcal{G}_s. \qquad (6.43)$$

Because this is not a particularly tractable condition, we will require a stronger formulation which implies (6.43).

We say that $\{F_{s,t},\ 0 \leqslant s \leqslant t\}$, the random field of instantaneous forward rates, satisfies the **first Markov** property if we have

$$F_{s_1,t_1} \perp F_{s_3,t_2} \mid F_{s_2,t_2} \quad \text{for all} \quad 0 \leqslant s_1 \leqslant s_2 < s_3,\ s_1 \leqslant t_1,\ s_3 \leqslant t_2. \tag{6.44}$$

In particular, the first Markov property implies that, for each t, the stochastic process $\{F_{s,t},\ 0 \leqslant s \leqslant t\}$ of instantaneous forward rates at t forms a Markov process in the usual sense.

Note that if A, B, C, D and E are random variables having a joint normal distribution then the conditions

$$A \perp D \mid C, \quad A \perp E \mid C, \quad B \perp C \mid D \quad \text{and} \quad B \perp E \mid D \tag{6.45}$$

imply that $(A, B) \perp E \mid (C, D)$; furthermore if in addition to the conditions (6.45) either $A \perp B \mid C$ or $A \perp B \mid D$ then both hold and it follows that $A \perp B \mid (C, D)$. The conditions (6.45) imply that the joint distribution of (A, B) conditional on (C, D, E) is the same as the joint distribution of (A, B) conditional on (C, D), which is equivalent to the statement that $(A, B) \perp E \mid (C, D)$. To see that the first Markov property in the form (6.44) implies (6.43), note that this observation shows that for $s < s_i \leqslant t_i$, $1 \leqslant i \leqslant k$, the conditional joint distribution of $\{F_{s_i,t_i}, 1 \leqslant i \leqslant k\}$ given $\mathcal{F}_s$ is the same as the conditional joint distribution of $\{F_{s_i,t_i}, 1 \leqslant i \leqslant k\}$ given $\{F_{s,t_i}, 1 \leqslant i \leqslant k\}$ which, in turn, is the same as the conditional joint distribution of $\{F_{s_i,t_i}, 1 \leqslant i \leqslant k\}$ given $\mathcal{G}_s$.

It should be observed that the independent-increments property implies the first Markov property, since for (6.44) to hold we require

$$\Gamma(s_1, t_1, s_3, t_2)\,\Gamma(s_2, t_2, s_2, t_2) = \Gamma(s_1, t_1, s_2, t_2)\,\Gamma(s_3, t_2, s_2, t_2),$$

but if (6.22) holds both sides of the equation equal $c(s_1, t_1, t_2)\,c(s_2, t_2, t_2)$. Note that while (6.43) would be a much stronger condition than (6.44) for general random fields, it might be expected not to be in the Gaussian context where pairwise independence implies independence. We say that the random field of instantaneous forward rates $\{F_{s,t},\ 0 \leqslant s \leqslant t\}$ satisfies the **second Markov** property when we have

$$F_{s_1,t_1} \perp F_{s_2,t_2} \mid F_{s_2,t_1} \quad \text{for all} \quad 0 \leqslant s_1 < s_2,\ t_1,\ t_2 \text{ with } s_2 \leqslant t_1 \wedge t_2. \tag{6.46}$$

The two definitions (6.44) and (6.46) are contrasted in Figure 6.1. We say that the random field of instantaneous forward rates $\{F_{s,t},\ 0 \leqslant s \leqslant t\}$ is **Markov** when it satisfies both the first and second Markov properties.

Theorem 6.4 *Suppose that the random field of forward rates $\{F_{s,t},\ 0 \leqslant s \leqslant t\}$ is Markov and satisfies the independent-increments property, then the covariance function c is of the form $c(s, t_1, t_2) = f(s)g(t_1, t_2)$ for some functions f and g*

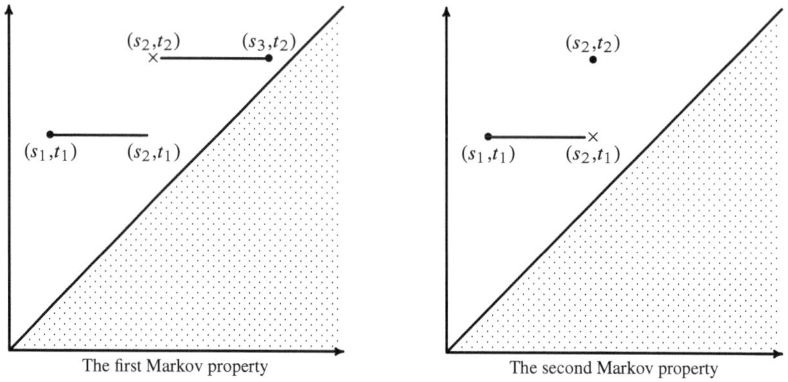

Figure 6.1: The first and second Markov properties

where f is non-decreasing and $g(t_1,t_2)$ is symmetric and non-negative definite in t_1 and t_2.

Proof. Since the Markov property holds, (6.46) is true. For $s_1 < s_2$, $s_i \leqslant t_i$, $i = 1, 2$ the conditional-independence condition $F_{s_1,t_1} \perp F_{s_2,t_2} \mid F_{s_2,t_1}$ holds so that

$$c(s_1,t_1,t_2)\,c(s_2,t_1,t_1) = c(s_1,t_1,t_1)\,c(s_2,t_1,t_2)\,.$$

A rearrangement of this relation gives

$$\frac{c(s_2,t_1,t_2)}{c(s_1,t_1,t_2)} = \frac{c(s_2,t_1,t_1)}{c(s_1,t_1,t_1)},$$

which demonstrates that the ratio $c(s_2,t_1,t_2)/c(s_1,t_1,t_2)$ does not depend on t_2; by symmetry in t_1 and t_2 the ratio cannot depend on t_1 either, from which it follows that the function c may be written in the form

$$c(s,t_1,t_2) = f(s)c(0,t_1,t_2) = f(s)g(t_1,t_2)\,, \quad \text{say}, \tag{6.47}$$

for some functions f and g, where $g(t_1,t_2)$ is necessarily symmetric and non-negative definite in t_1 and t_2 because c represents a covariance. It may be seen that the function f is non-decreasing since for $s_1 < s_2 \leqslant t$

$$\mathbb{V}ar\left(F_{s_1,t}\right)\mathbb{V}ar\left(F_{s_2,t}\right) \geqslant \left[\mathbb{C}ov\left(F_{s_1,t},F_{s_2,t}\right)\right]^2$$

which gives

$$c(s_1,t,t)\,c(s_2,t,t) \geqslant [c(s_1,t,t)]^2$$

implying that $f(s_2) \geqslant f(s_1)$. ∎

Remark Observe that when the random field is Markov then it is not necessarily the case that the spot-rate process $\{R_s,\ s \geqslant 0\}$ is a one-dimensional Markov process. To see this, note that Theorem 6.4 shows that the Markov property implies that $c(s, t_1, t_2) = f(s)g(t_1, t_2)$. The spot rate R_s is a one-dimensional Markov process when

$$R_{t_1} \perp R_{t_3} \mid R_{t_2} \quad \text{or} \quad F_{t_1,t_1} \perp F_{t_3,t_3} \mid F_{t_2,t_2} \quad \text{for} \quad t_1 < t_2 < t_3.$$

This is equivalent to the condition

$$\mathbb{C}ov\,(F_{t_1,t_1}, F_{t_3,t_3})\mathbb{V}ar(F_{t_2,t_2}) = \mathbb{C}ov\,(F_{t_1,t_1}, F_{t_2,t_2})\mathbb{C}ov\,(F_{t_3,t_3}, F_{t_2,t_2}),$$

which, in turn, is the same as $c(t_1, t_2, t_3)c(t_2, t_2, t_2) = c(t_1, t_1, t_2)c(t_2, t_2, t_3)$, or

$$g(t_1, t_3)g(t_2, t_2) = g(t_1, t_2)g(t_2, t_3), \tag{6.48}$$

for $t_1 < t_2 < t_3$. A further condition on the random field, which is equivalent to the relation (6.48), is considered below. □

We say that the random field is **stationary** if for each $t > 0$ the joint distributions of the set of random variables $\{F_{s,t},\ 0 \leqslant s \leqslant t\}$ are the same as those of the set $\{F_{s+u,t+u},\ 0 \leqslant s \leqslant t\}$, for any fixed $u > 0$.

Adding stationarity to the Markovian assumptions further restricts the possible forms of the covariance, as the next result demonstrates.

Theorem 6.5 *Suppose that the random field of forward rates $\{F_{s,t},\ 0 \leqslant s \leqslant t\}$ is Markov and stationary and satisfies the independent-increments property, then the covariance function c is of the form*

$$c(s, t_1, t_2) = e^{\lambda(s - t_1 \wedge t_2)} h(|t_1 - t_2|), \tag{6.49}$$

for some real $\lambda \geqslant 0$ and function h satisfying

$$|h(x)| \leqslant h(0)e^{-\lambda x/2}, \qquad x \geqslant 0. \tag{6.50}$$

Proof. From Theorem 6.4 it follows that $c(s, t_1, t_2) = f(s)g(t_1, t_2)$, so if the assumption that the random field is stationary is added, then this implies that for $s \leqslant t_1 \wedge t_2$,

$$f(0)g(t_1 - s, t_2 - s) = c(0, t_1 - s, t_2 - s) = c(s, t_1, t_2) = f(s)g(t_1, t_2),$$

after using (6.47). This holds for all $s \leqslant t_1 \wedge t_2$, so setting $s = t_1 \wedge t_2$ it follows that the function g may be written in the form

$$g(t_1, t_2) = \frac{h(|t_1 - t_2|)}{f(t_1 \wedge t_2)},$$

where $h(t) = g(0,t)$, if we normalize so that $f(0) = 1$. That is, the function c necessarily takes the form

$$c(s,t_1,t_2) = \frac{f(s)}{f(t_1 \wedge t_2)} h(|t_1 - t_2|). \tag{6.51}$$

Again use the fact that stationarity implies that

$$c(s,t_1,t_2) = c(s+u, t_1+u, t_2+u)$$

to see from (6.51) that

$$f(s) f((t_1 \wedge t_2) + u) = f(s+u) f(t_1 \wedge t_2);$$

setting $s = 0$, $v = t_1 \wedge t_2$ shows that f satisfies

$$f(u+v) = f(u) f(v),$$

the general solution of which is $f(s) = e^{\lambda s}$ for some constant λ. From Theorem 6.4, f is non-decreasing so it must be the case that $\lambda \geqslant 0$. Thus the most general form for the covariance function c with the assumptions of the Markov property and stationarity is given by an expression of the form (6.49). Recall that the function $g(t_1,t_2) = e^{-\lambda(t_1 \wedge t_2)} h(|t_1 - t_2|)$ must be non-negative definite so that

$$[g(t_1,t_2)]^2 \leqslant g(t_1,t_1)\, g(t_2,t_2),$$

hence (6.50) holds. □

Remark Notice that, in the case where the covariance is given by (6.49) and if the conditions of Theorem 6.2 hold, then substituting from (6.49) into (6.24), for $s \leqslant t$, the mean $\mu_{s,t} = \mathbb{E}\, F_{s,t}$ is given by

$$\begin{aligned} \mu_{s,t} = \mu_{0,t} &+ \int_0^t \left(1 - e^{-\lambda(t-v)}\right) h(v) dv \\ &- \int_0^{t-s} \left(1 - e^{-\lambda(t-s-v)}\right) h(v) dv. \end{aligned} \tag{6.52}$$

Stationarity gives that $\mu_{s+u,t+u} \equiv \mu_{s,t}$ which implies that

$$\mu_{0,t+u} + \int_0^{t+u} \left(1 - e^{-\lambda(t+u-v)}\right) h(v) dv = \mu_{0,t} + \int_0^t \left(1 - e^{-\lambda(t-v)}\right) h(v) dv,$$

for all $u \geqslant -t$; thus the right-hand side does not depend on t, whence

$$\mu_{0,t} = \nu - \int_0^t \left(1 - e^{-\lambda(t-v)}\right) h(v) dv \quad \text{for some constant} \quad \nu.$$

Substituting back into (6.52) the mean is given by

$$\mu_{s,t} = \nu - \int_0^{t-s} \left(1 - e^{-\lambda(t-s-v)}\right) h(v)dv.$$

Notice that the constant $\nu \equiv \mathbb{E} F_{s,s}$ is the mean spot rate under the martingale measure and thus in equilibrium the mean instantaneous forward rate for any time is less than the mean spot rate when h is non-negative. □

There is one final form of the Markov property that may be considered. We say that the random field of instantaneous forward rates $\{F_{s,t},\ 0 \leqslant s \leqslant t\}$ is **Markov in the t-direction**, that is in the maturity-time coordinate, if

$$\text{for all} \quad s \leqslant t_1 \leqslant t_2 \leqslant t_3 \quad \text{then} \quad F_{s,t_1} \perp F_{s,t_3} \mid F_{s,t_2}. \tag{6.53}$$

Although this property may seem intuitively less plausible than the previous Markovian assumptions it would mirror the Markov property in the real-time direction and furthermore, if the Markov property holds, so that (6.47) is true, then (6.53) is equivalent to the more intuitive condition that

$$\text{for all} \quad t_1 \leqslant t_2 \leqslant s \leqslant t_3 \leqslant t_4 \quad \text{then} \quad F_{t_1,t_2} \perp F_{t_3,t_4} \mid F_{s,s}. \tag{6.54}$$

That is, if the spot rate at s is known, then the past, (t_1, t_2), gives no further information about the future, (t_3, t_4). This is because if (6.47) is true then (6.53) holds if and only if

$$g(t_1, t_3)g(t_2, t_2) = g(t_1, t_2)g(t_2, t_3) \tag{6.55}$$

for all $t_1 \leqslant t_2 \leqslant t_3$ which is equivalent to (6.54). Combining the notions of the field being Markov and Markov in the t-direction gives the strongest form of the Markov property that we will consider.

We say that the random field $\{F_{s,t},\ 0 \leqslant s \leqslant t\}$ is **strictly Markov** if it is both Markov and Markov in the t-direction.

Note that the conditions (6.48) and (6.55) are the same so that we may see that when the random field has the Markov property and the independent-increments property then the spot-rate process, $\{R_s,\ s \geqslant 0\}$, is a one-dimensional Markov process if and only if the field is strictly Markov.

The principal conclusion of this section is contained in the next result which demonstrates that a stationary, strictly Markov random field with the independent-increments property necessarily has a covariance of a particular functional form and is determined by just three parameters. It shows that these structural conditions are extremely limiting to the forms of covariance that may be obtained; it also suggests that some, or all, of these conditions may have to be relaxed when attempting to fit models to data.

Theorem 6.6 *Suppose that the random field of forward rates $\{F_{s,t},\ 0 \leqslant s \leqslant t\}$ is stationary, strictly Markov and it satisfies the independent-increments property, then* *the covariance function c is of the form*

$$c(s,t_1,t_2) = \sigma^2 e^{\lambda(s-t_1\wedge t_2)-\mu|t_1-t_2|} = \sigma^2 e^{\lambda s+(2\mu-\lambda)(t_1\wedge t_2)-\mu(t_1+t_2)}, \qquad (6.56)$$

for some constants σ, $\lambda \geqslant 0$ *and* $\mu \geqslant \lambda/2$.

Proof. If the condition (6.53) is added to the form (6.49) then it gives

$$e^{\lambda(s-t_1)}h(t_3-t_1)\, e^{\lambda(s-t_2)}h(0) = e^{\lambda(s-t_1)}h(t_2-t_1)\, e^{\lambda(s-t_2)}h(t_3-t_2)\,.$$

Putting $u = t_2 - t_1$, $v = t_3 - t_2$ it follows that $h(u+v)h(0) = h(u)h(v)$, the general solution of which is of the form $h(u) = \sigma^2 e^{-\mu u}$, for some constants σ^2 and μ. In view of (6.50) it is necessary that $\mu \geqslant \lambda/2$. □

Remarks

1. When $W_{s,t}$ is the standard Brownian sheet, which is the Gaussian field with zero mean and with covariance specified by

$$\mathcal{C}ov\left(W_{s_1,t_1}, W_{s_2,t_2}\right) = (s_1 \wedge s_2)(t_1 \wedge t_2)\,,$$

then it is immediate that $\{F_{s,t}\}$ with covariance determined by (6.56) may be represented as

$$F_{s,t} = \mu_{s,t} + \sigma e^{-\mu t} W_{e^{\lambda s}, e^{(2\mu-\lambda)t}}\,.$$

2. In the model specified by (6.56), for any fixed $t \geqslant 0$, the one-dimensional process $\{F_{s,s+t},\ s \geqslant 0\}$ is a stationary Ornstein–Uhlenbeck process with covariance

$$\mathcal{C}ov\left(F_{s_1,s_1+t}, F_{s_2,s_2+t}\right) = \sigma^2 e^{-\lambda t-\mu|s_1-s_2|},$$

(see Exercise 6.5); this implies that the spot rate and all instantaneous forward rates of fixed duration are mean reverting. □

6.3.4 Finite-factor models and restricted information

Up to now it has been assumed that in our discussion of the random-field model that the information at time s is that contained in the σ-field

$$\mathcal{F}_s = \sigma\left\{F_{u,v},\ 0 \leqslant u \leqslant s, u \leqslant v\right\},$$

so that the whole yield curve is observed at each time. As noted above, most models of interest rates in the literature assume that there is a small number of bonds of particular maturities (or equivalently of interest rates of corresponding durations) for which the dynamics are specified and then all bond prices may be calculated in terms of these; for example, for $0 < t_1 < \ldots < t_k$ the evolution of $P_{s,s+t_1}, \ldots, P_{s,s+t_k}$ and the spot rate $R_s = F_{s,s}$, for $s \geqslant 0$, might be specified. When the rates are Gaussian we will see that typically these finite-factor models may be viewed as special cases of the independent-increments situation presented here but where there is 'restricted information' available. Suppose that $\mathcal{R}_s$ is a σ-field with $\mathcal{R}_s \subseteq \mathcal{F}_s$ and such that the conditional distributions of $\{F_{s,v},\ v \geqslant s\}$ given $\mathcal{R}_s$ are Gaussian. Then with this restricted information the bond prices will be

$$\hat{P}_{s,t} = \mathbb{E}\left[P_{s,t} \mid \mathcal{R}_s\right] = \mathbb{E}\left[e^{-\int_s^t F_{s,u}du} \mid \mathcal{R}_s\right] = e^{-\int_s^t \hat{F}_{s,u}du}, \quad \text{say,}$$

where

$$\int_s^t \hat{F}_{s,u}du = \int_s^t \mathbb{E}\left[F_{s,u} \mid \mathcal{R}_s\right] du - \frac{1}{2}\mathbb{V}ar\left(\int_s^t F_{s,u} \mid \mathcal{R}_s\right).$$

If we denote $\hat{c}(s,u,v) = \mathbb{C}ov\left(F_{s,u}, F_{s,v} \mid \mathcal{R}_s\right)$ then it follows that

$$\int_s^t \hat{F}_{s,u}du = \int_s^t \mathbb{E}\left[F_{s,u} \mid \mathcal{R}_s\right] du - \int_{u=s}^t \int_{v=s}^u \hat{c}(s,u,v)dudv,$$

which implies that

$$\hat{F}_{s,t} = \mathbb{E}\left[F_{s,t} \mid \mathcal{R}_s\right] - \int_s^t \hat{c}(s,u,t)du. \tag{6.57}$$

Consider the special case of the one-factor model when $\mathcal{R}_s = \sigma\{R_s\}$, so only the spot rate is observed and suppose that the field has the independent-increments property, with covariance specified by $c(\cdot,\cdot,\cdot)$; assume that $c(0,\cdot,\cdot) \equiv 0$ and hence $\mu_{0,s} \equiv F_{0,s}$. In this case

$$\mathbb{E}\left[F_{s,t} \mid \mathcal{R}_s\right] = \mu_{s,t} + \frac{c(s,s,t)}{c(s,s,s)}\left(R_s - \mu_{s,s}\right)$$

and

$$\hat{c}(s,u,v) = c(s,u,v) - \frac{c(s,s,u)c(s,s,v)}{c(s,s,s)}.$$

Recalling (6.24), we see that

$$\int_s^t \mu_{s,u}du = \int_s^t \mu_{0,u}du + \int_s^t \int_0^u c(s \wedge v, u, v)dvdu,$$

and it follows that $\hat{P}_{s,t}$ may be represented in the form

$$\hat{P}_{s,t} = e^{a_{s,t} - b_{s,t} R_s} \tag{6.58}$$

where

$$b_{s,t} = \int_s^t \frac{c(s,s,u)}{c(s,s,s)}du \quad \text{and} \quad a_{s,t} = \ln\left(\frac{P_{0,t}}{P_{0,s}}\right) + b_{s,t}F_{0,s} - \frac{c(s,s,s)}{2}b_{s,t}^2 + d_{s,t}$$

with

$$d_{s,t} = \int_{u=s}^{t} \int_{v=0}^{s} \left[\frac{c(v,v,s)c(s,s,u)}{c(s,s,s)} - c(v,v,u) \right] dv du.$$

If the field is assumed to be strictly Markov then $d_{s,t} \equiv 0$ and $\mathbb{E}\left[P_{s,t} \mid \mathcal{R}_s\right] \equiv \mathbb{E}\left[P_{s,t} \mid \mathcal{Q}_s\right]$ where $\mathcal{Q}_s = \sigma\{R_u, 0 \leqslant u \leqslant s\}$. When we take $c(s,u,v) = \sigma^2 s$ for all u and v then (6.58) gives the formula for bond prices of the model of Ho and Lee (see Exercise 6.3), while if we take $c(s,u,v) = \sigma^2 \sinh(\alpha s) e^{-\alpha(u \vee v)}/\alpha$ then (6.58) gives the formula for the extended Vasicek model (see Example 6.2).

In the general finite-factor model the information $\mathcal{R}_s$ is taken to be of the form

$$\mathcal{R}_s = \sigma\left\{R_s, P_{s,s+t_1}, \ldots, P_{s,s+t_k}\right\}. \tag{6.59}$$

Then, in principle, $\mathbb{E}\left[F_{s,t} \mid \mathcal{R}_s\right]$ may be calculated and $\hat{F}_{s,t}$ obtained from (6.57), although closed-form expressions are not easy to derive except in certain special cases.

6.4 Exercises

Exercise 6.1 Assume that the discounted zero-coupon bond prices $\{P_{s,t}\}$ are martingales, with the money-market account as numeraire, under the probability $\mathbb{P}$. For each fixed $t > 0$, define an equivalent probability $\mathbb{Q}_t$ (known as **the t-forward (risk-adjusted) measure**) by specifying the Radon–Nikodym derivative

$$\frac{d\mathbb{Q}_t}{d\mathbb{P}} = e^{-\int_0^t R_u du} / \mathbb{E}\, P_{0,t}.$$

(a) Show that, for each $t' \leqslant t$, the process $\{P_{s,t'}/P_{s,t}, \mathcal{F}_s, 0 \leqslant s \leqslant t'\}$ is a martingale under $\mathbb{Q}_t$; that is, with the zero-coupon bond maturing at time t as numeraire, the discounted bond prices are martingales under $\mathbb{Q}_t$.

(b) Show that, for $s \leqslant t$,

$$F_{s,t} = \mathbb{E}_{\mathbb{Q}_t}\left[R_t \mid \mathcal{F}_s\right];$$

that is, the instantaneous-forward-rate process $\{F_{s,t}, \mathcal{F}_s, 0 \leqslant s \leqslant t\}$ is a martingale under $\mathbb{Q}_t$.

(c) Suppose that a contingent claim pays an amount C at time t. Show that its price at time $s \leqslant t$ may be expressed as $P_{s,t}\mathbb{E}_{\mathbb{Q}_t}[C \mid \mathcal{F}_s]$.

Exercise 6.2 *Cox, Ingersoll and Ross model.* Consider the one-factor interest-rate model where the short rate satisfies the stochastic differential equation

$$dR_s = \alpha(\beta - R_s)\, ds + \sigma\sqrt{R_s}\, dW_s.$$

Show that bond prices have the form $P_{s,t} = \exp(a_{s,t} - b_{s,t} R_s)$ where

$$a_{s,t} = \frac{2\alpha\beta}{\sigma^2} \ln\left[\frac{2\gamma e^{(\alpha+\gamma)(t-s)/2}}{(\gamma+\alpha)\left(e^{\gamma(t-s)}-1\right)+2\gamma}\right] \quad \text{and}$$
$$b_{s,t} = \frac{2\left(e^{\gamma(t-s)}-1\right)}{(\gamma+\alpha)\left(e^{\gamma(t-s)}-1\right)+2\gamma}, \tag{6.60}$$

with $\gamma = \sqrt{\alpha^2 + 2\sigma^2}$.

Exercise 6.3 *Ho and Lee model.* The first one-factor model that was set up to fit the initial term structure (originally presented in a discrete-time version) supposes that

$$dR_s = \theta_s ds + \sigma dW_s,$$

where θ_s is a deterministic function of time s. Show that by choosing

$$\theta_s = \frac{dF_{0,s}}{ds} + \sigma^2 s$$

the bond prices will be of the form $P_{s,t} = \exp(a_{s,t} - b_{s,t} R_s)$ and will fit the term structure at time $s = 0$ when $b_{s,t} = t - s$ and

$$a_{s,t} = \ln(P_{0,t}/P_{0,s}) + (t-s)F_{0,s} - \sigma^2 s(t-s)^2/2. \tag{6.61}$$

Exercise 6.4 Consider a one-factor model in which the short rate evolves as

$$dR_s = \mu_s ds + \sigma_s dW_s.$$

Show that when the drift term and the volatility are of the form

$$\mu_s = c_1(s) + d_1(s)R_s \quad \text{and} \quad \sigma_s^2 = c_2(s) + d_2(s)R_s \tag{6.62}$$

for functions $c_i, d_i, i = 1, 2$, then the bond prices satisfy

$$P_{s,t} = e^{a(t-s) - b(t-s)R_s} \tag{6.63}$$

for appropriate functions a and b; that is, the model is affine.

Conversely, if $\mu_s = \mu(R_s, s)$ and $\sigma_s = \sigma(R_s, s)$ are functions of R_s and s, show that when the bond prices are of the form (6.63) then necessarily μ_s and σ_s are of the form (6.62).

Exercise 6.5 Consider the Ornstein–Uhlenbeck process satisfying

$$dX_t = -\alpha(X_t - \beta)\, dt + \sigma dW_t,$$

where it is assumed that the initial position X_0 has the $N\left(\beta, \sigma^2/2\alpha\right)$-distribution and is independent of the Brownian motion driving the stochastic differential equation. Verify that the process $\{X_t, t \geqslant 0\}$ is stationary and determine its covariance structure.

Consider the stationary Gaussian-random-field interest-rate model with covariance determined by

$$c(s, t_1, t_2) = \sigma^2 e^{\lambda(s - t_1 \wedge t_2) - \mu|t_1 - t_2|},$$

for some constants σ, $\lambda \geqslant 0$ and $\mu \geqslant \lambda/2$. Show that for any fixed $t \geqslant 0$, the one-dimensional process of forward interest rates $\{F_{s,s+t},\ s \geqslant 0\}$ is a stationary Ornstein–Uhlenbeck process.

Exercise 6.6 In the context of the Gaussian-random-field model, establish the formula (given in (6.42) on page 182) for the price at time s of a European call option, with expiry time t and strike price k, on the bond of maturity date $t + \Delta$ where $s < t < t + \Delta$.

Exercise 6.7 As in the proof of Theorem 6.3, suppose that, N_1 and N_2 are random variables with a joint normal distribution with means μ_1, μ_2 and variances σ_1^2, σ_2^2 respectively. Use the formula (6.37) to show that $\mathbb{E}\left[e^{-N_2}\left(e^{\gamma} - e^{N_1}\right)_+\right]$ equals

$$\begin{aligned} &e^{\gamma-\mu_2+\sigma_2^2/2}\Phi\left(\frac{\mathbb{C}ov\,(N_1, N_2) - \mu_1 + \gamma}{\sigma_1}\right) \\ &\qquad - e^{\mu_1-\mu_2+\mathbb{V}ar(N_1-N_2)/2}\Phi\left(\frac{\mathbb{C}ov\,(N_1, N_2) - \mu_1 + \gamma + \sigma_1^2}{\sigma_1}\right), \end{aligned} \tag{6.64}$$

where γ is a constant.

In the context of the Gaussian-random-field model, show that the price at time s of a European put option, with expiry time t and strike price k, on the bond of maturity date $t + \Delta$, where $s < t < t + \Delta$, is given by

$$\begin{aligned} &kP_{s,t}\,\Phi\left(\frac{\ln\left(P_{s,}/P_{s,t+\Delta}\right) + \ln(k)}{\sigma(s)} + \frac{\sigma(s)}{2}\right) \\ &\qquad - P_{s,t+\Delta}\,\Phi\left(\frac{\ln\left(P_{s,t}/P_{s,t+\Delta}\right) + \ln(k)}{\sigma(s)} - \frac{\sigma(s)}{2}\right), \end{aligned} \tag{6.65}$$

where again $\sigma(s)$ is given in (6.36).

Establish the form of the put-call parity relationship between the price of the put and the price of the call given in (6.42) (see Exercise 6.6).

Appendix A

MATHEMATICAL PRELIMINARIES

A.1 Probability background

A.1.1 Probability spaces

A probability triple $(\Omega, \mathcal{F}, \mathbb{P})$ consists of a non-empty set Ω, the sample space, a σ-field $\mathcal{F}$ of subsets of Ω and a probability (measure) $\mathbb{P}$ defined on $\mathcal{F}$. Recall that a σ-field $\mathcal{F}$ is a non-empty set of subsets (events) of Ω which is closed under taking complements:

$$A \in \mathcal{F} \quad \text{implies that} \quad A^c = \Omega \setminus A \in \mathcal{F},$$

and under taking countable unions:

$$A_i \in \mathcal{F}, i = 1, 2, \ldots, \quad \text{implies that} \quad \bigcup_{i=1}^{\infty} A_i \in \mathcal{F}.$$

A random variable X is a function $X : \Omega \to \mathbb{R}$ which is measurable with respect to $\mathcal{F}$; that is, the event $(X \leqslant c) = (\omega : X(\omega) \leqslant c) \in \mathcal{F}$ for all real numbers $c \in \mathbb{R}$. The smallest σ-field with respect to which a random variable X is measurable is the σ-field generated by X, usually denoted by $\sigma(X)$.

For a non-negative random variable X, the expectation is defined as

$$\mathbb{E}X = \int_{\Omega} X(\omega)\mathbb{P}(d\omega) = \int_{\Omega} X d\mathbb{P},$$

which may be $+\infty$. An arbitrary random variable X may be expressed as the difference of two non-negative random variables, $X = X_+ - X_-$ where $X_+ = \max(X, 0)$ and $X_- = \max(-X, 0)$. Provided at least one of $\mathbb{E}(X_+)$ and $\mathbb{E}(X_-)$ is finite then we define

$$\mathbb{E}X = \mathbb{E}(X_+) - \mathbb{E}(X_-).$$

Recall that the random variable X is said to be integrable when $\mathbb{E}|X| < \infty$, that is when both $\mathbb{E}(X_+)$ and $\mathbb{E}(X_-)$ are finite.

Throughout, we will adopt the usual convention that relations between random variables, X, Y say, such as $X \leqslant Y$, or $X = Y$, without any further qualification may be taken to hold with probability 1, or 'almost surely'.

A.1.2 Conditional expectations

Suppose that $\mathcal{G} \subseteq \mathcal{F}$ is a sub σ-field of $\mathcal{F}$. For a non-negative random variable X, the conditional expectation of X given $\mathcal{G}$, written $\mathbb{E}\,(X \mid \mathcal{G})$, is a $\mathcal{G}$-measurable random variable satisfying

$$\mathbb{E}\,[\mathbb{E}\,(X \mid \mathcal{G})\, I_A] = \mathbb{E}\,(X I_A) \quad \text{for all events} \quad A \in \mathcal{G}, \tag{A.1}$$

where I_A is the indicator of the event A; that is, $I_A(\omega) = 1$ or 0 according as $\omega \in A$ or $\omega \notin A$. Note that the conditional expectation $\mathbb{E}\,(X \mid \mathcal{G})$ is only defined up to sets of probability 0. For any random variable X for which $\mathbb{E}\,X$ is defined, we set

$$\mathbb{E}\,(X \mid \mathcal{G}) = \mathbb{E}\,(X_+ \mid \mathcal{G}) - \mathbb{E}\,(X_- \mid \mathcal{G})\,.$$

It follows that (A.1) holds for any X, for which $\mathbb{E}\,(X)$ is defined. It is immediate that $\mathbb{E}\,(\mathbb{E}\,(X \mid \mathcal{G})) = \mathbb{E}\,(X)$ by taking $A = \Omega$ in (A.1).

For a random variable X and $\mathcal{G}$-measurable random variable Y for which both $\mathbb{E}\,X$ and $\mathbb{E}\,(XY)$ are defined we have that

$$\mathbb{E}\,(XY \mid \mathcal{G}) = \mathbb{E}\,(X \mid \mathcal{G})Y. \tag{A.2}$$

The relation (A.2) shows that when Y is $\mathcal{G}$-measurable it may be treated effectively as a constant when conditioning on $\mathcal{G}$ and taken outside the conditional expectation; in other words, Y is known given $\mathcal{G}$.

For two σ-fields $\mathcal{G}$ and $\mathcal{H}$, with $\mathcal{G} \subseteq \mathcal{H} \subseteq \mathcal{F}$, we have that

$$\mathbb{E}\,(\mathbb{E}\,(X \mid \mathcal{H}) \mid \mathcal{G}) = \mathbb{E}\,(X \mid \mathcal{G})\,. \tag{A.3}$$

The relation (A.3) is often known as the **tower property** of conditional expectations.

Note also the conditional form of Jensen's inequality that

$$f(\mathbb{E}\,(X \mid \mathcal{G})) \leqslant \mathbb{E}\,(f(X) \mid \mathcal{G})$$

for a convex function $f : \mathbb{R} \to \mathbb{R}$ (see Section A.4), when $f(X)$ is integrable; the inequality is reversed when f is a concave function.

The random variable X is independent of the σ-field $\mathcal{G}$ when $\sigma(X)$, the σ-field generated by X, and $\mathcal{G}$ are independent σ-fields; in this case $\mathbb{E}\,(X \mid \mathcal{G}) = \mathbb{E}\,X$. Recall that σ-fields $\mathcal{G}$ and $\mathcal{H}$ are independent when $\mathbb{P}(G \cap H) = \mathbb{P}(G)\mathbb{P}(H)$ for all events $G \in \mathcal{G}$ and $H \in \mathcal{H}$.

A.1.3 Change of probability

A probability (measure) $\mathbb{Q}$ defined on $(\Omega, \mathcal{F})$ is **dominated** by $\mathbb{P}$ (or $\mathbb{Q}$ is **absolutely continuous** with respect to $\mathbb{P}$) if for any $A \in \mathcal{F}$, $\mathbb{P}(A) = 0$ implies that $\mathbb{Q}(A) = 0$; when $\mathbb{Q}$ is dominated by $\mathbb{P}$ we write $\mathbb{Q} \ll \mathbb{P}$. When $\mathbb{Q} \ll \mathbb{P}$ and $\mathbb{P} \ll \mathbb{Q}$ then $\mathbb{P}$ and $\mathbb{Q}$ are said to be **equivalent**; when $\mathbb{Q}$ and $\mathbb{P}$ are equivalent then for an event $A \in \mathcal{F}$, $\mathbb{P}(A) = 0$ if and only if $\mathbb{Q}(A) = 0$.

It is a consequence of the Radon–Nikodym theorem that when $\mathbb{Q} \ll \mathbb{P}$ there exists a non-negative random variable L so that the probability $\mathbb{Q}$ may be represented as

$$\mathbb{Q}(A) = \mathbb{E}\,(L I_A) \quad \text{for all events} \quad A \in \mathcal{F}. \tag{A.4}$$

Necessarily, $\mathbb{E}L = \mathbb{Q}\,(\Omega) = 1$. The random variable L is usually written as $L = d\mathbb{Q}/d\mathbb{P}$ and it is known as the **Radon–Nikodym derivative** of $\mathbb{Q}$ with respect to $\mathbb{P}$.

Conversely, one may start with any non-negative random variable L with $\mathbb{E}L = 1$ and define $\mathbb{Q}$ by (A.4); that is, set $\mathbb{Q}(A) = \mathbb{E}\,(LI_A)$ for all events $A \in \mathcal{F}$. This gives $\mathbb{Q}$ to be a probability which is absolutely continuous with respect to $\mathbb{P}$ and for which $L = d\mathbb{Q}/d\mathbb{P}$.

To illustrate the notion of the Radon–Nikodym derivative, suppose for example that $\Omega = \{\omega_1, \omega_2, \ldots\}$ is finite or countable, and that $\mathbb{P}(\{\omega_j\}) = p_j > 0$ and $\mathbb{Q}(\{\omega_j\}) = q_j > 0$ for each j. Then $\mathbb{P}$ and $\mathbb{Q}$ are equivalent, we will have

$$L(\omega_j) = \frac{d\mathbb{Q}}{d\mathbb{P}}(\omega_j) = \frac{q_j}{p_j}$$

and we may check that

$$\mathbb{Q}(A) = \sum_{j:\omega_j \in A} q_j = \sum_{j:\omega_j \in A} \left(\frac{q_j}{p_j}\right) p_j = \mathbb{E}\left(\frac{d\mathbb{Q}}{d\mathbb{P}} I_A\right).$$

It is important always to remember that $d\mathbb{Q}/d\mathbb{P}$ is a random variable and that, in statistical terminology, it is just the likelihood ratio of the two probabilities.

When $\mathbb{Q} \ll \mathbb{P}$, expectations with respect to the probability $\mathbb{Q}$ may be computed as $\mathbb{E}_{\mathbb{Q}}X = \mathbb{E}\,(LX)$ for any random variable X for which $\mathbb{E}\,(LX)$ is defined; here the unsubscripted expectation, $\mathbb{E}$, refers always to the expected value with respect to the original probability $\mathbb{P}$. For an event B with $\mathbb{Q}(B) > 0$, conditional probabilities and conditional expectations for $\mathbb{Q}$ are computed as

$$\mathbb{Q}\,(A \mid B) = \frac{\mathbb{E}\,(LI_A \mid B)}{\mathbb{E}\,(L \mid B)} \quad \text{and} \quad \mathbb{E}_{\mathbb{Q}}\,(X \mid B) = \frac{\mathbb{E}\,(LX \mid B)}{\mathbb{E}\,(L \mid B)}.$$

In terms of conditional expectations with respect to a sub σ-field $\mathcal{G} \subseteq \mathcal{F}$, the corresponding statement is

$$\mathbb{E}_{\mathbb{Q}}\,(X \mid \mathcal{G}) = \frac{\mathbb{E}\,(LX \mid \mathcal{G})}{\mathbb{E}\,(L \mid \mathcal{G})}. \tag{A.5}$$

When $\mathbb{P}$ and $\mathbb{Q}$ are equivalent, then the Radon–Nikodym derivative, $d\mathbb{Q}/d\mathbb{P}$, is strictly positive (with probability 1 under both $\mathbb{P}$ and $\mathbb{Q}$).

These notions may be generalized to the case where L is no longer non-negative. Consider an arbitrary random variable L with $\mathbb{E}|L| < \infty$ and $\mathbb{E}L = 1$, and set $\mathbb{Q}(A) = \mathbb{E}\,(LI_A)$ for events $A \in \mathcal{F}$ to define a signed measure $\mathbb{Q}$ on $(\Omega, \mathcal{F})$; a signed measure $\mathbb{Q}$ is a measure that may be expressed as $\mathbb{Q}(A) = \mathbb{Q}_+(A) - \mathbb{Q}_-(A)$, for events $A \in \mathcal{F}$, where $\mathbb{Q}_+$ and $\mathbb{Q}_-$ are non-negative measures for which not both $\mathbb{Q}_+(A)$ and $\mathbb{Q}_-(A)$ are positive for any $A \in \mathcal{F}$. Again write $L = d\mathbb{Q}/d\mathbb{P}$. The 'expectation' with respect to $\mathbb{Q}$ for random variables X is then given by $\mathbb{E}_{\mathbb{Q}}X = \mathbb{E}(LX)$ so long as the expectation $\mathbb{E}(LX)$ is well defined. Here $\mathbb{Q}$ will be a measure with total mass $\mathbb{Q}(\Omega) = 1$ but it will not be a probability measure unless $L \geqslant 0$. Conversely, one may start with a signed measure $\mathbb{Q}$, then $\mathbb{Q}$ is dominated by $\mathbb{P}$ when $\mathbb{P}(A) = 0$ implies that $\mathbb{Q}_+(A) = \mathbb{Q}_-(A) = 0$. As previously, by considering $\mathbb{Q}_+$ and $\mathbb{Q}_-$ separately, when $\mathbb{Q}$ is dominated by $\mathbb{P}$ and has total mass $\mathbb{Q}(\Omega) = 1$ then there exists a random variable $L\ (= d\mathbb{Q}/d\mathbb{P})$ with $\mathbb{E}L = 1$ for which $\mathbb{Q}(A) = \mathbb{E}(LI_A)$; furthermore 'expectations' with respect to $\mathbb{Q}$ may be written as $\mathbb{E}_{\mathbb{Q}}(X) = \mathbb{E}(LX)$, again with the proviso that $\mathbb{E}(LX)$ is well defined.

A.1.4 Essential supremum

For an uncountable collection of random variables $\{X_\gamma : \gamma \in \Gamma\}$, the quantity $\sup_{\gamma\in\Gamma} X_\gamma$ may not necessarily be a random variable; however, there exists a random variable Y, known as the **essential supremum** of the collection and written as $Y = \operatorname{ess\,sup}_{\gamma\in\Gamma} X_\gamma$, which has the properties

(i) $Y \geqslant X_\gamma$ for all $\gamma \in \Gamma$; and

(ii) if Z is a random variable with $Z \geqslant X_\gamma$ for all $\gamma \in \Gamma$ then $Z \geqslant Y$.

Recall that the inequalities $Y \geqslant X_\gamma$, $Z \geqslant X_\gamma$ and $Z \geqslant Y$ in (i) and (ii) are taken to hold with probability one.

The collection $\{X_\gamma : \gamma \in \Gamma\}$ is said to be **directed upwards** if for any two random variables X_{γ_1} and X_{γ_2} in the collection there is a third random variable X_{γ_3} in the collection such that $X_{\gamma_3} \geqslant \max(X_{\gamma_1}, X_{\gamma_2})$. It is always the case that there exists a sequence of random variables $\{X_{\gamma_i} : \gamma_i \in \Gamma,\ i \geqslant 1\}$ with

$$Y = \operatorname*{ess\,sup}_{\gamma\in\Gamma} = \sup_i X_{\gamma_i};$$

furthermore, when the collection of random variables is directed upwards we may take the sequence so that $X_{\gamma_i} \uparrow Y$ as $i \uparrow \infty$.

A.2 Martingales

We consider first the case of martingales in discrete time and then comment briefly below on the changes needed for continuous time. Suppose that $(\Omega, \mathcal{F}, \mathbb{P})$ is the underlying probability space. We define a **filtration** $\{\mathcal{F}_r, 0 \leqslant r \leqslant n\}$ to be an expanding sequence of σ-fields

$$\mathcal{F}_0 \subseteq \mathcal{F}_1 \subseteq \cdots \subseteq \mathcal{F}_n \subseteq \mathcal{F}.$$

Suppose that we make observations of a process over times $r = 0, 1, \ldots, n$, then we may think of $\mathcal{F}_r$ as the cumulative information available at time r; the σ-fields $\mathcal{F}_r$ correspond to increasingly finer partitions of the sample space Ω.

A sequence of random variables $\{X_r, 0 \leqslant r \leqslant n\}$ is **adapted** to the filtration $\{\mathcal{F}_r\}$ when X_r is $\mathcal{F}_r$-measurable for each $r = 0, 1, \ldots, n$; intuitively, this means that when the information in $\mathcal{F}_r$ has been observed the value of X_r is known.

A sequence of integrable random variables $\{X_r, 0 \leqslant r \leqslant n\}$ is a **martingale** (relative to a given filtration $\{\mathcal{F}_r, 0 \leqslant r \leqslant n\}$ and a probability $\mathbb{P}$) if the sequence is adapted and

$$\mathbb{E}\,(X_{r+1} \mid \mathcal{F}_r) = X_r \quad \text{for all} \quad 0 \leqslant r < n. \tag{A.6}$$

A **submartingale** is defined in a similar way except that the martingale property (A.6) is replaced by

$$\mathbb{E}\,(X_{r+1} \mid \mathcal{F}_r) \geqslant X_r \quad \text{for all} \quad 0 \leqslant r < n; \tag{A.7}$$

similarly, a **supermartingale** has (A.6) replaced by

$$\mathbb{E}\,(X_{r+1} \mid \mathcal{F}_r) \leqslant X_r \quad \text{for all} \quad 0 \leqslant r < n. \tag{A.8}$$

A martingale is both a submartingale and a supermartingale.

When $\{X_r, 0 \leqslant r \leqslant n\}$ is a martingale and $f : \mathbb{R} \to \mathbb{R}$ is a convex function for which $f(X_r)$ is integrable for each r, then by the conditional form of Jensen's inequality

$$\mathbb{E}\,(f(X_{r+1}) \mid \mathcal{F}_r) \geqslant f(\mathbb{E}\,(X_{r+1} \mid \mathcal{F}_r)) = f(X_r) \tag{A.9}$$

showing that $\{f(X_r), 0 \leqslant r \leqslant n\}$ is a submartingale; when f is concave instead of convex then the inequality in (A.9) is reversed so that $\{f(X_r), 0 \leqslant r \leqslant n\}$ is a supermartingale.

It follows immediately from (A.6), by induction on $s = r+1, r+2, \ldots$, that for a martingale we have

$$\mathbb{E}\,(X_s \mid \mathcal{F}_r) = X_r \quad \text{for all} \quad 0 \leqslant r < s \leqslant n, \tag{A.10}$$

and taking expectations we see that $\mathbb{E}\,X_0 = \mathbb{E}\,X_1 = \cdots = \mathbb{E}\,X_n$, showing that a martingale is constant in mean. The corresponding statement for a submartingale is that it is non-decreasing in mean, $\mathbb{E}\,X_0 \leqslant \mathbb{E}\,X_1 \leqslant \cdots \leqslant \mathbb{E}\,X_n$, and a supermartingale is non-increasing in mean, $\mathbb{E}\,X_0 \geqslant \mathbb{E}\,X_1 \geqslant \cdots \geqslant \mathbb{E}\,X_n$.

A **stopping time** (relative to the filtration $\{\mathcal{F}_r, 0 \leqslant r \leqslant n\}$) is a random variable taking values in the set $\{0, 1, \ldots, n\}$ such that the event

$$(T \leqslant r) \in \mathcal{F}_r \quad \text{for each} \quad r = 0, 1, \ldots, n. \tag{A.11}$$

The relation (A.11) is equivalent to requiring that the event $(T = r) \in \mathcal{F}_r$ for each r. Intuitively, a stopping time is a rule which tells us when to stop based only on knowing the history up to the instant of stopping; that is, it does not look into the future. Note that if S and T are stopping times then $S \wedge T = \min(S, T)$ and $S \vee T = \max(S, T)$ are also stopping times; in particular, $T \wedge r$ is a stopping time when r is a constant time.

For a stopping time T the σ-field $\mathcal{F}_T$ is defined to be the set of those events $A \in \mathcal{F}$ such that $A \cap (T \leqslant r) \in \mathcal{F}_r$ for each $r = 0, 1, \ldots, n$. It is straightforward to check that $\mathcal{F}_T$ is a σ-field and it should be noted that it represents the information available through observing the history up to the stopping time T. Furthermore, when a sequence of random variables $\{X_r, 0 \leqslant r \leqslant n\}$ is adapted then the random variable X_T is $\mathcal{F}_T$-measurable.

When S and T are stopping times with $S \leqslant T \leqslant n$ and $\{X_r, 0 \leqslant r \leqslant n\}$ is martingale then we have

$$\mathbb{E}\left(X_T \mid \mathcal{F}_S\right) = X_S, \tag{A.12}$$

which generalizes (A.10). The conclusion (A.12) which shows that the martingale property is preserved at stopping times is known as the Optional Sampling Theorem. We may deduce from a particular case of (A.12) that for any stopping time T, the sequence $\{X_{T \wedge r}, 0 \leqslant r \leqslant n\}$ is a martingale.

When $\{X_r, 0 \leqslant r \leqslant n\}$ is a submartingale, (A.12) is replaced by the inequality $\mathbb{E}\left(X_T \mid \mathcal{F}_S\right) \geqslant X_S$, and in the case of a supermartingale by $\mathbb{E}\left(X_T \mid \mathcal{F}_S\right) \leqslant X_S$.

Note that a sequence of random vectors $\{\boldsymbol{X}_r, 0 \leqslant r \leqslant n\}$ taking values in $\mathbb{R}^s$ with $\boldsymbol{X}_r = (X_{1,r}, \ldots, X_{s,r})^\top$, is a martingale relative to the fixed filtration and probability if each coordinate sequence $\{X_{i,r}, 0 \leqslant r \leqslant n\}$ is a martingale, $i = 1, \ldots, s$.

Turning to continuous time, only minor changes to the outline above are required. A collection $\{\mathcal{F}_t, 0 \leqslant t < \infty\}$ of sub σ-fields of $\mathcal{F}$ is a filtration when $\mathcal{F}_s \subseteq \mathcal{F}_t$ whenever $0 \leqslant s \leqslant t$; furthermore, a stochastic process $\{X_t, 0 \leqslant t < \infty\}$ (that is, a collection of random variables indexed by $t \geqslant 0$) which is adapted to the filtration $\{\mathcal{F}_t\}$ is a martingale when X_t is integrable for each t and

$$\mathbb{E}\left(X_t \mid \mathcal{F}_s\right) = X_s \quad \text{when} \quad 0 \leqslant s \leqslant t. \tag{A.13}$$

For a submartingale or a supermartingale we replace the equality in (A.13) by the inequality corresponding to (A.7) or (A.8) respectively. As in the discrete-time situation when $\{X_t, 0 \leqslant t < \infty\}$ is a martingale then $\{f(X_t), 0 \leqslant t < \infty\}$ is a submartingale when f is a convex function or a supermartingale when f is a concave function (assuming in each case that $f(X_t)$ is integrable for all t).

A.3 Gaussian random variables

A.3.1 Univariate normal distributions

We will use the terms 'Gaussian' and 'normally-distributed' interchangeably throughout. Recall that a real-valued random variable has the normal distribution with mean $\mathbb{E}X = \mu$, $-\infty < \mu < \infty$, and variance $\mathbb{V}ar(X) = \sigma^2$, $\sigma > 0$, when its probability density function is

$$\phi(x; \mu, \sigma) = \frac{1}{\sqrt{2\pi}\sigma} e^{-\frac{1}{2\sigma^2}(x-\mu)^2}, \quad \text{for} \quad -\infty < x < \infty;$$

as shorthand we say that X has the $N(\mu, \sigma^2)$-distribution. In the particular case when $\mu = 0$ and $\sigma^2 = 1$ we say that X has the standard normal distribution, $N(0, 1)$, in which case it has probability density function

$$\phi(x) = \frac{1}{\sqrt{2\pi}} e^{-\frac{1}{2}x^2} \quad \text{and distribution function} \quad \Phi(x) = \int_{-\infty}^{x} \frac{1}{\sqrt{2\pi}} e^{-\frac{1}{2}x^2} dx.$$

When X has the $N(\mu, \sigma^2)$-distribution and $a \neq 0$ and b are real constants then the random variable $Y = aX + b$ has the $N(a\mu + b, a^2\sigma^2)$-distribution. Furthermore

$$\mathbb{E}[f(X)(X - \mu)] = \sigma^2 \mathbb{E}[f'(X)] \tag{A.14}$$

for any differentiable function for which the expectations on both sides are finite. To see (A.14), write $X + \mu + \sigma Y$, so that Y has the standard $N(0, 1)$-distribution, and notice that for the standard normal density function ϕ, $\phi'(y) = -y\phi(y)$, then use integration-by-parts to see that

$$\begin{aligned}
\mathbb{E}[f(X)(X - \mu)] &= \sigma \mathbb{E}[f(\mu + \sigma Y) Y] \\
&= \sigma \int_{-\infty}^{\infty} y f(\mu + \sigma y)\phi(y) dy = -\sigma \int_{-\infty}^{\infty} f(\mu + \sigma y) d\phi(y) \\
&= \left[-\sigma f(\mu + \sigma y)\phi(y)\right]_{-\infty}^{\infty} + \sigma^2 \int_{-\infty}^{\infty} f'(\mu + \sigma y)\phi(y) dy \\
&= \sigma^2 \int_{-\infty}^{\infty} f'(\mu + \sigma y)\phi(y) dy = \sigma^2 \mathbb{E}\left[f'(X)\right].
\end{aligned}$$

Another important identity when X has the $N(\mu, \sigma^2)$-distribution is that, for any real number θ,

$$\mathbb{E}\left[e^{\theta X} f(X)\right] = e^{\mu\theta + \theta^2\sigma^2/2} \mathbb{E} f(X + \theta\sigma^2) \tag{A.15}$$

for all functions f for which the expectation on both sides of (A.15) is defined. This relation is true because the left-hand side is

$$\begin{aligned}
&\int_{-\infty}^{\infty} e^{\theta x} f(x) \frac{1}{\sqrt{2\pi}\sigma} e^{-\frac{1}{2\sigma^2}(x-\mu)^2} dx \\
&\qquad = e^{\mu\theta + \theta^2\sigma^2/2} \int_{-\infty}^{\infty} f(x) \frac{1}{\sqrt{2\pi}\sigma} e^{-\frac{1}{2\sigma^2}(x-\mu-\theta\sigma^2)^2} dx
\end{aligned}$$

which may be seen to be the right-hand side of (A.15) after substituting $x = y + \theta\sigma^2$ in the integral on the right-hand side; an alternative derivation of (A.15) is given in Exercise 3.3. One important special case of (A.15) gives the moment-generating function of X as

$$m(\theta) = \mathbb{E}\left[e^{\theta X}\right] = e^{\mu\theta + \theta^2\sigma^2/2}, \tag{A.16}$$

which may be derived by setting $f(\cdot) \equiv 1$. It is an immediate consequence of (A.16) that a linear combination of independent random variables each with the normal distribution has a normal distribution; suppose that $X_1, \ldots, X_n$ are independent, with the distribution of X_i being $N(\mu_i, \sigma_i^2)$, $1 \leqslant i \leqslant n$, then for real constants $a_1, \ldots, a_n$,

$$\begin{aligned}
\mathbb{E}\left[e^{\theta \sum_{i=1}^{n} a_i X_i}\right] &= \prod_{i=1}^{n} \mathbb{E}\left[e^{\theta a_i X_i}\right] = \prod_{i=1}^{n} e^{\theta a_i \mu_i + \theta^2 a_i^2 \sigma_i^2/2} \\
&= e^{\theta\left(\sum_{i=1}^{n} a_i \mu_i\right) + \theta^2\left(\sum_{i=1}^{n} a_i^2 \sigma_i^2\right)/2}
\end{aligned}$$

showing that the linear combination $\sum_{i=1}^{n} a_i X_i$ is $N\left(\sum_{i=1}^{n} a_i \mu_i, \sum_{i=1}^{n} a_i^2 \sigma_i^2\right)$.

One further useful identity when X has the $N(\mu, \sigma^2)$-distribution and $\Phi(x)$ is the standard normal distribution function is that

$$\mathbb{E}\,\Phi(X) = \Phi\left(\frac{\mu}{\sqrt{1+\sigma^2}}\right); \tag{A.17}$$

to see this, suppose that Z has the standard normal distribution and is independent of X, so that $\Phi(x) = \mathbb{P}(Z \leqslant x)$, we have

$$\mathbb{E}\,\Phi(X) = \mathbb{P}(Z \leqslant X) = \mathbb{P}\left(\frac{Z - X + \mu}{\sqrt{1+\sigma^2}} \leqslant \frac{\mu}{\sqrt{1+\sigma^2}}\right) = \Phi\left(\frac{\mu}{\sqrt{1+\sigma^2}}\right),$$

because $(Z - X + \mu)/\sqrt{1+\sigma^2}$ again has the standard normal distribution by the remarks above. When a and b are constants, (A.17) generalizes immediately to

$$\mathbb{E}\,\Phi(aX + b) = \Phi\left(\frac{a\mu + b}{\sqrt{1+a^2\sigma^2}}\right), \tag{A.18}$$

because $aX + b$ has the $N(a\mu + b, a^2\sigma^2)$-distribution.

A.3.2 Multivariate normal distributions

We say that random variables X and Y have a **bivariate normal distribution** (or **bivariate Gaussian distribution** or **joint normal distribution**) if their joint probability density function has the form

$$\phi_{X,Y}(x, y) = \frac{1}{2\pi\sigma\tau\sqrt{1-\rho^2}} \exp\left[-\frac{1}{2(1-\rho^2)}\left(\left(\frac{x-\mu}{\sigma}\right)^2 - 2\rho\left(\frac{x-\mu}{\sigma}\right)\left(\frac{y-\nu}{\tau}\right) + \left(\frac{y-\nu}{\tau}\right)^2\right)\right]$$

for $-\infty < x < \infty$ and $-\infty < y < \infty$ where the parameters satisfy $-\infty < \mu < \infty$, $-\infty < \nu < \infty$, $\sigma > 0$, $\tau > 0$ and $-1 < \rho < 1$.

First check that this expression is indeed a joint density function in that it integrates to 1. Make the substitutions $u = (x-\mu)/(\sigma\sqrt{1-\rho^2})$ and $v = (y-\nu)/(\tau\sqrt{1-\rho^2})$, then we have

$$I = \iint_{-\infty<x,y<\infty} \phi_{X,Y}(x, y)\,dx\,dy = \iint_{-\infty<u,v<\infty} \frac{\sqrt{1-\rho^2}}{2\pi} e^{-\frac{1}{2}(u^2 - 2\rho uv + v^2)}\,du\,dv$$

$$= \iint_{-\infty<u,v<\infty} \frac{\sqrt{1-\rho^2}}{2\pi} e^{-\frac{1}{2}((u-\rho v)^2 + (1-\rho^2)v^2)}\,du\,dv.$$

Now put $w = u - \rho v$ and $z = v\sqrt{1-\rho^2}$, or equivalently $u = w + \rho z/\sqrt{1-\rho^2}$ and $v = z/\sqrt{1-\rho^2}$, and calculate the Jacobian of this transformation

$$\frac{\partial(u,v)}{\partial(w,z)} = \begin{vmatrix} 1 & \dfrac{\rho}{\sqrt{1-\rho^2}} \\ 0 & \dfrac{1}{\sqrt{1-\rho^2}} \end{vmatrix} = \frac{1}{\sqrt{1-\rho^2}};$$

then we see that

$$I = \iint_{-\infty<w,z<\infty} \frac{1}{2\pi}\, e^{-(w^2+z^2)/2} dw dz = \left(\int_{-\infty}^{\infty} \frac{1}{\sqrt{2\pi}}\, e^{-w^2/2} dw\right)^2 = 1.$$

To see the relationship with the univariate normal distribution and to determine the marginal distributions of X and Y, consider the random variables

$$U = X, \quad V = Y - \nu - \rho\tau(X-\mu)/\sigma;$$

now put X and Y in terms of U and V to see that

$$X = U, \quad Y = V + \nu + \rho\tau(U-\mu)/\sigma,$$

and calculate the Jacobian of this transformation as

$$J = \begin{vmatrix} \dfrac{\partial x}{\partial u} & \dfrac{\partial x}{\partial v} \\ \dfrac{\partial y}{\partial u} & \dfrac{\partial y}{\partial v} \end{vmatrix} = \begin{vmatrix} 1 & 0 \\ \rho\tau/\sigma & 1 \end{vmatrix} = 1.$$

We may now calculate the joint density function of U and V, evaluated at (u, v), as

$$\left(\frac{1}{\sqrt{2\pi}\sigma}\, e^{-(u-\mu)^2/(2\sigma^2)}\right)\left(\frac{1}{\sqrt{2\pi(1-\rho^2)}\tau}\, e^{-v^2/(2\tau^2(1-\rho^2))}\right),$$

and we recognize these two expressions, the first in u is the density of the $N(\mu, \sigma^2)$ distribution, and the second in v is the density of the $N(0, \tau^2(1-\rho^2))$ distribution, and moreover, because the joint density factors into the product of these two densities, the random variables U and V are independent. We conclude that the marginal distribution of X is $N(\mu, \sigma^2)$ and, by the symmetry of the joint density of X and Y, we can see that the marginal density of Y is $N(\nu, \tau^2)$. To interpret the remaining parameter ρ, calculate

$$\begin{aligned} \mathbb{C}ov\,(X,Y) &= \mathbb{C}ov\,(U, V + \nu + \rho\tau(U-\mu)/\sigma) \\ &= \mathbb{C}ov\,(U,V) + \mathbb{C}ov\,(U, \rho\tau(U-\mu)/\sigma) = \mathbb{C}ov\,(U, \rho\tau(U-\mu)/\sigma), \end{aligned}$$

since ν is constant and U and V are independent, so that

$$\mathbb{C}ov\,(X,Y) = \rho\tau \mathbb{V}ar(U)/\sigma = \rho\sigma\tau = \rho\sqrt{\mathbb{V}ar(X)\mathbb{V}ar(Y)}.$$

Thus the parameter $\rho = \mathbb{C}orr\,(X, Y)$ is the correlation coefficient of the random variables X and Y. We may see immediately that

$$\phi_{X,Y}(x, y) = \left(\frac{1}{\sqrt{2\pi}\sigma} e^{-(x-\mu)^2/(2\sigma^2)}\right)\left(\frac{1}{\sqrt{2\pi}\tau} e^{-(y-\nu)^2/(2\tau^2)}\right) = \phi_X(x)\phi_Y(y),$$

for all x and y, if and only if $\rho = 0$, or equivalently if and only if $\mathbb{C}ov\,(X, Y) = 0$. Thus random variables which have a joint normal distribution are independent if and only if their covariance is zero. Recall that in general the covariance between two random variables being zero does not imply independence of the random variables, but it is an important and useful property that the covariance being zero is sufficient to show independence for normally distributed variables.

We may calculate the conditional density of one of the random variables Y, say, given the value of the other variable $X = x$, that is, the density $\phi_{Y|X}(y \mid x) = \phi_{X,Y}(x, y)/\phi_X(x)$, which equals

$$\frac{\exp\left[-\frac{1}{2(1-\rho^2)}\left(\left(\frac{x-\mu}{\sigma}\right)^2 - 2\rho\left(\frac{x-\mu}{\sigma}\right)\left(\frac{y-\nu}{\tau}\right) + \left(\frac{y-\nu}{\tau}\right)^2\right)\right]}{2\pi\sigma\tau\sqrt{1-\rho^2}} \Bigg/ \frac{\exp\left[-\frac{1}{2}\left(\frac{x-\mu}{\sigma}\right)^2\right]}{\sigma\sqrt{2\pi}}$$

$$= \exp\left[-\frac{1}{2(1-\rho^2)}\left(\rho^2\left(\frac{x-\mu}{\sigma}\right)^2 - 2\rho\left(\frac{x-\mu}{\sigma}\right)\left(\frac{y-\nu}{\tau}\right) + \left(\frac{y-\nu}{\tau}\right)^2\right)\right] \Big/ \tau\sqrt{2\pi(1-\rho^2)}$$

$$= \exp\left[-\frac{1}{2\tau^2(1-\rho^2)}\left(y - \nu - \rho\tau(x-\mu)/\sigma\right)^2\right] \Big/ \tau\sqrt{2\pi(1-\rho^2)}\,.$$

We recognize this last expression as being the density (in y) of the normal distribution with mean $\nu + \rho\tau(x - \mu)/\sigma$ and variance $\tau^2(1 - \rho^2)$, so that, in shorthand notation, we may express the conditional distribution of

$$Y \mid X \quad \text{is} \quad N\left(\nu + \rho\tau(X - \mu)/\sigma, \tau^2(1 - \rho^2)\right). \tag{A.19}$$

Notice that the conditional expectation of Y given X, which is

$$\begin{aligned} \mathbb{E}\left(Y \mid X\right) &= \nu + \rho\tau(X - \mu)/\sigma \\ &= \mathbb{E}Y + \mathbb{C}ov\,(X, Y)\,(X - \mathbb{E}X)\,/\,\mathbb{V}ar(X), \end{aligned} \tag{A.20}$$

depends on X, but the variance of Y conditional on X is the constant $\tau^2(1 - \rho^2)$, which is less than the unconditioned variance of Y, that is τ^2.

One consequence of (A.20) is that

$$\mathbb{C}ov\,(f(X), Y) = \mathbb{E}\left[f'(X)\right]\mathbb{C}ov\,(X, Y), \tag{A.21}$$

for any differentiable function f for which both sides of (A.21) are finite. To see this

$$\begin{aligned} \mathbb{C}ov\,(f(X), Y) &= \mathbb{E}\,[f(X)\,(Y - \mathbb{E}Y)] = \mathbb{E}\,[\mathbb{E}\,[f(X)\,(Y - \mathbb{E}Y) \mid X]] \\ &= \mathbb{E}\,[f(X)\mathbb{E}\,[Y - \mathbb{E}Y \mid X]]\,, \end{aligned}$$

because conditional on X, $f(X)$ may be treated as a constant and taken outside the conditional expectation, and then using (A.20) and (A.14), we see that this

$$= \frac{\mathbb{C}ov\,(X, Y)}{\mathbb{V}ar(X)}\mathbb{E}\,[f(X)\,(X - \mathbb{E}X)] = \mathbb{E}\left[f'(X)\right]\,\mathbb{C}ov\,(X, Y).$$

A further property that is straightforward to check is that when X and Y have a joint normal distribution and we define random variables R and S by

$$\begin{pmatrix} R \\ S \end{pmatrix} = \begin{pmatrix} a & b \\ c & d \end{pmatrix} \begin{pmatrix} X \\ Y \end{pmatrix} + \begin{pmatrix} g \\ h \end{pmatrix}$$

where a, b, c, d, g and h are constants with $ad \neq bc$, then R and S have a joint normal distribution; this shows that normal distributions are preserved under linear transformations. The condition $ad \neq bc$ is needed to ensure that $|\mathbb{C}orr\,(R, S)| \neq 1$; even if this condition does not hold, the random variables R and S will individually have normal distributions but their correlation coefficient will be 1 or -1.

Consider now the joint distribution of more than two normal random variables. Suppose that $Z_1, \ldots, Z_n$ are i.i.d. random variables each with the standard $N(0, 1)$ distribution. Suppose that $A = \left(a_{ij}\right)$ is an $n \times n$ invertible matrix and (using vector notation) suppose that

$$\boldsymbol{X} = \begin{pmatrix} X_1 \\ \vdots \\ X_n \end{pmatrix} = A \begin{pmatrix} Z_1 \\ \vdots \\ Z_n \end{pmatrix} + \begin{pmatrix} \mu_1 \\ \vdots \\ \mu_n \end{pmatrix} = A\boldsymbol{Z} + \boldsymbol{\mu},$$

where $\mu_1, \ldots, \mu_n$ are constants. Since each of the random variables $\{Z_j\}$ has mean zero, we see first that $\mathbb{E}\,X_i = \mu_i$, for each i. The joint probability density function of the components of $\boldsymbol{Z}$ at $\boldsymbol{z} = (z_1, \ldots, z_n)^\top$ is

$$\begin{aligned} \boldsymbol{\phi}(\boldsymbol{z}) &= \prod_{i=1}^{n} \phi(z_i) = \prod_{i=1}^{n} \frac{1}{\sqrt{2\pi}}\, e^{-z_i^2/2} \\ &= \left(\frac{1}{2\pi}\right)^{n/2} e^{-\sum_{i=1}^{n} z_i^2/2} = \left(\frac{1}{2\pi}\right)^{n/2} e^{-\boldsymbol{z}^\top \boldsymbol{z}/2}. \end{aligned}$$

Writing $\boldsymbol{z} = A^{-1}(\boldsymbol{x} - \boldsymbol{\mu})$, the Jacobian of the transformation is $|\det A|^{-1}$, so that the joint density for $\boldsymbol{X}$ is

$$\begin{aligned} \boldsymbol{\psi}(\boldsymbol{x}) &= \frac{1}{|\det A|}\, \boldsymbol{\phi}\left(A^{-1}(\boldsymbol{x} - \boldsymbol{\mu})\right) = \frac{1}{|\det A|} \left(\frac{1}{2\pi}\right)^{n/2} e^{-\frac{1}{2}\left(A^{-1}(\boldsymbol{x}-\boldsymbol{\mu})\right)^\top \left(A^{-1}(\boldsymbol{x}-\boldsymbol{\mu})\right)} \\ &= \frac{1}{|\det A|} \left(\frac{1}{2\pi}\right)^{n/2} e^{-\frac{1}{2}(\boldsymbol{x}-\boldsymbol{\mu})^\top (A^{-1})^\top A^{-1}(\boldsymbol{x}-\boldsymbol{\mu})} \\ &= \frac{1}{\sqrt{|\det V|}} \left(\frac{1}{2\pi}\right)^{n/2} e^{-\frac{1}{2}(\boldsymbol{x}-\boldsymbol{\mu})^\top V^{-1}(\boldsymbol{x}-\boldsymbol{\mu})}, \end{aligned} \qquad \text{(A.22)}$$

where $V = \left(v_{ij}\right) = AA^\top$. To interpret the matrix V we see that for any pair (i, j), $1 \leqslant i, j \leqslant n$,

$$\begin{aligned} \mathbb{C}ov\,\left(X_i, X_j\right) &= \mathbb{E}\left((X_i - \mu_i)(X_j - \mu_j)\right) = \mathbb{E}\left(\left(\sum_r a_{ir} Z_r\right)\left(\sum_s a_{js} Z_s\right)\right) \\ &= \sum_r a_{ir} a_{jr} = \left(AA^\top\right)_{ij} = v_{ij}, \end{aligned}$$

so that the entries of the matrix V are the covariances between the components of the random vector $\boldsymbol{X}$. Any joint density of the form (A.22) is a **multivariate normal distribution** with mean $\boldsymbol{\mu}$ and **covariance matrix** V, usually written $N(\boldsymbol{\mu}, V)$.

Notice that V is a symmetric matrix and it is positive definite in that $\boldsymbol{x}^\top V\boldsymbol{x} > 0$ for all vectors $\boldsymbol{x} \neq 0$; this follows because $\boldsymbol{x}^\top V\boldsymbol{x} = ||A^\top \boldsymbol{x}||^2 > 0$, since A is invertible. We may write $V = \mathbb{E}\left((\boldsymbol{X} - \boldsymbol{\mu})(\boldsymbol{X} - \boldsymbol{\mu})^\top\right)$.

We may also consider the multivariate counterpart of the result (A.19). Suppose that the random vectors $\boldsymbol{X} = \left(X_1, \dots, X_m\right)^\top$ and $\boldsymbol{Y} = \left(Y_1, \dots, Y_n\right)^\top$ are such that $\boldsymbol{X}$ and $\boldsymbol{Y}$ have a multivariate normal distribution in that

$$\begin{pmatrix} \boldsymbol{X} \\ \boldsymbol{Y} \end{pmatrix} = \left(X_1, \dots, X_m, Y_1, \dots, Y_n\right)^\top \quad \text{is} \quad N\left(\begin{pmatrix} \boldsymbol{\mu} \\ \boldsymbol{\nu} \end{pmatrix}, \begin{pmatrix} V_{11} & V_{12} \\ V_{21} & V_{22} \end{pmatrix}\right);$$

here V_{11} is the $m \times m$ covariance matrix of $\boldsymbol{X}$ and V_{22} the $n \times n$ covariance matrix of $\boldsymbol{Y}$, $\mathbb{E}\boldsymbol{X} = \boldsymbol{\mu}$, $\mathbb{E}\boldsymbol{Y} = \boldsymbol{\nu}$ and $V_{12} = V_{21}^\top$ is the $m \times n$ cross-covariance matrix

$$V_{12} = \mathbb{E}\left((\boldsymbol{X} - \boldsymbol{\mu})(\boldsymbol{Y} - \boldsymbol{\nu})^\top\right).$$

The conditional distribution of $\boldsymbol{X}$ given $\boldsymbol{Y}$ is also a multivariate normal with

$$\boldsymbol{X} \mid \boldsymbol{Y} \quad \text{being} \quad N\left(\boldsymbol{\mu} + V_{12}V_{22}^{-1}(\boldsymbol{Y} - \boldsymbol{\nu}), V_{11} - V_{12}V_{22}^{-1}V_{21}\right).$$

A.4 Convexity

A set $\mathcal{Z} \subseteq \mathbb{R}^n$ is **convex** if for all $\boldsymbol{z}_1, \boldsymbol{z}_2 \in \mathcal{Z}$, and $0 \leqslant \theta \leqslant 1$, the point

$$\theta \boldsymbol{z}_1 + (1 - \theta)\boldsymbol{z}_2 \in \mathcal{Z}.$$

That is, for any two points in the set the line segment joining the two points is contained in the set. Any set $\mathcal{H} \subset \mathbb{R}^n$ of the form $\mathcal{H} = \{\boldsymbol{z} \in \mathbb{R}^n : \boldsymbol{x}^\top \boldsymbol{z} = \beta\}$, for some fixed non-zero $\boldsymbol{x} \in \mathbb{R}^n$, and $\beta \in \mathbb{R}$ is known as a **hyperplane**. An important result is the following which we present without proof.

Theorem A.1 Separating Hyperplane Theorem *For any non-empty convex set $\mathcal{Z} \subset \mathbb{R}^n$ and point $\boldsymbol{y} \notin \mathcal{Z}$, there exists a hyperplane $\mathcal{H} = \{\boldsymbol{z} : \boldsymbol{x}^\top \boldsymbol{z} = \beta\}$ which separates $\boldsymbol{y}$ and $\mathcal{Z}$ in that $\boldsymbol{x}^\top \boldsymbol{y} \leqslant \beta \leqslant \boldsymbol{x}^\top \boldsymbol{z}$, for all $\boldsymbol{z} \in \mathcal{Z}$; furthermore, $\mathcal{H}$ may be chosen so that not both $\boldsymbol{y}$ and $\mathcal{Z}$ are contained in $\mathcal{H}$.*

Suppose that $\mathcal{Z} \subseteq \mathbb{R}^n$ is a convex set, then a function $f : \mathcal{Z} \to \mathbb{R}$ is **convex** if

$$f(\theta z_1 + (1 - \theta)z_2) \leqslant \theta f(z_1) + (1 - \theta) f(z_2) \tag{A.23}$$

for all $z_1, z_2 \in \mathcal{Z}$ and $0 \leqslant \theta \leqslant 1$. A function $f : \mathcal{Z} \to \mathbb{R}$ is **concave** if $-f$ is convex.

Note that when $n = 1$, a convex set $\mathcal{Z}$ is an interval or the whole of $\mathbb{R}$. When $n = 1$ an equivalent characterization of (A.23) is that

$$\frac{f(z_2) - f(z_1)}{z_2 - z_1} \leqslant \frac{f(z_3) - f(z_2)}{z_3 - z_2} \tag{A.24}$$

for all $z_i \in \mathcal{Z}$, $i = 1, 2, 3$, with $z_1 < z_2 < z_3$. When (A.23) holds, take

$$\theta = (z_3 - z_2)/(z_3 - z_1) \quad \text{and} \quad 1 - \theta = (z_2 - z_1)/(z_3 - z_2),$$

so that $z_2 = \theta z_1 + (1 - \theta) z_3$, and

$$f(z_2) \leqslant \left(\frac{z_3 - z_2}{z_3 - z_1}\right) f(z_1) + \left(\frac{z_2 - z_1}{z_3 - z_1}\right) f(z_3)$$

which may be rearranged to give (A.24); the argument may be reversed in a straightforward way to see that (A.24) implies (A.23).

A convex function $f : \mathbb{R} \to \mathbb{R}$ will be continuous but it may not necessarily be differentiable; however, from (A.24), it is immediate that it will have right-hand and left-hand derivatives

$$f'(x+) = \lim_{y \downarrow x} \frac{f(y) - f(x)}{y - x} \quad \text{and} \quad f'(x-) = \lim_{y \uparrow x} \frac{f(x) - f(y)}{x - y}$$

with $f'(x-) \leqslant f'(x+)$, for each $x \in \mathbb{R}$. When f is convex and differentiable, then $f'(x)$ is non-decreasing in x, and when f is twice differentiable then $f''(x) \geqslant 0$.

For a twice-differentiable convex function $f : \mathbb{R}^n \to \mathbb{R}$ the corresponding requirement is that its Hessian matrix

$$\mathcal{H}_f = \left(\frac{\partial^2 f}{\partial x_i \partial x_j}\right)_{i,j}$$

be non-negative definite.

Appendix B

SOLUTIONS TO THE EXERCISES

B.1 Portfolio Choice

Solution 1.1 With either distribution for X, the random variable $Y = X/\mu$ has a distribution not depending on μ so the equation $\mathbb{E}v(X+\alpha) = v(\mu)$ for the compensatory risk premium α reduces to $\mathbb{E}v(Y+c) = v(1)$ in both cases (a) and (b) when we substitute $\alpha = c\mu$. Similarly, for the insurance risk premium when we set $\beta = d\mu$, we see that d is the solution of the equation $\mathbb{E}v(Y) = v(1-d)$.

In the case of the uniform distribution and logarithmic utility, integrating by parts, we have

$$\begin{aligned}
\mathbb{E}\,v(Y+c) &= \int_0^2 \frac{\ln(y+c)}{2}\,dy = \left[\frac{y\ln(y+c)}{2}\right]_0^2 - \frac{1}{2}\int_0^2 \frac{y}{y+c}\,dy \\
&= \ln(2+c) - \frac{1}{2}\int_0^2 \left(1 - \frac{c}{y+c}\right) dy \\
&= \ln(2+c) - 1 + (c/2)\ln\left((2/c)+1\right). \qquad \text{(B.1)}
\end{aligned}$$

Equate the expression in (B.1) to $\ln(1) = 0$, and we may see that the compensatory risk premium is $\alpha = c\mu$ where c is the unique positive root of

$$\ln(2+c) + (c/2)\ln((2/c)+1) = 1;$$

this gives the value of $c \approx 0{\cdot}177$. For the insurance risk premium, $\beta = d\mu$, from the relation (B.1) with $c = 0$, we have that $\mathbb{E}\,\ln(Y) = \ln(2/e) = \ln(1-d)$, whence $d = 1 - 2/e \approx 0{\cdot}264$. □

Solution 1.2 First note that $\mathbb{E}X = a/2$, $\mathbb{V}ar(X) = a^2/12$ so that when $\mathbb{E}X = \mu$, $\mathbb{V}ar(X) = \mu^2/3$. We also have $\mathbb{E}Y = \gamma/\lambda = \mu$, $\mathbb{V}ar(Y) = \gamma/\lambda^2 = \mu^2/\gamma$, so that we need $\gamma = 3$. Use (A.16) to see that $\mathbb{E}Z = e^{\nu+\sigma^2/2} = \mu$ and $\mathbb{V}ar(Z) = e^{2\nu+\sigma^2}(e^{\sigma^2}-1) = \mu^2(e^{\sigma^2}-1)$, so that $e^{\sigma^2} - 1 = 1/3$ or $\sigma^2 = \ln(4/3)$. Now calculate

$$\mathbb{E}\sqrt{X} = \int_0^{2\mu} \frac{\sqrt{x}}{2\mu}\, dx = \frac{2^{3/2}}{3}\sqrt{\mu} \approx (0{\cdot}943)\sqrt{\mu}\,;$$

$$\mathbb{E}\sqrt{Y} = \int_0^{\infty} \sqrt{y}\,\frac{e^{-\lambda y}\lambda^3 y^2}{2}\, dy = \frac{\Gamma(\frac{7}{2})}{2\sqrt{\lambda}}\int_0^{\infty} \frac{e^{-\lambda y}\lambda^{7/2} y^{5/2}}{\Gamma(\frac{7}{2})}\, dy$$

and the last integral is 1 because the integrand is a probability density function, so this expectation

$$= \frac{\Gamma(\frac{7}{2})}{2\sqrt{\lambda}} = \frac{\Gamma(\frac{7}{2})}{2\sqrt{3}}\sqrt{\mu} = \frac{5\sqrt{3\pi}}{16}\sqrt{\mu} \approx (0{\cdot}959)\sqrt{\mu}\,;$$

and using (A.16) again

$$\mathbb{E}\sqrt{Z} = e^{\nu/2+\sigma^2/8} = e^{-\sigma^2/8}\sqrt{\mu} = (3/4)^{1/8}\sqrt{\mu} \approx (0{\cdot}965)\sqrt{\mu}.$$

We conclude that $Z \succ Y \succ X$. □

Solution 1.3 The compensatory risk premium α solves $\mathbb{E}v(\alpha + X) = v(\mu)$ while the insurance risk premium β solves $\mathbb{E}v(X) = v(\mu - \beta)$ giving the common value

$$\alpha = \beta = \mu + \frac{1}{a}\ln(\psi(a)).$$

The expansion for small a is straightforward; when $\alpha = a\mathbb{V}ar(X)/2$ for all $a > 0$ we have

$$\psi(a) = \mathbb{E}\big(e^{-aX}\big) = e^{-a\mu + a^2\mathbb{V}ar(X)/2}$$

which is true only when X has a normal distribution, using (A.16). For the final part

$$\psi''\psi - (\psi')^2 = \mathbb{E}\left(X^2 e^{-aX}\right)\mathbb{E}\left(e^{-aX}\right) - \left(\mathbb{E}\left(Xe^{-aX}\right)\right)^2 \geqslant 0$$

by the Cauchy–Schwarz inequality applied to the random variables $A = Xe^{-aX/2}$ and $B = e^{-aX/2}$. To see that α is increasing

$$\frac{d\alpha}{da} = \frac{1}{a^2}\left[\frac{a\psi'}{\psi} - \ln(\psi)\right] = \frac{1}{a^2}f(a), \quad \text{say.}$$

But $f(0) = 0$ and $f' = a\left[\psi''\psi - (\psi')^2\right]/\psi^2 \geqslant 0$ and the conclusion follows. □

Solution 1.4 The quadratic programming problem to be solved is

$$\text{minimize } 2x_1^2 + 4x_1x_2 + 3x_2^2 \quad \text{subject to } \begin{cases} x_1 + x_2 = 1 \\ 3x_1 + 4x_2 = \mu \end{cases}.$$

In this case the constraints have a unique solution given by $x_1 = 4-\mu$ and $x_2 = \mu-3$. The mean-variance efficient frontier is $\sigma^2 = \mu^2 - 6\mu + 11$, the global minimum-variance portfolio is when $d\sigma/d\mu = 0$ which leads to the mean return $\mu_g = 3$, while the diversified portfolio corresponds to $\lambda = 0$ which gives $\mu_d = 11/3$. Note that this is a situation where the set of attainable values of (σ, μ) coincides with the mean-variance frontier. Using the results in Section 1.3

$$V = \begin{pmatrix} 2 & 2 \\ 2 & 3 \end{pmatrix} \quad \text{with} \quad V^{-1} = \frac{1}{2}\begin{pmatrix} 3 & -2 \\ -2 & 2 \end{pmatrix}$$

to give $\alpha = 1/2$, $\beta = 3/2$, $\gamma = 11/2$ and $\delta = 1/2$. When the riskless asset is added the problem to be solved is

$$\text{minimize} \quad 2x_1^2 + 4x_1x_2 + 3x_2^2 \qquad \text{subject to} \quad \begin{cases} x_0 + \ x_1 + \ x_2 = 1 \\ \frac{3}{2}x_0 + 3x_1 + 4x_2 = \mu \end{cases} ;$$

minimizing the Lagrangian

$$\mathcal{L} = 2x_1^2 + 4x_1x_2 + 3x_2^2 + \lambda(1 - x_0 - x_1 - x_2) + \nu(\mu - \tfrac{3}{2}x_0 - 3x_1 - 4x_2),$$

the term in x_0 implies that $\nu = -2\lambda/3$ while differentiating with respect to x_1 and x_2 leads to the equations

$$4x_1 + 4x_2 - \lambda - 3\nu = 0 \quad \text{and} \quad 4x_1 + 6x_2 - \lambda - 4\nu = 0.$$

Solving these gives $x_1 = \lambda/12$ and $x_2 = -\lambda/3$ and substituting into the constraints yields $\lambda = 12(3-2\mu)/17$ and the optimal portfolio is given by $x_0 = (26-6\mu)/17$, $x_1 = (3-2\mu)/17$ and $x_2 = -4(3-2\mu)/17$. The tangency portfolio corresponds to $x_0 = 0$ or $\mu_t = 13/3$. ∎

Solution 1.5 Write $X = \mu + \sigma Y$ where Y has the standard normal distribution, $N(0, 1)$. Then it follows that

$$\frac{\partial f}{\partial \mu} = \mathbb{E}\, v'(\mu + \sigma Y) > 0 \quad \text{when } v' > 0,$$

and using the relation (A.14)

$$\frac{\partial f}{\partial \sigma} = \mathbb{E}\left[Y v'(\mu + \sigma Y)\right] = \sigma \mathbb{E}\, v''(\mu + \sigma Y) \leqslant 0,$$

by the concavity of v. Now when returns are normally distributed then the wealth created by each portfolio has a normal distribution; this argument shows that maximizing in σ for fixed μ gives a value of (σ, μ) on the efficient frontier. To see the concavity of f, note that

$$\frac{\partial^2 f}{\partial \mu^2} = \mathbb{E}\, v''(\mu + \sigma Y) \leqslant 0 \quad \text{and} \quad \frac{\partial^2 f}{\partial \sigma^2} = \mathbb{E}\left[Y^2 v''(\mu + \sigma Y)\right] \leqslant 0,$$

because $v'' \leqslant 0$ since v is concave, while

$$\frac{\partial^2 f}{\partial\mu\partial\sigma} = \mathbb{E}\left[Y v''(\mu + \sigma Y)\right], \quad \text{and then} \quad \frac{\partial^2 f}{\partial\mu^2}\frac{\partial^2 f}{\partial\sigma^2} \geqslant \left(\frac{\partial^2 f}{\partial\mu\partial\sigma}\right)^2$$

follows by applying the Cauchy–Schwarz inequality to the random variables $A = Y\sqrt{-v''(\mu + \sigma Y)}$ and $B = \sqrt{-v''(\mu + \sigma Y)}$; this shows that the 2×2 matrix of second derivatives has non-positive diagonal entries and a non-negative determinant which is sufficient for the matrix to be negative semi-definite. The fact that f is concave means that sets of the form $\{(\sigma, \mu) : f(\sigma, \mu) \geqslant c\}$ are convex which gives the last statement. □

Solution 1.6 The objective function to maximize is

$$f(\boldsymbol{x}) = \mathbb{E}\, v\Big(w\Big(r_0 + \sum_{j=1}^{s} x_j\,(R_j - r_0)\Big)\Big)$$

where $\boldsymbol{x} = (x_1, \ldots, x_s)^\top$ and we have used the condition that $x_0 + \sum_{j=1}^{s} x_j = 1$. The first-order conditions give

$$\frac{\partial f}{\partial x_j} = w\mathbb{E}\left[v'\left(\overline{W}\right)\left(R_j - r_0\right)\right] = 0, \quad \text{for} \quad 1 \leqslant j \leqslant s.$$

Since $r_j = \mathbb{E}\, R_j$ and the fact that $\overline{W}$ and R_j have a joint normal distribution we have that

$$\begin{aligned} 0 = \mathbb{E}\left[v'\left(\overline{W}\right)\left(R_j - r_0\right)\right] &= \mathbb{E}\left[v'\left(\overline{W}\right)\left(R_j - r_j\right)\right] + \mathbb{E}\left[v'\left(\overline{W}\right)\right]\left(r_j - r_0\right) \\ &= \mathbb{C}ov\left(v'\left(\overline{W}\right), R_j\right) + \mathbb{E}\left[v'\left(\overline{W}\right)\right]\left(r_j - r_0\right) \\ &= \mathbb{E}\left[v''\left(\overline{W}\right)\right]\mathbb{C}ov\left(\overline{W}, R_j\right) + \mathbb{E}\left[v'\left(\overline{W}\right)\right]\left(r_j - r_0\right), \end{aligned}$$

where the last equality uses (A.21), and this now gives the relation

$$r_j - r_0 = \alpha\, \mathbb{C}ov\left(\overline{W}, R_j\right),$$

as required.

For the final part, recall that for random variables X and Y and a a constant $\mathbb{C}ov\,(X, Y + a) = \mathbb{C}ov\,(X, Y)$ and $\mathbb{C}ov\,(aX, Y) = a\mathbb{C}ov\,(X, Y)$. Now for each i

$$\alpha_i^{-1}\left(r_j - r_0\right) = \mathbb{C}ov\left(\overline{W}_i, R_j\right)$$

and summing this on i yields

$$\Big(\sum_{i=1}^{n} \alpha_i^{-1}\Big)\left(r_j - r_0\right) = \Big(\sum_{i=1}^{n} w_i\Big)\mathbb{C}ov\left(M, R_j\right).$$

Divide through this relation by n and multiply by $\overline{\alpha}$, where $(\overline{\alpha})^{-1} = \sum_{i=1}^{n} \alpha_i^{-1}/n$, to obtain

$$\mathbb{E}\, R_j - r_0 = \overline{w}\,\overline{\alpha}\, \mathbb{C}ov\,(M, R_j). \tag{B.2}$$

When $\overline{x}_{ij}$ is the optimal proportion invested by investor i in asset j then

$$\overline{W}_i = w_i \left[r_0 + \sum_{j=1}^{s} \overline{x}_{ij} \left(R_j - r_0 \right) \right]$$

which when summed on i gives

$$(M - r_0) \left(\sum_{i=1}^{n} w_i \right) = \sum_{i=1}^{n} \sum_{j=1}^{s} w_i \overline{x}_{ij} \left(R_j - r_0 \right). \tag{B.3}$$

Take the expectation in (B.3), multiply (B.2) by $w_i \overline{x}_{ij}$, sum on i and j, rearrange the expression using the two properties of covariance mentioned above and the result (1.21) follows. This shows that the risk premium for the market is proportional to $\overline{\alpha}$ which is a measure of the risk aversion in the economy. □

Solution 1.7 Suppose that the investor's initial wealth is $w > 0$ and that he wishes to minimize $\mathbb{E}\left(e^{-aW}\right)$ where

$$W = w\left(r_0 + \sum_{j=1}^{s} x_j \left(R_j - r_0 \right) \right) = w \left[r_0 \left(1 - \boldsymbol{x}^\top \boldsymbol{e} \right) + \boldsymbol{x}^\top \boldsymbol{R} \right],$$

where $\boldsymbol{x} = (x_1, \ldots, x_s)^\top$ and $\boldsymbol{e} = (1, \ldots, 1)^\top$ as usual; the proportion of his wealth in the riskless asset is $x_0 = 1 - \boldsymbol{x}^\top \boldsymbol{e}$. Note that the linear combination $\boldsymbol{x}^\top \boldsymbol{R}$ has the $N\left(\boldsymbol{r}^\top \boldsymbol{x}, \boldsymbol{x}^\top V \boldsymbol{x}\right)$-distribution, then use the expression (A.16) for the moment-generating function of a normally distributed random variable to see that

$$\mathbb{E}\left(e^{-aW}\right) = \exp\left(-awr_0 \left(1 - \boldsymbol{x}^\top \boldsymbol{e} \right) - aw\boldsymbol{r}^\top \boldsymbol{x} + \tfrac{1}{2} a^2 w^2 \boldsymbol{x}^\top V \boldsymbol{x} \right).$$

It is necessary to minimize the expression

$$\tfrac{1}{2} aw \boldsymbol{x}^\top V \boldsymbol{x} - \boldsymbol{x}^\top \left(\boldsymbol{r} - r_0 \boldsymbol{e} \right),$$

for which the minimum occurs when $\boldsymbol{x} = (1/aw) V^{-1} (\boldsymbol{r} - r_0 \boldsymbol{e})$, and the conclusion follows from (1.17). The amount of his wealth invested in the risky assets is $(\boldsymbol{x}^\top \boldsymbol{e}) w = (\beta - \alpha r_0)/a$, which decreases in $a > 0$ when $\beta > \alpha r_0$. □

Solution 1.8 When R_i has the gamma distribution $\Gamma(\gamma_i, \lambda_i)$ we have that $\mathbb{E}\, R_i = r_i = \gamma_i / \lambda_i$ and $\mathbb{V}ar(R_i) = \gamma_i / \lambda_i^2$, from which it follows that $\gamma_i = r_i^2 / \sigma_i^2$ and

$\lambda_i = r_i/\sigma_i^2$. For $\phi + \lambda_i > 0$, note that

$$\mathbb{E}\left(e^{-\phi R_i}\right) = \int_0^\infty \frac{e^{-\phi x} e^{-\lambda_i x} \lambda_i^{\gamma_i} x^{\gamma_i - 1}}{\Gamma(\gamma_i)} dx$$
$$= \left(\frac{\lambda_i}{\phi + \lambda_i}\right)^{\gamma_i} \int_0^\infty \frac{e^{-\phi x} e^{-\lambda_i x} (\phi + \lambda_i)^{\gamma_i} x^{\gamma_i - 1}}{\Gamma(\gamma_i)} dx = \left(\frac{\lambda_i}{\phi + \lambda_i}\right)^{\gamma_i},$$

because the integrand in the latter integral is a probability density function, and so the value of the integral is 1. The investor wishes to solve the constrained optimization problem:

$$\text{maximize } \mathbb{E}\left(1 - e^{-aw(\boldsymbol{x}^\top \boldsymbol{R})}\right) \quad \text{subject to } \boldsymbol{x}^\top \boldsymbol{e} = 1,$$

but this is equivalent to minimizing

$$\mathbb{E}\left(e^{-aw(\boldsymbol{x}^\top \boldsymbol{R})}\right) = \prod_{i=1}^{s} \mathbb{E}\left(e^{-awx_i R_i}\right) = \prod_{i=1}^{s} \left(\frac{\lambda_i}{awx_i + \lambda_i}\right)^{\gamma_i},$$

subject to the constraint. Taking logarithms, we need to

$$\text{maximize } \sum_{i=1}^{s} \gamma_i \ln(awx_i + \lambda_i) \quad \text{subject to } \sum_{i=1}^{s} x_i = 1.$$

Maximizing the Lagrangian

$$\mathcal{L} = \sum_{i=1}^{s} \gamma_i \ln(awx_i + \lambda_i) + \theta\left(1 - \sum_{i=1}^{s} x_i\right)$$

in x_i gives $x_i = (\gamma_i/\theta) - \lambda_i/(aw)$. Substituting back into the constraint shows that the Lagrange multiplier is given as

$$\theta = \frac{\sum_{j=1}^{s} \gamma_j}{1 + (aw)^{-1} \sum_{j=1}^{s} \lambda_j},$$

from which it follows that the optimal portfolio may be expressed as

$$\boldsymbol{x} = \left(1 + (aw)^{-1} \sum_{j=1}^{s} \lambda_j\right) \overline{\boldsymbol{x}} - \left((aw)^{-1} \sum_{j=1}^{s} \lambda_j\right) \boldsymbol{x}_d,$$

where the two portfolios $\overline{\boldsymbol{x}}$ and $\boldsymbol{x}_d$ are

$$(\overline{\boldsymbol{x}})_i = \frac{\gamma_i}{\sum_j \gamma_j} = \frac{r_i^2/\sigma_i^2}{\sum_j r_j^2/\sigma_j^2} \quad \text{and} \quad (\boldsymbol{x}_d)_i = \frac{\lambda_i}{\sum_j \lambda_j} = \frac{r_i/\sigma_i^2}{\sum_j r_j/\sigma_j^2},$$

with the latter portfolio being the diversified portfolio (see Example 1.1 on page 12). As his initial wealth is w, the investor invests the amount $w + \sum_j \lambda_j/a$ in $\overline{\boldsymbol{x}}$ and the amount $-\sum_j \lambda_j/a$ in the diversified portfolio; that is, he is long in $\overline{\boldsymbol{x}}$ and short in the diversified portfolio.

Note that in the case when the random variables R_i have exponential distributions, then $\gamma_i = 1$, or $r_i^2 = \sigma_i^2$, for each $1 \leqslant i \leqslant s$, so that the portfolio $\overline{\boldsymbol{x}}$ is just the **uniform** portfolio $\overline{\boldsymbol{x}} = (1/s, \dots, 1/s)^\top$ which apportions wealth equally between the s risky assets.

For the final part, when there is a riskless asset and we set $x_0 = 1 - \boldsymbol{x}^\top \boldsymbol{e}$, we see that we wish to minimize the expression

$$\begin{aligned}\mathbb{E}\left(e^{-aw\left(r_0(1-\boldsymbol{x}^\top\boldsymbol{e})+\boldsymbol{x}^\top\boldsymbol{R}\right)}\right) &= e^{awr_0(\sum_j x_j - 1)} \prod_{i=1}^{s} \mathbb{E}\left(e^{-awx_iR_i}\right)\\ &= e^{awr_0(\sum_j x_j - 1)} \prod_{i=1}^{s} \left(\frac{\lambda_i}{awx_i + \lambda_i}\right)^{\gamma_i},\end{aligned}$$

in $\boldsymbol{x} = (x_1, \dots, x_s)^\top$, which is equivalent to maximizing

$$\sum_{i=1}^{s} \gamma_i \ln(awx_i + \lambda_i) - awr_0 \sum_{i=1}^{s} x_i.$$

Deduce that for $1 \leqslant i \leqslant s$, the optimal $x_i = (aw)^{-1}((\gamma_i/r_0) - \lambda_i)$, and the optimal investment in the risky assets is determined by

$$\boldsymbol{x} = \left((awr_0)^{-1}\sum_{j=1}^{s}\gamma_j\right)\overline{\boldsymbol{x}} - \left((aw)^{-1}\sum_{j=1}^{s}\lambda_j\right)\boldsymbol{x}_d.$$

The investor is long in the particular risky asset i when $x_i > 0$, which is true if and only if $r_i > r_0$; he is long overall in risky assets if and only if $\sum_{j=1}^{s} x_j > 0$ which is equivalent to the condition that $r_0^{-1} > \sum_{j=1}^{s}(r_j/\sigma_j^2)/\sum_{j=1}^{s}(r_j^2/\sigma_j^2)$. □

B.2 The Binomial Model

Solution 2.1 Let W be the final wealth, then after setting $w_i = W(\omega_i)$, $i = 1, 2$, the problem is

$$\text{maximize} \quad p_1\sqrt{w_1} + p_2\sqrt{w_2} \quad \text{subject to} \quad \alpha\,(q_1w_1 + q_2w_2) = w_0,$$

where q_1, q_2 are the martingale probabilities. Maximizing the Lagrangian for this gives $1/(2\sqrt{w_i}) = \lambda\alpha(q_i/p_i)$ so that $w_i = p_i^2/(2\lambda\alpha q_i)^2$. This yields $\o\lambda =$

$\sqrt{\gamma/(4w_0\alpha)}$ where $\gamma = \sum_{j=1}^{2}(p_j^2/q_j)$, and hence $w_i = w_0(p_i/q_i)^2/(\alpha\gamma)$. The holding in stock is given by

$$\begin{aligned}
x = \frac{1}{S_0}\left(\frac{w_1 - w_2}{u-d}\right) &= \frac{w_0}{\alpha\gamma S_0}\left(\frac{p_1^2}{q_1^2} - \frac{p_2^2}{q_2^2}\right) \\
&= \frac{w_0(u-d)}{\alpha\gamma S_0}\left(\frac{p_1^2}{(1/\alpha - d)^2} - \frac{p_2^2}{(u-1/\alpha)^2}\right) \\
&= \frac{w_0(u-d)}{\alpha\gamma S_0}\left[\frac{(p_1 u + p_2 d - 1/\alpha)(p_1(u-1/\alpha) + p_2(1/\alpha - d))}{(1/\alpha - d)^2(u-1/\alpha)^2}\right] \\
&= \frac{w_0(u-d)}{\alpha^2\gamma S_0^2}\left[\frac{(\mathbb{E}(\alpha S_1) - S_0)(p_1(u-1/\alpha) + p_2(1/\alpha - d))}{(1/\alpha - d)^2(u-1/\alpha)^2}\right]
\end{aligned}$$

from which the result follows. □

Solution 2.2 For (i), we have $u = \frac{3}{2}$, $d = \frac{1}{2}$ and the martingale probability of an up jump is $q = \frac{5}{6}$. The calculations are essentially the same as those made in Example 2.2 and yield the following values:

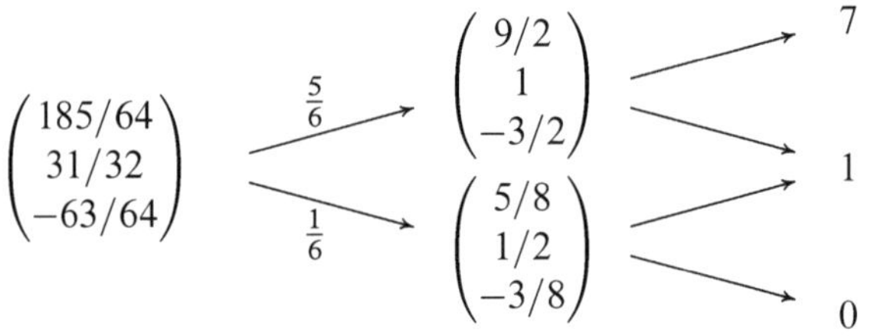

The triples on the nodes are made up of, respectively, the price of the option, the number of units of stock held and the holding in the bank account in the hedging portfolio.

For (ii) we need the condition $\rho < \frac{1}{2}$ to ensure no arbitrage. The martingale probabilities are $q = \rho + \frac{1}{2}$ and $1 - q = \frac{1}{2} - \rho$, which give the calculations on the tree if the American put is not to be exercised before expiry as:

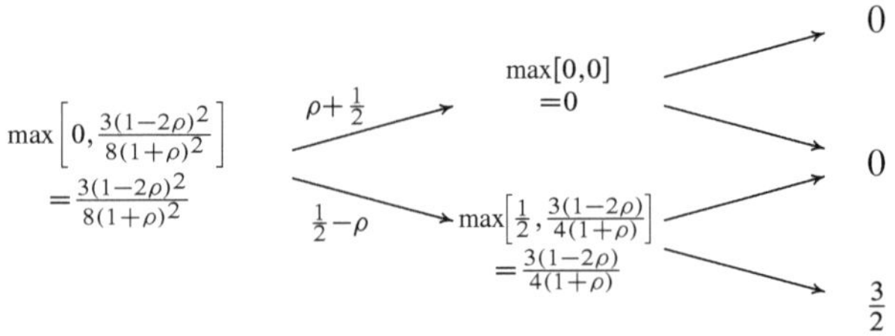

We obtain the condition that $1/2 \leqslant 3(1-2\rho)/4(1+\rho)$ which holds when the interest rate is in the range $0 \leqslant \rho < 1/8$. □

Solution 2.3 In the case of (i), from (2.29) on page 37,

$$
\begin{aligned}
C_r &= \alpha^{n-r} \mathbb{E}_Q\left(S_n^2 \mid \mathcal{F}_r\right) \\
&= \alpha^{n-r} \sum_{i=0}^{n-r} u^{2i} d^{2(n-r-i)} S_r^2 \binom{n-r}{i} q^r (1-q)^{n-r-i} \\
&= \alpha^{n-r} S_r^2 \left[u^2 q + (1-q) d^2\right]^{n-r} = \alpha^{n-r} S_r^2 \left[(1+\rho)(u+d) - ud\right]^{n-r},
\end{aligned}
$$

after substituting $q = (1+\rho-d)/(u-d)$. Using (2.18), it follows that

$$
\begin{aligned}
X_r &= \frac{\alpha^{n-r-1}\left[(1+\rho)(u+d) - ud\right]^{n-r-1}\left((uS_r)^2 - (dS_r)^2\right)}{S_r(u-d)} \\
&= (u+d)\alpha^{n-r-1} S_r \left[(1+\rho)(u+d) - ud\right]^{n-r-1}.
\end{aligned}
$$

In the case of (ii), a similar calculation yields

$$
\begin{aligned}
C_r &= \alpha^{n-r}\left[\ln(S_r d^n) + (n-r) q \ln(u/d)\right] \quad \text{and} \\
X_r &= \alpha^{n-r-1} \ln(u/d) / \left[S_r(u-d)\right].
\end{aligned}
$$

Note that in this latter case, the value of the holding in stock discounted to time-0 values, $\alpha^r X_r S_r$, is held constant. □

Solution 2.4 Recall equation (2.30) satisfied by the holding in the bank account, Y_r at time r,

$$
Y_r = \mathbb{E}_Q(\alpha Y_{r+1} \mid \mathcal{F}_r), \quad \text{for} \quad 0 \leqslant r < n-1,
$$

so that $Y_{r+1} \leqslant 0$ (respectively, $Y_{r+1} \geqslant 0$) shows that $Y_r \leqslant 0$ (respectively, $Y_r \geqslant 0$), so it is only necessary to prove that $Y_{n-1} \leqslant 0$ in the case of (i), or $Y_{n-1} \geqslant 0$ in the case of (ii), and the result follows in each case by backward induction on r. In the case of (i), using (2.73), we have

$$
\begin{aligned}
Y_{n-1} &= \frac{\alpha}{u-d}\left[u f(dS_{n-1}) - d f(uS_{n-1})\right] \\
&= \frac{\alpha d}{u-d}\left[\frac{u}{d} f(dS_{n-1}) - f(uS_{n-1})\right] \\
&\leqslant \frac{\alpha d}{u-d}\left[f(uS_{n-1}) - f(uS_{n-1})\right] = 0.
\end{aligned}
$$

The inequality is reversed in the case of (ii). For the European put option at strike c, $f(x) = (c-x)_+$; for $\lambda \geqslant 1$,

$$
\begin{aligned}
f(\lambda x) &= (c - \lambda x)_+ = (\lambda(c-x) - (\lambda-1)c)_+ \\
&\leqslant (\lambda(x-c))_+ = \lambda(x-c)_+ = \lambda f(x),
\end{aligned}
$$

so that the inequality (2.74) holds. □

Solution 2.5 Set $S_r = x$ say, then from (2.29)

$$f_r(x) = \sum_{i=0}^{n-r} \alpha^{n-r} \binom{n-r}{i} f\left(xu^i d^{n-r-i}\right) q^i (1-q)^{n-r-i}. \tag{B.4}$$

Now when f is convex, so that

$$f(\lambda x + (1-\lambda)\, y) \leqslant \lambda f(x) + (1-\lambda)\, f(y) \tag{B.5}$$

for $0 \leqslant \lambda \leqslant 1$, the same property carries over immediately to f_r from the representation (B.4); the inequality in (B.5) is reversed for concavity. Note that an alternative characterization of convexity is that for $x_1 < x_2 < x_3 < x_4$,

$$\frac{f(x_2) - f(x_1)}{x_2 - x_1} \leqslant \frac{f(x_4) - f(x_3)}{x_4 - x_3},$$

with strict inequality in the case of strict convexity. For the second part of the exercise, suppose that the stock price moves between times r and $r+2$ on the nodes on the left-hand tree with the corresponding values of the claim given on the right-hand side; thus in the notation of the question, $w_1 = f_{r+2}\left(u^2 S_r\right)$ for example.

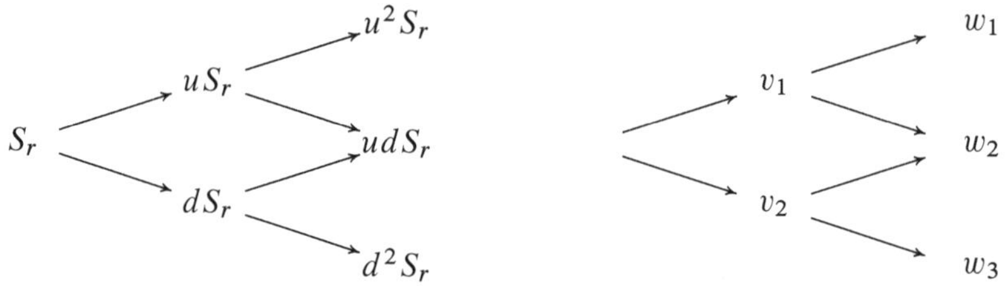

Remember that $v_1 = \alpha\,[qw_1 + (1-q)w_2]$ and $v_2 = \alpha\,[qw_2 + (1-q)w_3]$. The holding in stock in the hedging portfolio at the node where the stock price is S_r is $(v_1 - v_2) / (S_r(u-d))$; when the stock is uS_r it is $(w_1 - w_2) / (S_r u(u-d))$, thus the holding increases if and only if $u\,(v_1 - v_2) \leqslant w_1 - w_2$. The convexity of f_{r+2} implies that

$$\frac{w_2 - w_3}{S_r d(u-d)} \leqslant \frac{w_1 - w_2}{S_r u(u-d)}$$

which gives $w_2 - w_3 \leqslant d\,(w_1 - w_2)\,/u$. Then

$$\begin{aligned} v_1 - v_2 &= \alpha\,[q\,(w_1 - w_2) + (1-q)\,(w_2 - w_3)] \\ &\leqslant \alpha\,[q + (1-q)d/u]\,(w_1 - w_2) \\ &= \alpha\,[qu + (1-q)d]\,(w_1 - w_2)\,/u = (w_1 - w_2)\,/u, \end{aligned}$$

which is the required inequality. □

Solution 2.6 For (i), from (2.42) the optimal wealth satisfies

$$C^{(1/\gamma)-1} = \lambda\alpha^n \frac{d\mathbb{Q}}{d\mathbb{P}} \quad \text{where} \quad \frac{d\mathbb{Q}}{d\mathbb{P}} = L_n = \left(\frac{1-q}{1-p}\right)^n \left[\frac{S_n}{S_0 d^n}\right]^\theta$$

with θ given in (2.37). Let δ satisfy $\gamma^{-1} + \delta^{-1} = 1$, then $C = 1/\left(\lambda^{\delta}\alpha^{n\delta}L_n^{\delta}\right)$. Substituting into the constraint $w_0 = \mathbb{E}_{\mathbb{Q}}(\alpha^n C) = \alpha^n \mathbb{E}(L_n C)$ to determine the Lagrange multiplier λ gives

$$w_0 = \lambda^{-\delta}\alpha^{n(1-\delta)}\mathbb{E}\left(1/L_n^{\delta-1}\right) \quad \text{whence } C = \frac{w_0}{\alpha^n}\frac{1/L_n^{\delta}}{\mathbb{E}\left(1/L_n^{\delta-1}\right)}.$$

Notice that, for real β,

$$\mathbb{E}\left[(S_0/S_n)^{\beta}\right] = \mathbb{E}\left[1/\prod_1^n Z_j^{\beta}\right] = \prod_1^n \mathbb{E}\left(1/Z_j^{\beta}\right) = \left[pu^{-\beta} + (1-p)d^{-\beta}\right]^n,$$

and so

$$C = w_0\left[\frac{d^{\theta}(1-p)}{\alpha(1-q)\left(pu^{\theta(1-\delta)} + (1-p)d^{\theta(1-\delta)}\right)}\right]^n \left(\frac{S_0}{S_n}\right)^{\theta\delta}$$

gives the expression for the optimal C.

For (ii), in this case $v'(C) = e^{-aC} = \lambda\alpha^n L_n$; solving for C and substituting into the constraint to determine λ gives

$$C = \frac{w_0}{\alpha^n} - \frac{\theta}{a}\left(\ln S_n - \mathbb{E}_{\mathbb{Q}}(\ln(S_n))\right),$$

where $\mathbb{E}_{\mathbb{Q}}(\ln(S_n)) = \ln S_0 + n\,(q\ln u + (1-q)\ln d)$ (see also Exercise 2.3). ☐

Solution 2.7 As in Example 2.6 on page 57, this is a non-standard binomial tree in which there are three one-period binomial models embedded, A, B and C with the

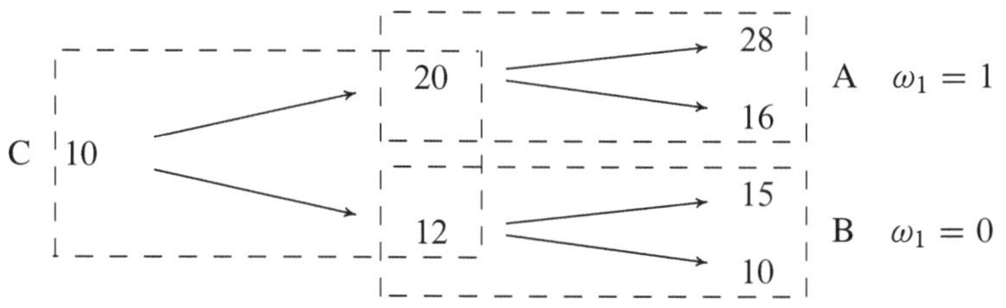

proportional up jumps, u, and down jumps, d, different in each case. The no-arbitrage condition in each submodel is $u > 1 + \rho > d$, which reduces for A to $\frac{2}{5} > \rho > -\frac{1}{5}$, for B to $\frac{1}{4} > \rho > -\frac{1}{6}$ and for C to $1 > \rho > \frac{1}{5}$; we conclude that for (i) we must have $\frac{1}{4} > \rho > \frac{1}{5}$. For (ii) the figures for each of the submodels are summarized in the table below.

	u	d	ρ	α	q
A	$\frac{7}{5}$	$\frac{4}{5}$	$\frac{1}{5}$	$\frac{5}{6}$	$\frac{2}{3}$
B	$\frac{5}{4}$	$\frac{5}{6}$	$\frac{1}{6}$	$\frac{6}{7}$	$\frac{4}{5}$
C	2	$\frac{6}{5}$	$\frac{2}{5}$	$\frac{5}{7}$	$\frac{1}{4}$

At the start of each of the relevant periods, the calculations for the European call of (ii) (a) give the value of the option, the amount of stock held and the holding in the bank account as

$$A: \begin{pmatrix} 10 \\ 1 \\ -10 \end{pmatrix}, \quad B: \begin{pmatrix} \frac{72}{35} \\ \frac{3}{5} \\ -\frac{36}{7} \end{pmatrix}, \quad C: \begin{pmatrix} \frac{283}{98} \\ \frac{139}{140} \\ -\frac{345}{49} \end{pmatrix}.$$

For part (ii) (b), the value of the claim at the start of A is 10, at the start of B it is $\frac{12}{5}$ and at the start of C is $\frac{43}{14}$; it is never optimal to exercise before the expiry time. □

Solution 2.8 The problem is to determine the initial amount of stock, $X_0 = x$, and the initial amount held in the bank account, $Y_0 = y$, so as to

$$\begin{array}{ll} \text{minimize} & x s_{0,0} + y \\ \text{subject to} & x s_{i,n} + (1+\rho)^n y \geqslant f(s_{i,n}), \quad 0 \leqslant i \leqslant n. \end{array} \tag{B.6}$$

Note that (B.6) is a linear programming problem, for which the dual problem is to find $\lambda_0, \ldots, \lambda_n$ so as to

$$\begin{array}{ll} \text{maximize} & \sum_{i=0}^{n} \lambda_i f(s_{i,n}) \\ \text{subject to} & \sum_{i=0}^{n} \lambda_i s_{i,n} = s_{0,0}, \\ & (1+\rho)^n \sum_{i=1}^{n} \lambda_i = 1 \quad \text{and} \quad \lambda_i \geqslant 0, \quad 0 \leqslant i \leqslant n. \end{array} \tag{B.7}$$

Observe that we may write the first constraint in (B.7) as

$$\sum_{i=0}^{n} \lambda_i u^i d^{n-i} = 1.$$

The complementary slackness conditions linking the primal and dual problems, (B.6) and (B.7), are

$$\lambda_i \left[x s_{i,n} + (1+\rho)^n y - f(s_{i,n}) \right] = 0, \quad 0 \leqslant i \leqslant n. \tag{B.8}$$

To find the optimal solution of (B.6), we seek x, y, $\lambda_0, \ldots, \lambda_n$ so that x and y are feasible for (B.6), $\lambda_0, \ldots, \lambda_n$ are feasible for (B.7) and the complementary slackness condition (B.8) is satisfied. The convexity of f suggests that we solve the constraints in (B.6) for $i = 0$ and $i = n$ with equality to get

$$x = \frac{f(s_{n,n}) - f(s_{0,n})}{s_{n,n} - s_{0,n}} \quad \text{and} \quad y = \alpha^n \left[\frac{s_{n,n} f(s_{0,n}) - s_{0,n} f(s_{n,n})}{s_{n,n} - s_{0,n}} \right]. \tag{B.9}$$

Check that x and y given in (B.9) satisfy the constraints in (B.6), which follows because

$$\begin{aligned} x s_{i,n} + (1+\rho)^n y &= \left[\frac{f(s_{n,n}) - f(s_{0,n})}{s_{n,n} - s_{0,n}} \right] s_{i,n} + \frac{s_{n,n} f(s_{0,n}) - s_{0,n} f(s_{n,n})}{s_{n,n} - s_{0,n}} \\ &= \left(\frac{s_{n,n} - s_{i,n}}{s_{n,n} - s_{0,n}} \right) f(s_{0,n}) + \left(\frac{s_{i,n} - s_{0,n}}{s_{n,n} - s_{0,n}} \right) f(s_{n,n}) \\ &\geqslant f\left(\left(\frac{s_{n,n} - s_{i,n}}{s_{n,n} - s_{0,n}} \right) s_{0,n} + \left(\frac{s_{i,n} - s_{0,n}}{s_{n,n} - s_{0,n}} \right) s_{n,n} \right) = f(s_{i,n}), \end{aligned}$$

by the convexity of f. Choose $\lambda_1 = \cdots = \lambda_{n-1} = 0$ and solve the constraints in (B.7) to obtain

$$\lambda_0 = \frac{\alpha^n u^n - 1}{u^n - d^n} > 0 \quad \text{and} \quad \lambda_n = \frac{1 - \alpha^n d^n}{u^n - d^n} > 0. \tag{B.10}$$

It follows that x and y given in (B.9) are optimal for (B.6) because the $\{\lambda_i\}$ defined in (B.10) satisfy dual feasibility and complementary slackness.

To see why the initial value of this optimal static hedging portfolio must be at least as large as the price of the claim, suppose that the price of the claim exceeds the price of the portfolio then an arbitrage may be formed by buying the portfolio and selling the claim; at time 2 the portfolio is worth at least the amount of the claim for all values of S_n and thus a positive profit is made with certainty. Alternatively, we may see this conclusion directly from the dual problem (B.7) without making a direct appeal to arbitrage considerations. Suppose that we set $r_i = \lambda_i (1+\rho)^n$, then we may reformulate (B.7) as

$$\begin{aligned} &\text{maximize} \quad \alpha^n \sum_{i=0}^{n} r_i f(s_{i,n}) \\ &\text{subject to} \quad \alpha^n \sum_{i=0}^{n} r_i s_{i,n} = s_{0,0}, \\ &\qquad \sum_{i=1}^{n} r_i = 1 \quad \text{and} \quad r_i \geqslant 0, \quad 0 \leqslant i \leqslant n. \end{aligned} \tag{B.11}$$

Any feasible solution (r_i) for (B.11) may be thought of as defining a probability distribution $\mathcal{Q}'$ for the stock price S_n, where $r_i = \mathcal{Q}'(S_n = s_{i,n})$, in which case (B.11) is the problem of determining a probability distribution $\mathcal{Q}'$ so as to maximize $\alpha^n \mathbb{E}_{\mathcal{Q}'}(f(S_n))$ over all probability distributions $\mathcal{Q}'$ satisfying $\alpha^n \mathbb{E}_{\mathcal{Q}'}(S_n) = S_0$. The martingale probability $\mathcal{Q}$ satisfies $\alpha^n \mathbb{E}_{\mathcal{Q}}(S_n) = S_0$ and is therefore feasible for the dual problem in this formulation; its dual value is $\alpha^n \mathbb{E}_{\mathbb{Q}}(f(S_n))$ which is the price of the claim at time 0. The initial price of the optimal static hedging portfolio is the optimal value of the primal linear programming problem (B.6) and thus is equal to the optimal value for the dual problem, which in turn dominates the value for any feasible solution to the dual. □

Solution 2.9 We may compute the amounts that the exact hedging portfolio must hold between times 1 and 2 in the usual way, the vectors on the nodes in the diagram below indicate the number of units of stock and the amount in the bank held in each case; for example when the quoted stock price at time 1 is 6, the portfolio between time 1 and time 2 consists of 1 unit of stock and $-\frac{3}{2}$ in the bank. Now suppose that

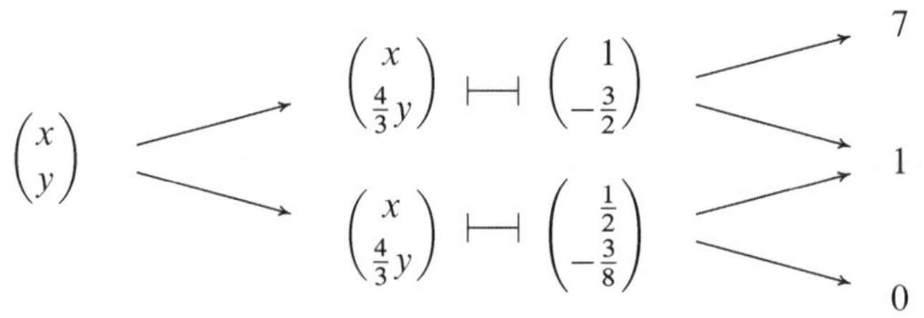

initially the number of units of stock is x and the amount in the bank at time 0 is y; from the discussion in the solution to Exercise 2.5, we would expect to have to buy more stock at time 1 when the stock price goes up and sell stock when the price goes down, so we take $\frac{1}{2} < x < 1$. When the stock price goes up, at time 1 we have to rearrange the portfolio $\left(x, \frac{4}{3}y\right)$ to $\left(1, -\frac{3}{2}\right)$ by buying $(1-x)$ units of stock which gives the condition

$$\tfrac{4}{3}y - (6+\epsilon)(1-x) = -\tfrac{3}{2}. \tag{B.12}$$

Similarly when the stock price goes down, to rearrange $\left(x, \frac{4}{3}y\right)$ to $\left(\frac{1}{2}, -\frac{3}{8}\right)$, we must sell $\left(x - \frac{1}{2}\right)$ units of stock which gives the condition

$$\tfrac{4}{3}y + (2-\delta)\left(x - \tfrac{1}{2}\right) = -\tfrac{3}{8}. \tag{B.13}$$

Solving (B.12) and (B.13) for x and y, yields

$$x = \frac{4\delta + 8\epsilon + 31}{8\,(4+\delta+\epsilon)}, \quad y = \frac{3\,(4\delta\epsilon + 12\delta - 11\epsilon - 42)}{32\,(4+\delta+\epsilon)},$$

which has initial cost

$$4x + y = \frac{12\delta\epsilon + 100\delta + 95\epsilon + 370}{32\,(4+\delta+\epsilon)}. \tag{B.14}$$

The initial cost of the portfolio set up with $\frac{7}{8}$ units of stock and $-\frac{63}{128}$ in the bank costs $\frac{385}{128}$ which is less than the expression in (B.14) if and only if the condition (2.75) holds. For example, the condition holds when $\delta = 1$ and $\epsilon = \frac{3}{2}$. Note that the portfolio with initial holding $\left(\frac{7}{8}, -\frac{63}{128}\right)$ is the optimal static hedging portfolio from Exercise 2.8 in this case. □

B.3 A General Discrete-Time Model

Solution 3.1 For part (i), note that $\mathcal{R}(C) = 0$ implies that $C = aS_1 + b$ for constants a and b (at least with probability one); the conclusion follows immediately. For the second part notice that in the case where there is just one risky asset we may write

$$\mathcal{R}(C) = \mathbb{V}ar(C) - (\mathbb{C}ov(C, S_1))^2 / \mathbb{V}ar(S_1),$$

and then calculate that

$$\mathcal{R}(C_1 + C_2) = \mathcal{R}(C_1) + \mathcal{R}(C_2) + 2\left[\mathbb{C}ov(C_1, C_2) - \frac{\mathbb{C}ov(C_1, S_1)\,\mathbb{C}ov(C_2, S_1)}{\mathbb{V}ar(S_1)}\right].$$

But when the random variables (C_1, C_2, S_1) are jointly normal we have that

$$\mathbb{C}ov\left(C_1, C_2 \mid S_1\right) = \mathbb{C}ov(C_1, C_2) - \mathbb{C}ov(C_1, S_1)\,\mathbb{C}ov(C_2, S_1) / \mathbb{V}ar(S_1);$$

this follows because the random variables $C_i - \mathbb{C}ov(C_i, S_1)\, S_1 / \mathbb{V}ar(S_1)$, $i = 1, 2$, are independent of S_1; recall that in the case of normally-distributed variables the covariance being equal to zero implies independence and thus it follows that

$$\begin{aligned}
\mathbb{C}ov\left(C_1, C_2 \mid S_1\right) &= \mathbb{C}ov\left(C_1 - \frac{\mathbb{C}ov(C_1, S_1)\, S_1}{\mathbb{V}ar(S_1)}, C_2 - \frac{\mathbb{C}ov(C_2, S_1)\, S_1}{\mathbb{V}ar(S_1)} \;\middle|\; S_1\right) \\
&= \mathbb{C}ov\left(C_1 - \frac{\mathbb{C}ov(C_1, S_1)\, S_1}{\mathbb{V}ar(S_1)}, C_2 - \frac{\mathbb{C}ov(C_2, S_1)\, S_1}{\mathbb{V}ar(S_1)}\right) \\
&= \mathbb{C}ov(C_1, C_2) - \mathbb{C}ov(C_1, S_1)\,\mathbb{C}ov(C_2, S_1) / \mathbb{V}ar(S_1),
\end{aligned}$$

which completes the argument. □

Solution 3.2 There are at least a couple of ways to approach this problem:
Either: Argue from first principles, set $L = d\mathbb{Q}/d\mathbb{P}$ for a dominated martingale measure $\mathbb{Q}$, then the problem of minimizing $\mathbb{E}\left[(d\mathbb{Q}/d\mathbb{P})^2\right]$ becomes the constrained optimization problem of finding a random variable L which

$$\text{minimizes} \quad \mathbb{E}(L^2) \quad \text{subject to} \quad \mathbb{E}L = 1 \quad \text{and} \quad S_0 = \mathbb{E}(B_1 L S_1).$$

To simplify the calculation slightly, minimize $\frac{1}{2}\mathbb{E}(L^2)$ rather than $\mathbb{E}(L^2)$, and to do this form a Lagrangian $\mathcal{L}$ (with Lagrange multipliers x_0 and $\boldsymbol{x}$ for the constraints in the problem) given by

$$\begin{aligned}\mathcal{L} &= \tfrac{1}{2}\mathbb{E}(L^2) + x_0(1 - \mathbb{E}L) + \boldsymbol{x}^\top(S_0 - \mathbb{E}(B_1 L S_1)) \\ &= \mathbb{E}\left[\tfrac{1}{2}L^2 - x_0 L - \left(B_1 \boldsymbol{x}^\top S_1\right) L\right] + x_0 + \boldsymbol{x}^\top S_0.\end{aligned}$$

Minimize inside the expectation in $\mathcal{L}$ to obtain $L = x_0 + B_1 \boldsymbol{x}^\top S_1$, and then substituting back into the constraints to determine x_0 and $\boldsymbol{x}$ yields

$$x_0 + B_1 \mathbb{E}\left(\boldsymbol{x}^\top S_1\right) = 1 \quad \text{and} \quad S_0 = \mathbb{E}\left[B_1\left(x_0 + B_1 \boldsymbol{x}^\top S_1\right) S_1\right].$$

Use the relation

$$\mathbb{E}\left[\left(\boldsymbol{x}^\top S_1\right) S_1\right] = V\boldsymbol{x} + \mathbb{E}\left(\boldsymbol{x}^\top S_1\right) \mathbb{E} S_1,$$

to see that

$$S_0 = B_1 x_0 \mathbb{E} S_1 + B_1^2\left(V\boldsymbol{x} + \mathbb{E}\left(\boldsymbol{x}^\top S_1\right) \mathbb{E} S_1\right) = B_1 \mathbb{E} S_1 + B_1^2 V\boldsymbol{x},$$

so that $\boldsymbol{x} = B_1^{-2} V^{-1}(S_0 - B_1 \mathbb{E} S_1)$; this then shows that the minimizing L gives the minimal martingale measure, since we obtain the expression in (3.11)

$$\begin{aligned}L &= 1 - B_1 \mathbb{E}\left(\boldsymbol{x}^\top S_1\right) + B_1 \boldsymbol{x}^\top S_1 = 1 + B_1 \boldsymbol{x}^\top (S_1 - \mathbb{E} S_1) \\ &= 1 - (\mathbb{E} S_1 - r_0 S_0)^\top V^{-1}(S_1 - \mathbb{E} S_1),\end{aligned}$$

after recalling that $B_1 = 1/r_0$.
Or: Alternatively, from (3.11)

$$L = 1 - (\mathbb{E} S_1 - r_0 S_0)^\top V^{-1}(S_1 - \mathbb{E} S_1)$$

defines the Radon–Nikodym derivative of the minimal martingale measure and note that it shows that L $\in \mathcal{A}$, the space of attainable claims; then let $L' = d\mathbb{Q}'/d\mathbb{P}$ for any other dominated martingale measure $\mathbb{Q}'$; without loss of generality assume that $\mathbb{E}\left[(L')^2\right] < \infty$. Now write $L' = L + (L' - L)$ and observe that $L' - L \in \mathcal{U}$, the space of unattainable claims, since $\mathbb{E}(L' - L) = 1 - 1 = 0$ and

$$\mathbb{E}\left[(L' - L) S_1\right] = B_1^{-1}(S_0 - S_0) = 0.$$

It follows that $\mathbb{E}[L(L' - L)] = 0$ and so

$$\mathbb{E}\left[(L')^2\right] = \mathbb{E}(L^2) + \mathbb{E}\left[(L' - L)^2\right] \geqslant \mathbb{E}(L^2),$$

which gives the conclusion. This latter argument shows that the Radon–Nikodym derivative of the minimal martingale measure is the projection onto the space of attainable claims of the Radon–Nikodym derivative of any other dominated martingale measure. □

Solution 3.3 First recall that $\mathbb{E}\left[e^{\theta X_1}\right] = e^{\theta\mu + \theta^2\sigma^2/2}$ so that $d\mathbb{Q}/d\mathbb{P}$ is a strictly positive random variable with $\mathbb{E}\left[d\mathbb{Q}/d\mathbb{P}\right] = 1$ and hence $\mathbb{Q}$ is an equivalent probability. For any real $\{\phi_i\}$, the joint moment-generating function of the $\{X_i\}$ under $\mathbb{Q}$ is

$$\mathbb{E}_{\mathbb{Q}}\left[e^{\sum_1^n \phi_i X_i}\right] = \mathbb{E}\left[\left(\frac{d\mathbb{Q}}{d\mathbb{P}}\right)e^{\sum_1^n \phi_i X_i}\right] = e^{-n\theta\mu - n\theta^2\sigma^2/2}\,\mathbb{E}\left[e^{\sum_1^n (\phi_i+\theta)X_i}\right]$$
$$= \prod_1^n e^{-\theta\mu-\theta^2\sigma^2/2}\,\mathbb{E}\left[e^{(\phi_i+\theta)X_i}\right]$$

by independence, and use (A.16) to show that

$$\mathbb{E}_{\mathbb{Q}}\left[e^{\sum_1^n \phi_i X_i}\right] = \prod_1^n \left[e^{(\phi_i+\theta)\mu + (\phi_i+\theta)^2\sigma^2/2 - \theta\mu - \theta^2\sigma^2/2}\right]$$
$$= \prod_1^n \left[e^{\phi_i(\mu+\theta\sigma^2) + \phi_i^2\sigma^2/2}\right];$$

thus the joint moment-generating function factors into the product of the appropriate individual moment-generating functions as required. The last identity can be rewritten as $\mathbb{E}_{\mathbb{Q}}\left[f(X_1)\right] = \mathbb{E}\left[f\left(X_1+\theta\sigma^2\right)\right]$, which is just a restatement of the previous conclusion that X_1 has the $N(\mu+\theta\sigma^2, \sigma^2)$-distribution under $\mathbb{Q}$. ∎

Solution 3.4 For (i), a claim $U = f(S_1)$ is unattainable if $\mathbb{E}U = 0$ and $\mathbb{E}(US_1) = 0$; from Exercise 3.3, this second condition implies that

$$\mathbb{E}\left[e^X f\left(S_0 e^X\right)\right] = e^{\mu+\sigma^2/2}\,\mathbb{E}\left[f\left(S_0 e^{X+\sigma^2}\right)\right] = 0.$$

Now when we set $X = \sigma Y + \mu$, so that Y has the standard normal distribution, and put $g(y) = f\left(S_0 e^{\sigma y+\mu}\right)$, the two conditions we require are that

$$\mathbb{E}g(Y) = 0 \quad \text{and} \quad \mathbb{E}g(Y+\sigma) = 0.$$

Because the distribution of Y is symmetric about 0, the first condition will be satisfied if, for example, g is an odd function, $g(y) = -g(-y)$, and the second if $g(y+\sigma)$ is odd so that $g(y+\sigma) = -g(-y+\sigma)$; replacing y by $y-\sigma$ in this second condition, we see that the two conditions will be satisfied if $g(y) = -g(-y)$ and $g(y) = g(y+2\sigma)$. An example of a function that will do the trick is

$$g(y) = \begin{cases} 1, & \text{when } 2r\sigma \leqslant y < (2r+1)\sigma, \\ -1, & \text{when } (2r-1)\sigma \leqslant y < 2r\sigma, \end{cases}$$

for integer r, $-\infty < r < \infty$. Of course, if one is not starting from the definition, one could take C to be some claim that is not attainable (some non-linear function of S_1, say $C = S_1^2$) and project it onto the space of unattainable claims to get

$$U = C - \mathbb{E}C - \mathbb{C}ov\,(C, S_1)\,(S_1 - \mathbb{E}S_1)\,/\,\mathbb{V}ar\,(S_1), \quad \text{etc.}$$

For (ii), there is no arbitrage because there exists an equivalent martingale probability $\mathbb{Q}$ for which $\mathbb{E}_{\mathbb{Q}}\left[(1/r_0)\, S_1\right] = S_0$; this $\mathbb{Q}$ must satisfy $\mathbb{E}_{\mathbb{Q}}\left(e^X\right) = r_0$. To find one, from Exercise 3.3, set $d\mathbb{Q}/d\mathbb{P} = e^{\theta X - \theta\mu - \theta^2\sigma^2/2}$, then under the probability $\mathbb{Q}$ the random variable X has the $N\left(\mu + \theta\sigma^2, \sigma^2\right)$-distribution, so that we need

$$\mathbb{E}_{\mathbb{Q}}\left(e^X\right) = e^{\mu + \theta\sigma^2 + \sigma^2/2} = r_0;$$

now choose $\theta = \left(\ln(r_0) - \mu - \sigma^2/2\right)/\sigma^2$. □

Solution 3.5 The steps of the proof of Lemma 3.1 may be followed making appropriate minor changes. Again, we will make no assumption about the integrability of the random vector $\boldsymbol{A}_1$; the proof may be simplified slightly when it is assumed that $\mathbb{E}\,\|\boldsymbol{A}_1\| < \infty$. Clearly both (a) and (b) cannot hold for if they did we would have

$$\begin{aligned} 0 < \mathbb{E}\left[\nu \boldsymbol{x}^\top(\boldsymbol{A}_1 - \boldsymbol{A}_0)\right] &= \mathbb{E}\left(\boldsymbol{x}^\top(\nu \boldsymbol{A}_1)\right) - \mathbb{E}\left(\nu \boldsymbol{x}^\top \boldsymbol{A}_0\right) \\ &= \boldsymbol{x}^\top \mathbb{E}\left(\nu \boldsymbol{A}_1\right) - \boldsymbol{x}^\top \boldsymbol{A}_0 = 0, \end{aligned}$$

giving a contradiction. Now let $\mathcal{Z}$ be the subset of $\mathbb{R}^r$ defined by

$$\begin{aligned} \mathcal{Z} = \{\boldsymbol{z} : \boldsymbol{z} = \mathbb{E}\left(\nu \boldsymbol{A}_1\right),\ &\text{for some random variable } \nu \\ &\text{with } \mathbb{P}\left(\nu > 0\right) = 1, \mathbb{E}\,\nu = 1 \text{ and } \mathbb{E}\,\|\nu \boldsymbol{A}_1\| < \infty\}. \end{aligned}$$

To see that $\mathcal{Z}$ is non-empty set $\overline{\nu} = 1/\left(1 + \max_i |(\boldsymbol{A}_1)_i|\right)$ so that $1 \geqslant \overline{\nu} > 0$ and $\mathbb{E}\,\|\overline{\nu}\boldsymbol{A}_1\| < \infty$, then take $\nu = \overline{\nu}/\mathbb{E}\,\overline{\nu}$ so that $\mathbb{E}\,\nu = 1$, $\mathbb{P}(\nu > 0) = 1$ and $\mathbb{E}\,\|\nu \boldsymbol{A}_1\| < \infty$; it is straightforward to check that $\mathcal{Z}$ is a convex set. Now when (b) does not hold then $\boldsymbol{A}_0 \notin \mathcal{Z}$ so by the Separating Hyperplane Theorem there exists a hyperplane $\mathcal{H} = \{\boldsymbol{z} : \boldsymbol{x}^\top \boldsymbol{z} = \beta\}$ not containing both $\boldsymbol{A}_0$ and $\mathcal{Z}$ with

$$\boldsymbol{x}^\top \boldsymbol{A}_0 \leqslant \beta \leqslant \boldsymbol{x}^\top \mathbb{E}\left(\nu \boldsymbol{A}_1\right)$$

for all ν, with $\mathbb{P}\left(\nu > 0\right) = 1$, $\mathbb{E}\,\nu = 1$ and $\mathbb{E}\,\|\nu \boldsymbol{A}_1\| < \infty$. For all such random variables ν it follows that

$$\mathbb{E}\left[\nu \boldsymbol{x}^\top(\boldsymbol{A}_1 - \boldsymbol{A}_0)\right] \geqslant 0. \tag{B.15}$$

Suppose that $\mathbb{P}\left(\boldsymbol{x}^\top(\boldsymbol{A}_1 - \boldsymbol{A}_0) < 0\right) > 0$, then take the random variable $\overline{\nu}$ as above, and set

$$c = \mathbb{E}\left(\overline{\nu} I_{(\boldsymbol{x}^\top(\boldsymbol{A}_1 - \boldsymbol{A}_0) < 0)}\right) > 0 \quad \text{and} \quad d = \mathbb{E}\left(\overline{\nu} I_{(\boldsymbol{x}^\top(\boldsymbol{A}_1 - \boldsymbol{A}_0) \geqslant 0)}\right) \geqslant 0.$$

Now, for $\epsilon > 0$, consider the random variable

$$\nu_\epsilon = \left(\frac{1 - \epsilon d}{c}\right) \overline{\nu} I_{(\boldsymbol{x}^\top(\boldsymbol{A}_1 - \boldsymbol{A}_0) < 0)} + \epsilon \overline{\nu} I_{(\boldsymbol{x}^\top(\boldsymbol{A}_1 - \boldsymbol{A}_0) \geqslant 0)};$$

this satisfies $\mathbb{E}\left(\nu_\epsilon\right) = 1$, $\mathbb{E}\,\|\nu_\epsilon \boldsymbol{A}_1\| < \infty$ and furthermore $\mathbb{P}\left(\nu_\epsilon > 0\right) = 1$ for all $\epsilon > 0$ sufficiently small. As $\epsilon \downarrow 0$,

$$\mathbb{E}\left[\nu_\epsilon \boldsymbol{x}^\top(\boldsymbol{A}_1 - \boldsymbol{A}_0)\right] \to -(1/c)\mathbb{E}\left[\overline{\nu}\left(\boldsymbol{x}^\top(\boldsymbol{A}_1 - \boldsymbol{A}_0)\right)_-\right] < 0$$

contradicting (B.15). Hence $\mathbb{P}\left(x^\top(A_1 - A_0) \geqslant 0\right) = 1$. We cannot have the situation where $x^\top(A_1 - A_0) \equiv 0$ or else the hyperplane $\mathscr{H}$ would contain both the point A_0 and the set $\mathcal{Z}$. □

Solution 3.6 To see that (a) implies (b), suppose that (x, y) is an arbitrage then $c = x^\top S_0 + y \leqslant 0$ and $x^\top S_1 + yr_1 \geqslant 0$ with $\mathbb{P}\left(x^\top S_1 + yr_1 > 0\right) > 0$. It follows that

$$x^\top(S_1 - r_1 S_0) = x^\top S_1 + (y - c)r_1 \geqslant x^\top S_1 + yr_1 \geqslant 0,$$

so that x satisfies the conditions in (b). Conversely, suppose that x satisfies the conditions in (b), then set $y = -x^\top S_0$ so that

$$x^\top S_1 + yr_1 = x^\top(S_1 - r_1 S_0) \geqslant 0,$$

from which it follows that (x, y) is an arbitrage so that (a) holds. To prove Theorem 3.3, just take $A_0 = r_1 S_0$, $A_1 = S_1$ and see that no arbitrage corresponds to the non-occurrence of case (a) of Exercise 3.5 so that there exists a strictly positive random variable ν with $\mathbb{E}(\nu) = 1$ and $r_1 S_0 = \mathbb{E}(\nu S_1)$; when $\nu = d\mathbb{Q}/d\mathbb{P}$, $\mathbb{Q}$ is an equivalent martingale probability. □

Solution 3.7 The investor chooses x, y to maximize $\mathbb{E}\,v(W)$ subject to the constraint $x^\top S_0 + y = w$, where $W = x^\top S_1 + yr_1$. Eliminate y to see that he needs to choose x to maximize the function

$$f(x) = \mathbb{E}\,v(x^\top(S_1 - r_1 S_0) + wr_1).$$

Set the gradient of f, with respect to x, equal to zero at the optimum value $\overline{W} = \overline{x}^\top(S_1 - r_1 S_0) + wr_1$

$$\frac{\partial f}{\partial x} = \mathbb{E}\left[v'(\overline{W})(S_1 - r_1 S_0)\right] = 0. \tag{B.16}$$

This gives a maximum since the function f is a concave function, which follows by showing that $f(\lambda x_1 + (1-\lambda)x_2) \geqslant \lambda f(x_1) + (1-\lambda)f(x_2)$, for $0 \leqslant \lambda \leqslant 1$, using the corresponding inequality from the concavity of v. Now rearrange (B.16), using the relation $B_1 = 1/r_1$, as

$$\mathbb{E}\left[v'(\overline{W})B_1 S_1\right] = \mathbb{E}(v'(\overline{W}))S_0$$

which in turn may be rewritten as the martingale property $\mathbb{E}_{\mathbb{Q}}(B_1 S_1) = S_0$ when we take the equivalent probability $\mathbb{Q}$ to be specified by $d\mathbb{Q}/d\mathbb{P} = v'(\overline{W})/\mathbb{E}\,v'(\overline{W})$. Note that we have $d\mathbb{Q}/d\mathbb{P} > 0$ because $v' > 0$ since v is strictly increasing. The lack of arbitrage is then a consequence of Theorem 3.3. We are assuming implicitly here that the random variables $v'(W)$ and $v'(W)S_{i,1}$, $1 \leqslant i \leqslant s$, are integrable for all W. □

B.4 Brownian Motion

Solution 4.1 Use the fact that $\{tW_{1/t}\}$ is again a standard Brownian motion, so that for a fixed time $t > 0$,

$$\begin{aligned} \mathbb{P}(M > t) &= \mathbb{P}(W_s = as, \text{ for some } s > t) = \mathbb{P}\left(sW_{1/s} = as, \text{ for some } s > t\right) \\ &= \mathbb{P}(W_u = a, \text{ for some } u < 1/t) = \mathbb{P}(T_a < 1/t); \end{aligned}$$

that is, M has the same distribution as $1/T_a$ and it follows the probability density function of M is $f_{T_a}(1/t)/t^2$, where $f_{T_a}(t) = ae^{-a^2/2t}/\sqrt{2\pi t^3}$, for $t > 0$, from (4.14). Furthermore,

$$\mathbb{E}M = \int_0^\infty (1/t)\, f_{T_a}(1/t)\, dt = 1/a^2,$$

which may be seen by making the substitution $u = a\sqrt{t}$ in the integral and using the identity $\left(\sqrt{2\pi}\right)^{-1}\int_0^\infty u^2 e^{-u^2/2}du = 1/2$. □

Solution 4.2 For the first suggested method in the hint, take expectations in the given relation to obtain

$$\begin{aligned} e^{-b\sqrt{2\theta}} &= \mathbb{E}e^{-\theta T_b} = \mathbb{E}\left[e^{-\theta T} I_{(T_b<T_a)}\right] + \mathbb{E}\left[e^{-\theta(T_b - T_a + T)} I_{(T_a<T_b)}\right] \\ &= \mathbb{E}\left[e^{-\theta T} I_{(T_b<T_a)}\right] + e^{-(b-a)\sqrt{2\theta}}\,\mathbb{E}\left[e^{-\theta T} I_{(T_a<T_b)}\right]. \end{aligned}$$

Interchange the roles of a and b and remember that $a < 0$ to see that

$$e^{a\sqrt{2\theta}} = \mathbb{E}\left[e^{-\theta T} I_{(T_a<T_b)}\right] + e^{-(b-a)\sqrt{2\theta}}\,\mathbb{E}\left[e^{-\theta T} I_{(T_b<T_a)}\right].$$

This gives two simultaneous equations for

$$x = \mathbb{E}\left[e^{-\theta T} I_{(T_b<T_a)}\right] \quad \text{and} \quad y = \mathbb{E}\left[e^{-\theta T} I_{(T_a<T_b)}\right],$$

where $x + y = \mathbb{E}\left[e^{-\theta T}\right]$. Solve these two equations to obtain

$$x = -\frac{\sinh\left(a\sqrt{2\theta}\right)}{\sinh\left((b-a)\sqrt{2\theta}\right)} \quad \text{and} \quad y = \frac{\sinh\left(b\sqrt{2\theta}\right)}{\sinh\left((b-a)\sqrt{2\theta}\right)},$$

and the result follows after using the identities

$$\begin{aligned} \sinh 2z &= 2\sinh z \cosh z \quad \text{and} \\ \sinh z_1 - \sinh z_2 &= 2\sinh((z_1 - z_2)/2)\cosh((z_1 + z_2)/2). \end{aligned}$$

The alternative argument to obtain the two simultaneous equations for x and y above is to observe that

$$X_t = e^{\sqrt{2\theta} W_t - \theta t} \quad \text{and} \quad Y_t = e^{-\sqrt{2\theta} W_t - \theta t}$$

are martingales (both of which are bounded on the event $(T > t)$ for all t) and then the fact that $\mathbb{E} X_T = 1 = \mathbb{E} Y_T$, which follows since $\mathbb{P}(T > t) \to 0$ as $t \to \infty$, gives the two equations after observing that $W_{T_a} = a$ and $W_{T_b} = b$. □

Solution 4.3 Let $T = \min(T_x, T_{-x})$, then $\mathbb{P}(\max_{0 \leqslant s \leqslant t} |W_s| \leqslant x)$ may be expressed as

$$\begin{aligned}
&\mathbb{P}\big(|W_t| \leqslant x\big) - \mathbb{P}\big(|W_t| \leqslant x, T \leqslant t\big) \\
&\quad = \mathbb{P}\big(|W_t| \leqslant x\big) - \mathbb{P}\big(|W_t| \leqslant x, T = T_x \leqslant t\big) - \mathbb{P}\big(|W_t| \leqslant x, T = T_{-x} \leqslant t\big) \\
&\quad = \mathbb{P}\big(|W_t| \leqslant x\big) - 2\mathbb{P}\big(|W_t| \leqslant x, T = T_x \leqslant t\big),
\end{aligned}$$

where the second equality follows from symmetry. Using the reflection principle

$$\begin{aligned}
\mathbb{P}\big(|W_t| \leqslant x, T = T_x \leqslant t\big) &= \mathbb{P}\big(-x \leqslant W_t \leqslant x, T = T_x \leqslant t\big) \\
&= \mathbb{P}\big(x \leqslant W_t \leqslant 3x, T = T_x\big);
\end{aligned}$$

this probability equals

$$\begin{aligned}
&\mathbb{P}\big(x \leqslant W_t \leqslant 3x\big) - \mathbb{P}\big(x \leqslant W_t \leqslant 3x, T = T_{-x}\big) \\
&\qquad = \mathbb{P}\big(x \leqslant W_t \leqslant 3x\big) - \mathbb{P}\big(-5x \leqslant W_t \leqslant -3x, T = T_{-x}\big)
\end{aligned}$$

after another application of the reflection principle, and in turn this

$$= \mathbb{P}\big(x \leqslant W_t \leqslant 3x\big) - \mathbb{P}\big(3x \leqslant W_t \leqslant 5x, T = T_x\big),$$

using the symmetry of the distribution of the process $\{W_t\}$ about 0. The same symmetry implies that

$$2\mathbb{P}\big((2r-1)x \leqslant W_t \leqslant (2r+1)x\big) = \mathbb{P}\big((2r-1)x \leqslant |W_t| \leqslant (2r+1)x\big),$$

and then an inductive argument on n yields

$$\begin{aligned}
\mathbb{P}\Big(\max_{0 \leqslant s \leqslant t} |W_s| \leqslant x\Big) = \mathbb{P}\big(|W_t| \leqslant x\big) + \sum_{r=1}^{n-1} (-1)^r \mathbb{P}\big((2r-1)x \leqslant |W_t| \leqslant (2r+1)x\big) \\
+ (-1)^n 2\mathbb{P}\big((2n-1)x \leqslant W_t \leqslant (2n+1)x, T = T_x\big);
\end{aligned}$$

but the last term $\to 0$ as $n \to \infty$ giving the result. □

Solution 4.4 Let $\{\widetilde{W}_s\}$ be the Brownian motion reflected in the level a after the time T_a, so that $\widetilde{W}_s = W_s$ for $s \leqslant T_a$ and $\widetilde{W}_s = 2a - W_s$ for $s \geqslant T_a$; let $\widetilde{T}_a = T_a$

and $\widetilde{T}_b^a = \inf\{t \geqslant \widetilde{T}_a : \widetilde{W}_t = b\}$ denote the corresponding times for the process $\{\widetilde{W}_s\}$. Then for $x \geqslant b$, by using the reflection principle twice, first reflecting in the level a,

$$\mathbb{E}\left[e^{\nu W_t} I_{(T_b^a \leqslant t,\, W_t \geqslant x)}\right] = \mathbb{E}\left[e^{\nu \widetilde{W}_t} I_{\left(\widetilde{T}_b^a \leqslant t,\, \widetilde{W}_t \geqslant x\right)}\right]$$
$$= \mathbb{E}\left[e^{\nu(2a - W_t)} I_{(T_{2a-b} \leqslant t,\, W_t \leqslant 2a - x)}\right];$$

and then using reflection in the level $2a - b$ to see that this expression

$$= e^{2a\nu}\, \mathbb{E}\left[e^{-\nu(4a-2b-W_t)} I_{(W_t \geqslant 2(a-b)+x)}\right]$$
$$= e^{-2\nu(a-b)+\nu^2 t/2}\, \mathbb{P}\big(W_t + \nu t \geqslant 2(a-b) + x\big),$$

with the last equality coming from Girsanov's Theorem, and then the first result follows because W_t has the $N(0,t)$-distribution. For the case $x \leqslant b$,

$$\mathbb{E}\left[e^{\nu W_t} I_{(T_b^a \leqslant t,\, W_t \leqslant x)}\right] = \mathbb{E}\left[e^{\nu \widetilde{W}_t} I_{\left(\widetilde{T}_b^a \leqslant t,\, \widetilde{W}_t \leqslant x\right)}\right]$$
$$= \mathbb{E}\left[e^{\nu(2a - W_t)} I_{(T_{2a-b} \leqslant t,\, W_t \geqslant 2a - x)}\right]$$
$$= e^{2a\nu}\, \mathbb{E}\left[e^{-\nu W_t} I_{(W_t \geqslant 2a - x)}\right],$$

since $2a - x \geqslant 2a - b$; it follows from Girsanov again that this expression is

$$e^{2a\nu + \nu^2 t/2}\, \mathbb{P}\big(W_t - \nu t \geqslant 2a - x\big) = e^{2a\nu + \nu^2 t/2}\, \Phi\left(\frac{x - 2a - \nu t}{\sqrt{t}}\right)$$

as required. For the final part, when A is the event that the Brownian motion with drift ν hits a and then b before t, then

$$\mathbb{P}\big(A, W_t^\nu \leqslant x\big) = \mathbb{E}\left[e^{\nu W_t - \nu^2 t/2} I_{(T_b^a \leqslant t,\, W_t \leqslant x)}\right],$$

from Girsanov. □

Solution 4.5 Use Girsanov's Theorem to see that

$$\mathbb{E}\left[e^{\theta W_t - \theta^2 t/2} I_{(T_{a,b} > t)}\right] = \mathbb{E}\left[e^{\theta W_t - \theta^2 t/2} I_{\left(\sup_{0 \leqslant s \leqslant t}(W_s - bs) < a\right)}\right]$$
$$= \mathbb{P}\left(\sup_{0 \leqslant s \leqslant t} \big(W_s - (b - \theta)\, s\big) < a\right)$$
$$= \Phi\left(\frac{a + (b-\theta)t}{\sqrt{t}}\right) - e^{-2a(b-\theta)}\, \Phi\left(\frac{-a + (b-\theta)t}{\sqrt{t}}\right).$$

Next, recall that $\{X_t,\, t \geqslant 0\}$ is a martingale where $X_t = e^{\theta W_t - \theta^2 t/2}$, and so by the Optional Sampling Theorem $\mathbb{E}\, X_{T_{a,b} \wedge t} = X_0 = 1$, which gives

$$1 = \mathbb{E}\left[e^{\theta W_t - \theta^2 t/2} I_{(T_{a,b} > t)}\right] + \mathbb{E}\left[e^{\theta(a + bT_{a,b}) - \theta^2 T_{a,b}/2} I_{(T_{a,b} \leqslant t)}\right],$$

after observing that $W_{T_{a,b}} = a + bT_{a,b}$, when $T_{a,b} < \infty$. Set $u = \theta b - \theta^2/2$ and solve for $\theta = b \pm \sqrt{b^2 - 2u}$, for $2u \leqslant b^2$, to see that

$$\mathbb{E}\left[e^{uT_{a,b}} I_{(T_{a,b} \leqslant t)}\right] = e^{-\theta a}\, \mathbb{P}\Big(\sup_{0 \leqslant s \leqslant t} (W_s - (b - \theta)\, s) \geqslant a\Big),$$

and evaluating this expression using the above shows that this equals

$$e^{-ab}\left[e^{-a\sqrt{b^2-2u}}\,\Phi\left(\frac{-a + t\sqrt{b^2 - 2u}}{\sqrt{t}}\right) + e^{a\sqrt{b^2-2u}}\,\Phi\left(\frac{-a - t\sqrt{b^2 - 2u}}{\sqrt{t}}\right)\right];$$

now replace u by θ to get the result. It should be noted that, for $t < \infty$, it does not matter which of the roots for θ in terms of u is used as both lead to the same expression. □

Solution 4.6 Since $d\mathbb{Q}/d\mathbb{P} > 0$, to see that $\mathbb{Q}$ is a probability, it is only necessary to check that $\mathbb{E}\,[d\mathbb{Q}/d\mathbb{P}] = 1$ which follows since

$$\mathbb{E}\left(\theta^{N_t}\right) = \sum_{k=0}^{\infty} \theta^k e^{-\lambda t} \frac{(\lambda t)^k}{k!} = e^{-\lambda t(1-\theta)}.$$

To see that under the probability $\mathbb{Q}$ the process is Poisson with rate $\theta\lambda$, consider fixed times, $0 = t_0 \leqslant t_1 \leqslant \cdots \leqslant t_k = t$, and non-negative integers n_i, $1 \leqslant i \leqslant k$, with $\sum_{i=1}^{k} n_i = n$, then we have

$$\begin{aligned}
\mathbb{Q}\left(N_{t_i} - N_{t_{i-1}} = n_i,\ 1 \leqslant i \leqslant k\right) &= \mathbb{E}\left[\theta^{N_t} e^{\lambda t(1-\theta)} I_{(N_{t_i} - N_{t_{i-1}} = n_i,\, 1 \leqslant i \leqslant k)}\right] \\
&= \theta^n e^{\lambda t(1-\theta)} \prod_{i=1}^{k} e^{-\lambda(t_i - t_{i-1})} \frac{[\lambda\,(t_i - t_{i-1})]^{n_i}}{n_i!} \\
&= \prod_{i=1}^{k} e^{-\theta\lambda(t_i - t_{i-1})} \frac{[\theta\lambda\,(t_i - t_{i-1})]^{n_i}}{n_i!},
\end{aligned}$$

after observing that $\sum_{i=1}^{k} (t_i - t_{i-1}) = t$, which gives the result. □

Solution 4.7 By considering $a_i > 0$, $i = 1, 2$, and keeping b and c fixed it is clear that

$$\mathbb{E}\left[e^{-\theta\tau_{a_1+a_2,b}}\right] = \mathbb{E}\left[e^{-\theta\tau_{a_1,b}}\right]\mathbb{E}\left[e^{-\theta\tau_{a_2,b}}\right],$$

so that $\mathbb{E}\left[e^{-\theta\tau_{a,b}}\right] = e^{-ax}$, for some $x = x(b, \lambda, \theta) \geqslant 0$. Let $T_{a,b}$ denote the first hitting time of the line $a + bt$ by the Brownian motion, then

$$\begin{aligned}
\mathbb{E}\left[e^{-\theta\tau_{a,b}}\right] &= \mathbb{E}\left[\mathbb{E}\left[e^{-\theta\tau_{a,b}} \mid T_{a,b}, \{W_t, N_t, t \leqslant T_{a,b}\}\right]\right] \\
&= \mathbb{E}\left[e^{-\theta T_{a,b}}\, \mathbb{E}\left[e^{-\theta(\tau_{a,b} - T_{a,b})} \mid T_{a,b}, \{W_t, N_t, t \leqslant T_{a,b}\}\right]\right] \\
&= \mathbb{E}\left[e^{-\theta T_{a,b} - cxN_{T_{a,b}}}\right],
\end{aligned}$$

since conditional on $T_{a,b}$ and $N_{T_{a,b}} = n$, say, the difference $\tau_{a,b} - T_{a,b}$ has the distribution of $\tau_{cn,b}$. Now observe that $\mathbb{E}\left[e^{-xN_t}\right] = \exp\left(-\lambda t\left(1-e^{-x}\right)\right)$ and that $T_{a,b}$ and $\{N_t\}$ are independent, then it follows that

$$\mathbb{E}\left[e^{-\theta\tau_{a,b}}\right] = \mathbb{E}\left[e^{-(\theta+\lambda(1-e^{-cx}))T_{a,b}}\right] = e^{-a\left(b+\sqrt{b^2+2\theta+2\lambda(1-e^{-cx})}\right)};$$

equating the two expressions for $\mathbb{E}\left[e^{-\theta\tau_{a,b}}\right]$ shows that

$$x = b + \sqrt{b^2 + 2\theta + 2\lambda\left(1 - e^{-cx}\right)}\,.$$

To see that there is a unique positive root of this equation, note that the right-hand side increases from $b + \sqrt{b^2 + 2\theta}$ to $b + \sqrt{b^2 + 2\theta + 2\lambda}$ as x goes from 0 to ∞ and is concave in x, since if we set $f(x) = \sqrt{b^2 + 2\theta + 2\lambda\left(1 - e^{-cx}\right)}$, then

$$f' = c\lambda e^{-cx}/f \quad \text{and} \quad f'' = -\left(c\lambda e^{-cx}/f\right) - \left(c^2\lambda^2 e^{-2cx}/f^2\right) \leqslant 0.$$

For the final part, take $a = \ln\left(h/S_0\right)/\sigma$, $b = -\mu/\sigma$ and $c = \nu/\sigma$ in the above.

An alternative approach is to use a martingale argument, via the Optional Sampling Theorem, since for any real θ and c the process

$$X_t = \exp\left[\theta\left(W_t - cN_t\right) - \theta^2 t/2 + \lambda t\left(1 - e^{-\theta c}\right)\right],$$

is a martingale relative to the filtration generated by the processes $\{W_t\}$ and $\{N_t\}$. □

Solution 4.8 Set $F = fg$, then apply Itô's Lemma to F to obtain

$$dF\left(X_t, t\right) = \left(Y_t\frac{\partial F}{\partial x} + \frac{\partial F}{\partial t} + \frac{1}{2}Z_t^2\frac{\partial^2 F}{\partial x^2}\right)dt + Z_t\frac{\partial F}{\partial x}\,dW_t.$$

Now substitute in

$$\frac{\partial F}{\partial x} = f\frac{\partial g}{\partial x} + g\frac{\partial f}{\partial x},\ \frac{\partial F}{\partial t} = f\frac{\partial g}{\partial t} + g\frac{\partial f}{\partial t} \ \text{and}\ \frac{\partial^2 F}{\partial x^2} = f\frac{\partial^2 g}{\partial x^2} + g\frac{\partial^2 f}{\partial x^2} + 2\frac{\partial f}{\partial x}\frac{\partial g}{\partial x}$$

to get the result. □

Solution 4.9 Suppose that the integral equals $f\left(W_t, t\right)$; by Itô's Lemma

$$W_t dW_t = df\left(W_t, t\right) = \left(\frac{\partial f}{\partial t} + \frac{1}{2}\frac{\partial^2 f}{\partial x^2}\right)dt + \frac{\partial f}{\partial x}\,dW_t;$$

which gives

$$\frac{\partial f}{\partial t} + \frac{1}{2}\frac{\partial^2 f}{\partial x^2} = 0 \quad \text{and} \quad \frac{\partial f}{\partial x} = x,$$

from which we see that

$$\frac{\partial^2 f}{\partial x^2} = 1 \quad \text{and} \quad \frac{\partial f}{\partial t} = -\frac{1}{2}.$$

Integrate to get $f = -t/2 + g(x)$, with $g'' = 1$ so that $g(x) = x^2/2 + c$; because $f(x, 0) = 0$ we see that $c = 0$. It follows that $\int_0^t W_s dW_s = \left(W_t^2 - t\right)/2$. □

B.5 The Black–Scholes Model

Solution 5.1 To see that q gives the price of the put, calculate exactly as for the Black–Scholes formula. Then with $p(S_t, t)$ as the price of the European call, with p defined in (5.6), put-call parity would imply that

$$q(x,t) = p(x,t) + ce^{-\rho(t_0-t)} - x, \tag{B.17}$$

which may be checked using the identity, $\Phi(x) = 1 - \Phi(-x)$. Recalling results for p from Section 5.2.2, we have from (B.17) that

$$\begin{aligned}\frac{\partial q}{\partial x} &= \frac{\partial p}{\partial x} - 1 = \Phi(d_1) + \left[x\phi(d_1) - ce^{-\rho(t_0-t)}\phi(d_2)\right]\frac{\partial d_1}{\partial x} - 1 \\ &= \Phi(d_1) - 1 < 0,\end{aligned}$$

after using the fundamental identity (5.7), while

$$\frac{\partial^2 q}{\partial x^2} = \frac{\partial^2 p}{\partial x^2} = \phi(d_1)\frac{\partial d_1}{\partial x} = \frac{\phi(d_1)}{\sigma x\sqrt{t_0 - t}} > 0.$$

Similarly

$$\begin{aligned}\frac{\partial q}{\partial c} &= \frac{\partial p}{\partial c} + e^{-\rho(t_0-t)} \\ &= x\phi(d_1)\frac{\partial d_1}{\partial c} + e^{-\rho(t_0-t)}(1 - \Phi(d_2)) - ce^{-\rho(t_0-t)}\phi(d_2)\frac{\partial d_2}{\partial c} \\ &= e^{-\rho(t_0-t)}(1 - \Phi(d_2)) > 0; \\ \frac{\partial^2 q}{\partial c^2} &= \frac{\partial^2 p}{\partial c^2} = -e^{-\rho(t_0-t)}\phi(d_2)\frac{\partial d_2}{\partial c} = \frac{e^{-\rho(t_0-t)}\phi(d_2)}{\sigma c\sqrt{t_0 - t}} > 0;\end{aligned}$$

and

$$\frac{\partial q}{\partial \rho} = \frac{\partial p}{\partial \rho} - c(t_0 - t)e^{-\rho(t_0-t)} = c(t_0 - t)e^{-\rho(t_0-t)}(\Phi(d_2) - 1) < 0.$$

Finally,

$$\frac{\partial q}{\partial t} = \frac{\partial p}{\partial t} + c\rho e^{-\rho(t_0-t)} = ce^{-\rho(t_0-t)}\left[\rho(1 - \Phi(d_2)) - \frac{\sigma\phi(d_2)}{2\sqrt{t_0 - t}}\right];$$

and it is easy to see that this expression may take both positive and negative values. □

Solution 5.2 In the case of (i), recall that $e^{-\rho t} S_t$ has the martingale property under the probability $\mathbb{Q}$, so that $\mathbb{E}_{\mathbb{Q}}\left(e^{-\rho t} S_t\right) = S_0$, and hence the price is

$$\mathbb{E}_{\mathbb{Q}}\left[e^{-\rho t_0}\int_0^{t_0} S_t dt\right] = e^{-\rho t_0}\int_0^{t_0} \mathbb{E}_{\mathbb{Q}}\left(S_t\right) dt$$

$$= e^{-\rho t_0} S_0 \int_0^{t_0} e^{\rho t} dt = S_0\left(1-e^{-\rho t_0}\right)/\rho.$$

For (ii), because calculations under $\mathbb{Q}$ correspond to setting $\mu = \rho$ and performing calculations under $\mathbb{P}$, we see that

$$e^{-\rho t_0}\mathbb{E}_{\mathbb{Q}}\left[\left(\ln S_{t_0}\right)^2\right] = e^{-\rho t_0}\mathbb{E}_{\mathbb{Q}}\left[\left(\ln S_0 + \sigma W_{t_0} + \left(\mu - \sigma^2/2\right) t_0\right)^2\right]$$

$$= e^{-\rho t_0}\mathbb{E}\left[\left(\ln S_0 + \sigma W_{t_0} + \left(\rho - \sigma^2/2\right) t_0\right)^2\right]$$

$$= e^{-\rho t_0}\left[\left(\ln S_0 + \left(\rho - \sigma^2/2\right) t_0\right)^2 + \sigma^2 t_0\right],$$

gives the price, after using the facts that $\mathbb{E}\left[W_{t_0}\right] = 0$ and $\mathbb{E}\left[W_{t_0}^2\right] = t_0$. □

Solution 5.3 If the claim is held between time t and time t_0, it is equivalent to receiving the amount

$$C = \int_t^{t_0} e^{\rho(t_0-u)}\theta \ln\left(S_u\right) du$$

at time t_0, so its price at time t is

$$\mathbb{E}_{\mathbb{Q}}\left[e^{-\rho(t_0-t)}\int_t^{t_0} e^{\rho(t_0-u)}\theta \ln\left(S_u\right) du \,\middle|\, \mathcal{F}_t\right]$$

$$= \theta\,\mathbb{E}_{\mathbb{Q}}\left[\int_t^{t_0} e^{\rho(t-u)} \ln\left(S_u\right) du \,\middle|\, \mathcal{F}_t\right],$$

which reduces to

$$\theta\,\mathbb{E}\left[\int_t^{t_0} e^{\rho(t-u)}\left(\ln\left(S_t\right) + \sigma\left(W_u - W_t\right) + \left(\rho - \sigma^2/2\right)(u-t)\right) du \,\middle|\, \mathcal{F}_t\right].$$

When $\rho \neq 0$, this shows that the price $p(S_t, t)$ is given by

$$\theta\left[\left(1-e^{-\rho(t_0-t)}\right)\left(\rho\ln\left(S_t\right) + \rho - \sigma^2/2\right) - \rho\left(\rho - \sigma^2/2\right)\left(t_0 - t\right)e^{-\rho(t_0-t)}\right]/\rho^2,$$

since $\mathbb{E}\left[\int_t^{t_0}\left(W_u - W_t\right) \,\middle|\, \mathcal{F}_t\right] = \int_t^{t_0}\mathbb{E}\left(W_u - W_t \mid \mathcal{F}_t\right) du = 0$. When $\rho = 0$, the expression for $p(S_t, t)$ is

$$\theta\left[\left(t_0 - t\right)\ln\left(S_t\right) - \sigma^2\left(t_0 - t\right)^2/4\right].$$

The holding in stock is then

$$\left.\frac{\partial p}{\partial x}\right|_{(S_t,t)} = \theta\left(1 - e^{-\rho(t_0-t)}\right) / (\rho S_t), \quad \text{when} \quad \rho \neq 0,$$

or $\theta\,(t_0 - t)\,/\,S_t$, when $\rho = 0$. Verifying that the partial differential equation is satisfied involves a straightforward calculation. □

Solution 5.4 The function $p(\rho, \sigma)$ may be represented as

$$p(\rho, \sigma) = e^{-\rho t_0}\,\mathbb{E}\left[f\left(S_0 e^{\sigma Z\sqrt{t_0} + (\rho - \sigma^2/2)t_0}\right)\right],$$

where Z has the standard normal distribution. Set $\overline{\rho} = \rho_0 - \tau^2 t_0/2, \overline{\sigma} = \sqrt{\sigma^2 + \tau^2 t_0}$ and recall that when X is a random variable with the $N(\mu, \sigma^2)$-distribution then

$$\mathbb{E}\left[e^{\theta X} g(X)\right] = e^{\mu\theta + \theta^2\sigma^2/2}\,\mathbb{E}\,g(X + \theta\sigma^2), \quad \text{for all real} \quad \theta.$$

It follows that

$$\mathbb{E}\,p(\rho, \sigma) = e^{-\rho_0 t_0 + \tau^2 t_0^2/2}\,\mathbb{E}\left[f\left(S_0 e^{\sigma Z\sqrt{t_0} + (\rho - \tau^2 t_0 - \sigma^2/2)t_0}\right)\right],$$

where the expectation on the right-hand side extends over values of Z and ρ, which may be taken to be independent, so this expression

$$= e^{-\overline{\rho} t_0}\,\mathbb{E}\left[f\left(S_0 e^{\overline{\sigma}\overline{Z}\sqrt{t_0} + (\overline{\rho} - \overline{\sigma}^2/2)t_0}\right)\right] = p(\overline{\rho}, \overline{\sigma}),$$

since $\overline{Z} = \left[\sigma Z\sqrt{t_0} + (\rho - \rho_0)\,t_0\right] / \left(\overline{\sigma}\sqrt{t_0}\right)$ has the standard normal distribution. □

Solution 5.5 The problem is to find the optimal final wealth C, which will satisfy

$$v'(C) = C^{-1/q} = \lambda e^{-\rho t_0}\frac{d\mathbb{Q}}{d\mathbb{P}},$$

for suitable Lagrange multiplier λ, where q satisfies $p^{-1} + q^{-1} = 1$. Since

$$\frac{d\mathbb{Q}}{d\mathbb{P}} \propto \left(S_{t_0}/S_0\right)^{(\rho-\mu)/\sigma^2},$$

we may write $C = a\left(S_{t_0}/S_0\right)^{q(\mu-\rho)/\sigma^2}/\lambda^q$ for some constant a and substitute into the constraint $w_0 = e^{-\rho t_0}\,\mathbb{E}_{\mathbb{Q}}\,(C)$, to determine λ, and hence C. Now set $\gamma = q\,(\mu - \rho)\,/\sigma^2$, then it follows that $C = w_0 e^{\rho t_0}\left(S_{t_0}/S_0\right)^\gamma / \mathbb{E}_{\mathbb{Q}}\left(S_{t_0}/S_0\right)^\gamma$, and

$$\begin{aligned}
\mathbb{E}_{\mathbb{Q}}\left(S_{t_0}/S_0\right)^\gamma &= \mathbb{E}\left[\exp\left(\gamma\sigma W_{t_0} + \gamma\left(\rho - \sigma^2/2\right)t_0\right)\right] \\
&= \exp\left(\gamma^2\sigma^2 t_0/2 + \gamma\left(\rho - \sigma^2/2\right)t_0\right) \\
&= \exp\left(\gamma\rho t_0 - \gamma\,(1 - \gamma)\,\sigma^2 t_0/2\right),
\end{aligned}$$

which gives

$$C = w_0 e^{(1-\gamma)(\rho+\gamma\sigma^2/2)t_0} \left(S_{t_0}/S_0\right)^{\gamma}.$$

The conclusion about the proportion of wealth held in stock follows from the discussion in Example 5.1. □

Solution 5.6 To establish the identities, first note that when Z has the standard normal distribution, then for $\theta < 1/2$,

$$\mathbb{E}\left[e^{\theta Z^2} I_{(Z \leqslant x)}\right] = \Phi\left(x\sqrt{1-2\theta}\right)/\sqrt{1-2\theta},$$

while in the case $\theta = 1/2$, $\mathbb{E}\left[e^{Z^2/2} I_{(0 \leqslant Z \leqslant a)}\right] = a/\sqrt{2\pi}$. Observe also that

$$\mathbb{E}\left[Z^2 I_{(0 \leqslant Z \leqslant a)}\right] = \Phi(a) - a\phi(a) - 1/2.$$

For the first identity consider the case $c < 0$ and $c^2 > 2b$, then

$$\begin{aligned}
I &= \int_0^a e^{bx} \Phi\left(c\sqrt{x}\right) dx \\
&= \mathbb{E}\left[\int_0^a e^{bx} I_{(Z \leqslant c\sqrt{x})} dx\right] = \mathbb{E}\left[\int_0^{a \wedge (Z^2/c^2)} e^{bx} I_{(Z \leqslant 0)} dx\right] \\
&= \left[\mathbb{E}\left(e^{b[a \wedge (Z^2/c^2)]} I_{(Z \leqslant 0)}\right) - 1/2\right]/b \\
&= \left[e^{ab}\,\mathbb{P}\left(Z \leqslant c\sqrt{a}\right) + \mathbb{E}\left(e^{b(Z^2/c^2)} I_{(c\sqrt{a} \leqslant Z \leqslant 0)}\right) - 1/2\right]/b,
\end{aligned}$$

which gives the result after using the above expressions. For the case $c > 0$, we see that

$$\begin{aligned}
I = \int_0^a e^{bx} \Phi\left(c\sqrt{x}\right) dx &= \int_0^a e^{bx}\left[1 - \Phi(-c\sqrt{x})\right] dx \\
&= (e^{ab} - 1)/b - \int_0^a e^{bx} \Phi(-c\sqrt{x}) dx,
\end{aligned}$$

and the result for the case $c < 0$ may be used to finish the derivation. The other identities follow in a similar fashion.

Now turn to the average-expiry call (i), the price is (up to the factor $(1/t_0)$),

$$\mathbb{E}_{\mathbb{Q}}\left[e^{-\rho t_0} \int_0^{t_0} (S_t - S_0)_+ \, dt\right] = \int_0^{t_0} e^{-\rho(t_0 - t)} \mathbb{E}_{\mathbb{Q}}\left[e^{-\rho t} (S_t - S_0)_+\right] dt$$

which equals

$$e^{-\rho t_0} S_0 \int_0^{t_0} \left[e^{\rho t} \Phi\left(\frac{2\rho + \sigma^2}{2\sigma}\sqrt{t}\right) - \Phi\left(\frac{2\rho - \sigma^2}{2\sigma}\sqrt{t}\right)\right] dt, \qquad \text{(B.18)}$$

as $\mathbb{E}_{\mathbb{Q}}\left[e^{-\rho t}\left(S_t - S_0\right)_+\right]$ is just the time-0 price of a European call at strike price $c = S_0$ and expiry time t. The integral in (B.18) may now be evaluated using the identities; however, the expression it gives is somewhat unwieldy and it does not simplify significantly. Note that in evaluating $\sqrt{c^2 - 2b}$ when $c = \left(2\rho + \sigma^2\right)/(2\sigma)$ and $b = \rho$, we have to take the absolute value, so that

$$\sqrt{c^2 - 2b} = \left|\left(2\rho - \sigma^2\right)/(2\sigma)\right|.$$

For the average-forward-start case the price is (again up to the factor $(1/t_0)$),

$$\begin{aligned} &\mathbb{E}_{\mathbb{Q}}\left[e^{-\rho t_0}\int_0^{t_0}\left(S_{t_0} - S_t\right)_+ dt\right] \\ &\qquad = \int_0^{t_0} \mathbb{E}_{\mathbb{Q}}\left[e^{-\rho t}\,\mathbb{E}_{\mathbb{Q}}\left[e^{-\rho(t_0 - t)}\left(S_{t_0} - S_t\right)_+ \mid \mathcal{F}_t\right]\right] dt, \end{aligned}$$

which, after seeing that the inner conditional expectation is the time-t price of the call with expiry t_0 and strike equal to S_t, becomes

$$\int_0^{t_0} \mathbb{E}_{\mathbb{Q}}\left[e^{-\rho t}\left[S_t\Phi\left(\frac{2\rho+\sigma^2}{2\sigma}\sqrt{t_0 - t}\right) - S_t e^{-\rho(t_0-t)}\Phi\left(\frac{2\rho-\sigma^2}{2\sigma}\sqrt{t_0 - t}\right)\right]\right] dt,$$

which, in turn is

$$S_0\int_0^{t_0}\left[\Phi\left(\frac{2\rho+\sigma^2}{2\sigma}\sqrt{t}\right) - e^{-\rho t}\Phi\left(\frac{2\rho-\sigma^2}{2\sigma}\sqrt{t}\right)\right] dt,$$

after using the fact that $\mathbb{E}_{\mathbb{Q}}\left[e^{-\rho t}S_t\right] = S_0$ and replacing the integrating variable t by $t_0 - t$. This expression may be evaluated using the identities. □

Solution 5.7 When the barrier has already been hit, so that $\tau_b \leqslant t \leqslant t_0$, the claim is just the standard digital call; write

$$d(x, c) = \frac{\ln(x/c) + \left(\rho - \sigma^2/2\right)(t_0 - t)}{\sigma\sqrt{t_0 - t}},$$

then, with $S_t = x$, the price is given by $e^{-\rho(t_0-t)}\Phi\left(d(x,c)\right)$. Differentiating with respect to x gives the holding in stock to be

$$e^{-\rho(t_0-t)}\phi\left(d(x,c)\right)/\left(x\sigma\sqrt{t_0 - t}\right). \tag{B.19}$$

When $\tau_b > t$, recalling that $\nu = \left(2\rho - \sigma^2\right)/(2\sigma)$, it may be seen from the discussion of up-and-in options that the price is

$$e^{-\rho(t_0-t)}\left[\Phi\left(d(x,b)\right) + (b/x)^{2\nu/\sigma}\left(\Phi\left(d(b^2, xc)\right) - \Phi\left(d(b,x)\right)\right)\right];$$

differentiating with respect to x, to obtain the holding in stock, yields

$$e^{-\rho(t_0-t)}\left[\phi\left(d(x,b)\right)+(b/x)^{2\nu/\sigma}\left(-\phi\left(d(b^2,xc)\right)+\phi\left(d(b,x)\right)\right)\right]/\left(x\sigma\sqrt{t_0-t}\right)$$
$$-\left(2\nu/(x\sigma)\right)(b/x)^{2\nu/\sigma}\left[\Phi\left(d(b^2,xc)\right)-\Phi\left(d(b,x)\right)\right]. \tag{B.20}$$

When $t\to\tau_b$, first from above and then from below, we have $x\to b$ in (B.19) and (B.20) respectively, and subtracting (B.20) from (B.19) gives the expression

$$\frac{2e^{-\rho(t_0-t)}}{b\sigma\sqrt{t_0-t}}\left[\phi\left(d(b,c)\right)-\phi\left(d(b,b)\right)+\nu\sqrt{t_0-t}\left(\Phi\left(d(b,c)\right)-\Phi\left(d(b,b)\right)\right)\right],$$

which is <0 from the hint (5.86); to see this, take $a=d(b,b)=\nu\sqrt{t_0-t}$ and $d(b,c)=x+a$, where $x=\ln(b/c)/\left(\sigma\sqrt{t_0-t}\right)>0$ because $b>c$. To establish the hint (5.86), the difference between the left-hand side and the right-hand side is zero when $x=0$ and its derivative with respect to x is $a\phi(x+a)-(x+a)\phi(x+a)=-x\phi(x+a)<0$ for $x>0$. □

Solution 5.8 Recall from Chapter 4 that when $T_{a,b}$ denotes the first hitting time of the line $a+bt$ by a standard Brownian motion then $\mathbb{E}\left[e^{\theta T_{a,b}}I_{(T_{a,b}\leqslant t)}\right]$ equals

$$e^{-ab}\left[e^{-a\sqrt{b^2-2\theta}}\,\Phi\left(\frac{-a+t\sqrt{b^2-2\theta}}{\sqrt{t}}\right)+e^{a\sqrt{b^2-2\theta}}\,\Phi\left(\frac{-a-t\sqrt{b^2-2\theta}}{\sqrt{t}}\right)\right],$$

for $2\theta\leqslant b^2$. In terms of τ_b, the first hitting time of the barrier b by the stock-price process $\{S_t\}$, the time-0 price of the option is $\mathbb{E}_{\mathbb{Q}}\left[e^{-\rho\tau_b}I_{(\tau_b\leqslant t_0)}\right]$, where $\mathbb{Q}$ is the usual martingale probability. Since $S_t=S_0e^{\sigma W_t+(\mu-\sigma^2/2)t}$, we have $S_t\geqslant b$ if and only if $W_t\geqslant(1/\sigma)\ln\left(b/S_0\right)-\left(\mu-\sigma^2/2\right)t/\sigma$, and under $\mathbb{Q}$ we make calculations by setting $\mu=\rho$. Thus, letting $a=(1/\sigma)\ln\left(b/S_0\right)$ and $\nu=\left(2\rho-\sigma^2\right)/(2\sigma)$, the price is $\mathbb{E}\left[e^{-\rho T_{a,-\nu}}I_{(T_{a,-\nu}\leqslant t_0)}\right]$. Note that $\sqrt{\nu^2+2\rho}=\left(2\rho+\sigma^2\right)/(2\sigma)$. Substituting into the above expression gives

$$\left(\frac{S_0}{b}\right)\Phi\left(\frac{\ln\left(S_0/b\right)+\left(\rho+\sigma^2/2\right)t_0}{\sigma\sqrt{t_0}}\right)$$
$$+\left(\frac{b}{S_0}\right)^{2\rho/\sigma^2}\Phi\left(\frac{\ln\left(S_0/b\right)-\left(\rho+\sigma^2/2\right)t_0}{\sigma\sqrt{t_0}}\right),$$

as the required price. □

Solution 5.9 As usual, $\tau_{b_1}=\inf\{t:S_t\geqslant b_1\}$ and let $\tau_{b_2}^{b_1}=\inf\left\{t\geqslant\tau_{b_1}:S_t\leqslant b_2\right\}$ so that the payoff of this claim is $f\left(S_{t_0}\right)I_{\left(\tau_{b_2}^{b_1}\leqslant t_0\right)}$. Under the martingale probability, $\mathbb{Q}$, the stock-price process $\{S_t\}$ is distributed as $\left\{S_0e^{\sigma W_t^\nu}\right\}$, where $\{W_t^\nu\}$ is a Brownian motion with drift ν; under $\mathbb{Q}$, τ_{b_1} has the same distribution as the first hitting time of the level $a = (1/\sigma)\ln(b_1/S_0)$ by $\{W_t^\nu\}$ and $\tau_{b_2}^{b_1}$ is the subsequent first hitting time of the level $b = (1/\sigma)\ln(b_2/S_0)$. Recall that

$$\mathbb{Q}(S_t \geqslant x) = \mathbb{P}\left(W_t^\nu \geqslant (1/\sigma)\ln(x/S_0)\right) = \Phi\left(\frac{\ln(S_0/x) + \sigma\nu t}{\sigma\sqrt{t}}\right).$$

Using the result from Exercise 4.4, for $x \geqslant b_2$,

$$\begin{aligned}\mathbb{Q}\left(\tau_{b_2}^{b_1} \leqslant t, S_t \geqslant x\right) &= (\kappa_1/\kappa_2)^{\nu/\sigma}\,\Phi\left(\frac{2\ln(b_2/b_1) + \ln(S_0/x) + \sigma\nu t}{\sigma\sqrt{t}}\right)\\ &= (\kappa_1/\kappa_2)^{\nu/\sigma}\,\mathbb{Q}(S_t \geqslant (\kappa_2/\kappa_1)\,x),\end{aligned}$$

while for $x \leqslant b_2$

$$\begin{aligned}\mathbb{Q}\left(\tau_{b_2}^{b_1} \leqslant t, S_t \leqslant x\right) &= (1/\kappa_1)^{\nu/\sigma}\,\Phi\left(\frac{\ln(x/S_0) + \ln(\kappa_1) - \sigma\nu t}{\sigma\sqrt{t}}\right)\\ &= (1/\kappa_1)^{\nu/\sigma}\,\mathbb{Q}(S_t \leqslant \kappa_1 x).\end{aligned}$$

The conclusion follows immediately. □

Solution 5.10 Let $p(\rho, \sigma)$ denote the price of the claim when the stock pays no dividend and the interest rate is ρ (and the volatility is σ). Then from the discussion in Section 5.5, $q(\theta, \sigma) = e^{-\theta t_0} p(\rho - \theta, \sigma)$, so that

$$\begin{aligned}\mathbb{E}\,q(\theta,\sigma) = \mathbb{E}\left[e^{-\theta t_0} p(\rho - \theta, \sigma)\right] &= e^{-\theta_0 t_0 + \tau^2 t_0^2/2}\,\mathbb{E}\,p(\rho - \theta + \tau^2 t_0, \sigma)\\ &= e^{-\theta_0 t_0 + \tau^2 t_0^2/2}\,p\left(\rho - \theta_0 + \tau^2 t_0/2, \sqrt{\sigma^2 + \tau^2 t_0}\right)\end{aligned}$$

using Exercise 5.4, since $\rho - \theta + \tau^2 t_0$ has the $N\left(\rho - \theta_0 + \tau^2 t_0, \tau^2\right)$-distribution, and this

$$= q\left(\theta_0 - \tau^2 t_0/2, \sqrt{\sigma^2 + \tau^2 t_0}\right),$$

as required. □

Solution 5.11 The price is $e^{-\rho t_0}\mathbb{E}_{\mathbb{Q}}\left[f\left(\left(\prod_{i=1}^n S_{t_i}\right)^{1/n}\right)\right]$ where $\mathbb{Q}$ is the usual martingale probability under which the product $\left(\prod_{i=1}^n S_{t_i}\right)^{1/n}$ has the same distribution as

$$S_0 \exp\left(\frac{\sigma}{n}\sum_{i=1}^n W_{t_i} + (\rho - \sigma^2/2)\sum_{i=1}^n t_i/n\right);$$

here $\{W_t, 0 \leqslant t \leqslant t_0\}$ is a standard Brownian motion under $\mathbb{Q}$. For a standard Brownian motion

$$\mathbb{V}ar\left(\sum_{i=1}^{n} W_{t_i}\right) = \sum_{i=1}^{n}\sum_{j=1}^{n} t_i \wedge t_j = \sum_{i=1}^{n}\left(\sum_{j=1}^{i} t_j + \sum_{j=i+1}^{n} t_i\right)$$
$$= \sum_{i=1}^{n}\left(\sum_{j=i}^{n} t_i + \sum_{j=i+1}^{n} t_i\right) = \sum_{i=1}^{n} (2n - 2i + 1)\, t_i,$$

whence $(\sigma/n)\sum_{i=1}^{n} W_{t_i}$ has the same distribution as $\overline{\sigma} W_{t_0}$. To complete the argument note that

$$\left(\rho - \sigma^2/2\right)\sum_{i=1}^{n} t_i/n = \left(\rho - \theta - \overline{\sigma}^2/2\right) t_0.$$

When $t_i = t_0 - (n-i)\delta/n$, it may be seen that

$$\sum_{i=1}^{n} (2n - 2i + 1)\, t_i = n^2 t_0 - \delta\,(n-1)\,(4n+1)\,/6,$$

from which it may be calculated that

$$\overline{\sigma}^2 = \sigma^2\left(1 - \frac{\delta\,(n-1)\,(4n+1)}{6n^2 t_0}\right) \quad \text{and} \quad \theta t_0 = \frac{\delta\,(n-1)}{2n}\left(\rho + \frac{\sigma^2\,(n+1)}{6n}\right),$$

which, in turn, gives the last result. □

Solution 5.12 Conditional on $\mathcal{F}_u$, under the probability $\mathbb{Q}$, we know that S_t has the same distribution as

$$S_u e^{(\rho - \sigma^2/2)(t-u) + \sigma(W_t - W_u)}$$

where $W_t - W_u$ is independent of S_u and has the $N(0, t-u)$-distribution. We then see that $d_2\,(S_t, t)$ has the same distribution as

$$\frac{W_t - W_u}{\sqrt{t_0 - t}} + \sqrt{\frac{t_0 - u}{t_0 - t}}\, d_2\,(S_u, u) = X, \quad \text{say},$$

where X has the $N(\nu, \tau^2)$-distribution, with

$$\nu = \sqrt{\frac{t_0 - u}{t_0 - t}}\, d_2\,(S_u, u) \quad \text{and} \quad \tau^2 = \frac{t-u}{t_0 - t};$$

apply the hint to see that

$$\mathbb{E}_{\mathbb{Q}}\left[\Phi\left(d_2\,(S_t, t)\right) \,\middle|\, \mathcal{F}_u\right] = \Phi\left(\sqrt{\frac{t_0 - u}{t_0 - t}}\, d_2\,(S_u, u) \Big/ \sqrt{1 + \frac{t-u}{t_0 - t}}\right)$$
$$= \Phi\left(d_2\,(S_u, u)\right),$$

as required. For the last part, in place of the hint use the fact that for a random variable $X \sim N(\nu, \tau^2)$,

$$\begin{aligned}\mathbb{E}\left[e^{\theta X}\Phi(X)\right] &= e^{\nu\theta+\theta^2\tau^2/2}\,\mathbb{E}\,\Phi(X+\theta\tau^2)\\ &= e^{\nu\theta+\theta^2\tau^2/2}\Phi\left(\left(\nu+\theta\tau^2\right)/\sqrt{1+\tau^2}\right),\end{aligned}$$

and argue in a similar way to the above. □

Solution 5.13 Use Itô's Lemma and (5.28) to see that

$$\begin{aligned}d\left(e^{-\rho t}p(S_t,t)\right) &= e^{-\rho t}\left[\frac{\partial p}{\partial t}-\rho p+\frac{1}{2}\sigma^2x^2\frac{\partial^2 p}{\partial x^2}\right]_{(S_t,t)}dt+e^{-\rho t}\left.\frac{\partial p}{\partial x}\right|_{(S_t,t)}dS_t\\ &= -e^{-\rho t}\left[\rho x\frac{\partial p}{\partial x}\right]_{(S_t,t)}dt+e^{-\rho t}\left.\frac{\partial p}{\partial x}\right|_{(S_t,t)}dS_t\\ &= \left.\frac{\partial p}{\partial x}\right|_{(S_t,t)}d\left(e^{-\rho t}S_t\right)\end{aligned}$$

Thus $e^{-\rho t}p(S_t,t)$ is a martingale under the probability $\mathbb{Q}$ since it is a stochastic integral with respect to the martingale $e^{-\rho t}S_t$ (provided that the integrand satisfies the condition to ensure the existence of the stochastic integral; a sufficient condition for this would be that $\partial p/\partial x$ is bounded, for example). It follows that

$$e^{-\rho t}p(S_t,t)=\mathbb{E}_{\mathbb{Q}}\left[e^{-\rho t_0}p\left(S_{t_0},t_0\right)\,\middle|\,\mathcal{F}_t\right]=\mathbb{E}_{\mathbb{Q}}\left[e^{-\rho t_0}f\left(S_{t_0}\right)\,\middle|\,\mathcal{F}_t\right],$$

as required. Next, write

$$k(x,t)=xg(x,t)=x\frac{\partial p}{\partial x},$$

and obtain

$$d\left(e^{-\rho t}k(S_t,t)\right)=e^{-\rho t}\left[\frac{\partial k}{\partial t}-\rho k+\frac{1}{2}\sigma^2x^2\frac{\partial^2 k}{\partial x^2}\right]_{(S_t,t)}dt+e^{-\rho t}\left.\frac{\partial k}{\partial x}\right|_{(S_t,t)}dS_t;$$

substitute in

$$\frac{\partial k}{\partial x}=\frac{\partial p}{\partial x}+x\frac{\partial^2 p}{\partial x^2},\quad \frac{\partial k}{\partial t}=x\frac{\partial^2 p}{\partial x\partial t}\quad\text{and}\quad \frac{\partial^2 k}{\partial x^2}=2\frac{\partial^2 p}{\partial x^2}+x\frac{\partial^2 p}{\partial x^3}$$

and use (5.33) to give

$$\begin{aligned}d\left(e^{-\rho t}S_t\right) &= -\rho e^{-\rho t}\left[x\frac{\partial k}{\partial x}\right]_{(S_t,t)}dt+e^{-\rho t}\left.\frac{\partial k}{\partial x}\right|_{(S_t,t)}dS_t\\ &= \left.\frac{\partial k}{\partial x}\right|_{(S_t,t)}d\left(e^{-\rho t}S_t\right),\end{aligned}$$

from which the martingale property follows as previously. □

B.6 Interest-Rate Models

Solution 6.1 For (a), when $s_1 \leqslant s_2 \leqslant t' \leqslant t$, we have

$$\mathbb{E}_{\mathbb{Q}_t}\left[\frac{P_{s_2,t'}}{P_{s_2,t}} \,\Big|\, \mathcal{F}_{s_1}\right] = \mathbb{E}\left[e^{-\int_0^t R_u du}\frac{P_{s_2,t'}}{P_{s_2,t}} \,\Big|\, \mathcal{F}_{s_1}\right] \Big/ \mathbb{E}\left[e^{-\int_0^t R_u du} \,\Big|\, \mathcal{F}_{s_1}\right]$$

since $\mathbb{E}\, P_{0,t}$ cancels in the numerator and denominator; now this expression

$$= \mathbb{E}\left[e^{-\int_{s_1}^t R_u du}\frac{P_{s_2,t'}}{P_{s_2,t}} \,\Big|\, \mathcal{F}_{s_1}\right] \Big/ P_{s_1,t}$$

after dividing through in the numerator and denominator by $e^{-\int_0^{s_1} R_u du}$; now use the tower property of conditional expectations to see that this

$$= \mathbb{E}\left[\mathbb{E}\left[e^{-\int_{s_1}^t R_u du}\frac{P_{s_2,t'}}{P_{s_2,t}} \,\Big|\, \mathcal{F}_{s_2}\right] \Big|\, \mathcal{F}_{s_1}\right] \Big/ P_{s_1,t}$$

and then take the ratio $P_{s_2,t'}/P_{s_2,t}$ outside the inner conditional expectation to see that this

$$= \mathbb{E}\left[\mathbb{E}\left[e^{-\int_{s_1}^t R_u du} \,\Big|\, \mathcal{F}_{s_2}\right]\frac{P_{s_2,t'}}{P_{s_2,t}} \,\Big|\, \mathcal{F}_{s_1}\right] \Big/ P_{s_1,t}$$

and using (6.1), in turn this

$$= \mathbb{E}\left[e^{-\int_{s_1}^{s_2} R_u du} P_{s_2,t}\frac{P_{s_2,t'}}{P_{s_2,t}} \,\Big|\, \mathcal{F}_{s_1}\right] \Big/ P_{s_1,t}$$

$$= \mathbb{E}\left[e^{-\int_{s_1}^{s_2} R_u du} P_{s_2,t'} \,\Big|\, \mathcal{F}_{s_1}\right] \Big/ P_{s_1,t} = P_{s_1,t'}/P_{s_1,t},$$

which demonstrates the martingale property. The calculation for part (b) is similar, in that

$$\mathbb{E}_{\mathbb{Q}_t}\left[R_t \mid \mathcal{F}_s\right] = \mathbb{E}\left[e^{-\int_0^t R_u du} R_t \,\Big|\, \mathcal{F}_s\right] \Big/ \mathbb{E}\left[e^{-\int_0^t R_u du} \,\Big|\, \mathcal{F}_s\right]$$

$$= \mathbb{E}\left[e^{-\int_s^t R_u du} R_t \,\Big|\, \mathcal{F}_s\right] \Big/ P_{s,t}$$

$$= \left[-\frac{\partial}{\partial t}\mathbb{E}\left[e^{-\int_s^t R_u du} \,\Big|\, \mathcal{F}_s\right]\right] \Big/ P_{s,t}$$

$$= -\frac{1}{P_{s,t}}\frac{\partial P_{s,t}}{\partial t} = -\frac{\partial \ln(P_{s,t})}{\partial t} = F_{s,t}.$$

Note that in the second line we have assumed sufficient regularity to justify the interchange of the differentiation and the expectation. Lastly, for (c), the price at time s is

$$\begin{aligned}
\mathbb{E}\left[e^{-\int_s^t R_u du} C \;\middle|\; \mathcal{F}_s\right] &= e^{\int_0^s R_u du}\,\mathbb{E}\left[e^{-\int_0^t R_u du} C \;\middle|\; \mathcal{F}_s\right] \\
&= e^{\int_0^s R_u du}\,\mathbb{E}\left[e^{-\int_0^t R_u du} \;\middle|\; \mathcal{F}_s\right] \mathbb{E}_{\mathbb{Q}_t}\left[C \mid \mathcal{F}_s\right] \\
&= P_{s,t}\,\mathbb{E}_{\mathbb{Q}_t}\left[C \mid \mathcal{F}_s\right].
\end{aligned}$$

This last result may sometimes reduce the calculations involved in determining the price in circumstances where the distribution of C under $\mathbb{Q}_t$ is simpler to obtain than the joint distribution of C and $\int_s^t R_u du$ under $\mathbb{P}$. $\square$

Solution 6.2 Suppose that the bond price $P_{s,t} = f(R_s, s, t)$ for some appropriate function $f = f(r, s, t)$ then, as in the discussion of the Vasicek model, the function f must satisfy

$$\frac{1}{2}\sigma^2 r \frac{\partial^2 f}{\partial r^2} + \alpha(\beta - r)\frac{\partial f}{\partial r} + \frac{\partial f}{\partial s} - rf = 0.$$

Fix t, and try a solution of the form $f = \exp(a - br)$ where $a = a(s)$ and $b = b(s)$, then we obtain

$$\frac{1}{2}\sigma^2 r b^2 - \alpha(\beta - r)b + a' - rb' - r = 0.$$

Equating the coefficient of r to zero gives the two equations

$$\frac{1}{2}\sigma^2 b^2 + \alpha b - b' - 1 = 0 \quad \text{and} \quad a' - \alpha\beta b = 0;$$

the roots of $\frac{1}{2}\sigma^2 b^2 + \alpha b - 1 = 0$ are $b = -(\alpha \pm \gamma)/\sigma^2$, so using partial fractions on the first equation we have

$$\frac{b'}{\frac{1}{2}\sigma^2 b^2 + \alpha b - 1} = \frac{1}{\gamma}\left[\frac{1}{b + ((\alpha - \gamma)/\sigma^2)} - \frac{1}{b + ((\alpha + \gamma)/\sigma^2)}\right] b' = 1,$$

which may be integrated to give

$$\frac{1}{\gamma}\ln\left(\frac{\sigma^2 b + \alpha - \gamma}{\sigma^2 b + \alpha + \gamma}\right) = s + \text{ constant}.$$

Use the boundary condition that $b(t) = 0$ and then solve for b to see that this gives $b \equiv b_{s,t}$ given in (6.60). To derive the expression for a, note that if c is the denominator in the expression for b then, since $(\gamma + \alpha)(\gamma - \alpha) = 2\sigma^2$, the second equation may be written as

$$a' = \alpha\beta b = \frac{2\alpha\beta}{\sigma^2}\left(-\frac{\gamma + \alpha}{2} - \frac{c'}{c}\right),$$

which completes the verification of (6.60) after integrating and using the boundary condition $a(t) = 0$. ∎

Solution 6.3 For $u > s$,

$$R_u = R_s + \int_s^u \theta_v dv + \sigma\left(W_u - W_s\right),$$

and conditional on $\mathcal{F}_s = \sigma\left(R_v,\ v \leqslant s\right)$, $\int_s^t R_u du$ is normally distributed with mean

$$(t-s)R_s + \int_{u=s}^t \int_{v=s}^u \theta_v dv du = (t-s)R_s + \int_s^t (t-v)\theta_v dv,$$

and variance

$$\sigma^2 \mathbb{V}ar\left(\int_s^t \left(W_u - W_s\right) du\right) = \sigma^2 \int_{u=s}^t \int_{v=s}^t \left(u \wedge v - s\right) dv du = \sigma^2 (t-s)^3/3.$$

It follows that

$$\begin{aligned} P_{s,t} &= \mathbb{E}\left[e^{-\int_s^t R_u du} \,\middle|\, \mathcal{F}_s\right] \\ &= \exp\left(-(t-s)R_s - \int_s^t (t-v)\theta_v dv + \sigma^2 (t-s)^3/6\right); \end{aligned} \tag{B.21}$$

for this to agree with the initial term structure we require

$$tR_0 + \int_0^t (t-v)\theta_v dv - \sigma^2 t^3/6 = \int_0^t F_{0,u} du.$$

Differentiating twice with respect to t gives $\theta_t = \frac{dF_{0,t}}{dt} + \sigma^2 t$; calculating

$$\begin{aligned} \int_s^t (t-v)\theta_v dv &= \int_s^t (t-v)\left(\frac{dF_{0,v}}{dv} + \sigma^2 v\right) dv \\ &= -(t-s)F_{0,s} + \ln\left(P_{0,s}/P_{0,t}\right) + \sigma^2\left(t^3 - 3ts^2 + 2s^3\right)/6, \end{aligned}$$

and substituting this into (B.21) gives the expression in (6.61) for $a_{s,t}$. Notice that the expression for the bond prices here implies that the instantaneous forward rate for date t at time s is

$$\begin{aligned} F_{s,t} &= -\frac{\partial \ln P_{s,t}}{\partial t} = -\frac{\partial a_{s,t}}{\partial t} + R_s \\ &= F_{0,t} - F_{0,s} + \sigma^2 s(t-s) + R_s, \end{aligned}$$

so that all the forward rates are assumed to evolve probabilistically exactly as R_s since all the rates at time s differ from R_s by a deterministic quantity. ∎

Solution 6.4 Firstly, when μ_s, σ_s satisfy (6.62) then $P_{s,t} = f(R_s, s, t)$, where $f = f(r, s, t)$, satisfies

$$\frac{1}{2}(c_2 + d_2 r)\frac{\partial^2 f}{\partial r^2} + (c_1 + d_1 r)\frac{\partial f}{\partial r} + \frac{\partial f}{\partial s} - rf = 0.$$

Suppose that we seek a solution of the form $f = e^{a(t-s)-b(t-s)r}$, $f(r, t, t) \equiv 1$ then, equating the terms in r to zero, we obtain

$$\frac{1}{2}d_2 b^2 - d_1 b + b' - 1 = 0,$$

which, may, in principle be solved for b, with the boundary condition $b(0) = 0$, and the remaining equation is

$$a' = \frac{1}{2}c_2 b^2 - c_1 b$$

which, in turn, may be solved for a. Conversely, when $f = e^{a(t-s)-b(t-s)r}$, then

$$\frac{1}{2}\sigma_s^2\frac{\partial^2 f}{\partial r^2} + \mu_s\frac{\partial f}{\partial r} + \frac{\partial f}{\partial s} - rf = 0,$$

implies that

$$\frac{1}{2}b^2(t-s)\sigma^2(r,s) - b(t-s)\mu(r,s) = a'(t-s) + r\left(1 - b'(t-s)\right).$$

Fix s and choose two values of t so that $t - s = t_i$, $i = 1, 2$, then we get two linear equations for σ^2 and μ

$$\frac{1}{2}b^2(t_1)\sigma^2(r,s) - b(t_1)\mu(r,s) = a'(t_1) + r\left(1 - b'(t_1)\right)$$
$$\frac{1}{2}b^2(t_2)\sigma^2(r,s) - b(t_2)\mu(r,s) = a'(t_2) + r\left(1 - b'(t_2)\right),$$

which will yield solutions which are affine in r (provided choices of t_1, t_2 are made so that the equations are linearly independent, which will be the case when we have $b^2(t_1)b(t_2) \neq b^2(t_2)b(t_1)$, so that we choose t_1, t_2 such that $b(t_i) \neq 0$, $i = 1, 2$ and $b(t_1) \neq b(t_2)$). □

Solution 6.5 First recall that

$$X_t = \left(1 - e^{-\alpha t}\right)\beta + e^{-\alpha t}X_0 + \sigma e^{-\alpha t}\int_0^t e^{\alpha s}dW_s,$$

so that $\mathbb{E}X_t = \beta\left(1 - e^{-\alpha t}\right) + e^{-\alpha t}\mathbb{E}X_0 = \beta$, and for $t_1 \leqslant t_2$,

$$\begin{aligned}\mathbb{C}ov\left(X_{t_1}, X_{t_2}\right) &= \mathbb{C}ov\Bigg(e^{-\alpha t_1}X_0 + \sigma e^{-\alpha t_1}\int_0^{t_1} e^{\alpha s}dW_s, \\ &\qquad e^{-\alpha t_2}X_0 + \sigma e^{-\alpha t_2}\int_0^{t_2} e^{\alpha s}dW_s\Bigg) \\ &= e^{-\alpha(t_1+t_2)}\left[\mathbb{V}ar(X_0) + \sigma^2\mathbb{C}ov\left(\int_0^{t_1} e^{\alpha s}dW_s, \int_0^{t_2} e^{\alpha s}dW_s\right)\right].\end{aligned}$$

Because the random variables $\int_0^{t_1} e^{\alpha s} dW_s$ and $\int_{t_1}^{t_2} e^{\alpha s} dW_s$ are independent, and

$$\mathbb{V}ar\left(\int_0^{t_1} e^{\alpha s} dW_s\right) = \int_0^{t_1} e^{2\alpha s} ds = \left(e^{2\alpha t_1} - 1\right) / (2\alpha),$$

we deduce that $\mathbb{C}ov\left(X_{t_1}, X_{t_2}\right) = \sigma^2 e^{-\alpha(t_2 - t_1)} / (2\alpha)$ so that for all t_1 and t_2,

$$\mathbb{C}ov\left(X_{t_1}, X_{t_2}\right) = \frac{\sigma^2}{2\alpha} e^{-\alpha|t_1 - t_2|}.$$

Since a Gaussian process is determined by its mean and covariance structure, any Gaussian process with constant mean and covariance of this form is a stationary Ornstein–Uhlenbeck process.

For the second part,

$$\begin{aligned}\mathbb{C}ov\left(F_{s_1, s_1+t}, F_{s_2, s_2+t}\right) &= c\left(s_1 \wedge s_2, s_1 + t, s_2 + t\right) \\ &= \sigma^2 e^{-\lambda t} e^{-\mu|s_1 - s_2|},\end{aligned}$$

showing that the covariance is of the required form (and since it is stationary it must then be an Ornstein–Uhlenbeck process). Furthermore, after some calculation, from Theorem 6.2 we also have that

$$\begin{aligned}\mathbb{E}\, F_{s,s+t} = \mu_{s,s+t} &= \mu_{0,s+t} + \int_0^{s+t} \left[c(s \wedge v, v, s+t) - c(0, v, s+t)\right] dv \\ &= \mu_{0,s+t} + \frac{\sigma^2}{\mu - \lambda}\left[e^{-\lambda t}\left(1 - e^{-\lambda s}\right) - \frac{\lambda}{\mu} e^{-\mu t}\left(1 - e^{-\mu s}\right)\right].\end{aligned}$$

Since the forward-rate random field is stationary, we have $\mu_{s,s+t} \equiv \mu_{0,t}$ for all s, so that

$$\mu_{0,t} - \frac{\sigma^2}{\mu - \lambda}\left[e^{-\lambda t} - \frac{\lambda}{\mu} e^{-\mu t}\right] = \mu_{0,s+t} - \frac{\sigma^2}{\mu - \lambda}\left[e^{-\lambda(s+t)} - \frac{\lambda}{\mu} e^{-\mu(s+t)}\right],$$

showing that the expression on the left-hand side is independent of t, from which we may conclude that in this case

$$\mu_{0,t} = \frac{\sigma^2}{\mu - \lambda}\left[e^{-\lambda t} - \frac{\lambda}{\mu} e^{-\mu t}\right] + a,$$

for all $t \geqslant 0$, for some constant a. □

Solution 6.6 It is necessary to calculate the quantity given in (6.41) by an argument paralleling that given in the proof of Theorem 6.3; replace the N_1 and N_2 given in that calculation by

$$N_1 = -\int_t^{t+\Delta} F_{t,u} du \quad \text{and} \quad N_2 = \int_s^t R_u du.$$

Again, denote by μ_i and σ_i^2 the mean and variance of N_i conditional on $\mathcal{F}_s$ and use the previous calculations to see that

$$\mu_1 = -\int_t^{t+\Delta} F_{s,u} du - \int_{u=t}^{t+\Delta}\int_{v=s}^{t} (c(v,u,v) - c(s,u,v))\,dv du - \sigma^2(s)/2;$$

$$\mu_2 = \int_s^t F_{s,u} du + \int_{u=s}^{t}\int_{v=s}^{u} (c(v,u,v) - c(s,u,v))\,dv du; \quad \sigma_1^2 = \sigma^2(s);$$

$$\sigma_2^2 = 2\int_{u=s}^{t}\int_{v=s}^{u} (c(v,u,v) - c(s,u,v))\,dv du;$$

and

$$\mathbb{C}ov\,(N_1, N_2 \mid \mathcal{F}_s) = -\int_{u=t}^{t+\Delta}\int_{v=s}^{t} (c(v,u,v) - c(s,u,v))\,dv du.$$

It follows that, in place of (6.38)–(6.40), we have

$$\mu_1 - \mu_2 + \mathbb{V}ar(N_1 - N_2 \mid \mathcal{F}_s)/2 = -\int_s^{t+\Delta} F_{s,u} du = \ln(P_{s,t+\Delta});$$

$$-\mu_2 + \sigma_2^2/2 = -\int_s^t F_{s,u} du = \ln(P_{s,t}); \quad \text{and}$$

$$\begin{aligned}\mu_1 + \sigma_1^2 - \mathbb{C}ov\,(N_1, N_2 \mid \mathcal{F}_s) &= -\int_t^{t+\Delta} F_{s,u} du + \sigma^2(s)/2 \\ &= \ln(P_{s,t+\Delta}/P_{s,t}) + \sigma^2(s)/2.\end{aligned}$$

Substituting these expressions into (6.37), and recalling that here we have $\gamma = \ln k$, gives the formula (6.42). □

Solution 6.7 First note that

$$\left(e^{\gamma} - e^{N_1}\right)_+ - \left(e^{N_1} - e^{\gamma}\right)_+ = e^{\gamma} - e^{N_1}.$$

Mutiply by e^{-N_2} and take expectations to give

$$\begin{aligned}\mathbb{E}\left[e^{-N_2}\left(e^{\gamma} - e^{N_1}\right)_+\right] - \mathbb{E}\left[e^{-N_2}\left(e^{N_1} - e^{\gamma}\right)_+\right] &= \mathbb{E}\left(e^{\gamma - N_2}\right) - \mathbb{E}\left(e^{N_1 - N_2}\right) \\ &= e^{\gamma - \mu_2 + \sigma_2^2/2} - e^{\mu_1 - \mu_2 + \mathbb{V}ar(N_1 - N_2)/2},\end{aligned}$$

from the formula for the moment-generating function of a normally distributed random variable (see (A.16)). Now (6.64) follows from (6.37) using the fact that $\Phi(-x) = 1 - \Phi(x)$.

To obtain the price of the put use the same choice of N_1 and N_2 as in Exercise 6.6, and the values of μ_1, μ_2, etc. as in the solution to that exercise, and substitute into (6.64) to obtain (6.65).

For put-call parity, let $C(s)$ be the time-s price of the call, given in (6.42), and $P(s)$ be the time-s price of the put, given in (6.65), then, by taking the difference of the two expressions, we have

$$C(s) - P(s) = P_{s,t+\Delta} - kP_{s,t}. \tag{B.22}$$

That is, the relation (B.22) shows that the portfolio that holds the call and shorts the put has the same value as the portfolio which holds the bond maturing at time $t + \Delta$ and shorts k bonds maturing at time t. Alternatively, we could have seen that the two portfolios must have the same value at the expiry time t and use an arbitrage argument to say that their values at time s are the same. □

Further Reading

For background on the terminology, the basic ideas and many of the practicalities of pricing derivatives in a light mathematical setting, a good starting place is the classic book by Hull (2008).

The axiomatic treatment establishing the existence of a utility function presented in Chapter 1 follows the lines of Jarrow (1988) and Ingersoll (1987), where more details may be found. Both these books have treatments of the mean-variance approach to portfolio choice; as mentioned, this topic was originated by Markowitz and an account of that research is given in his book Markowitz (1987). The paper by Fama and French (2004) surveys the problems that may arise in applying the capital–asset pricing model.

The martingale approach to derivative pricing presented from Chapter 2 onwards stems from the work of Harrison and Kreps (1979) and Harrison and Pliska (1981, 1983). The binomial model dates back to Cox, Ross and Rubinstein (1979); a very full account of the model may be found in the book by Shreve (2004a), and it is also treated in Pliska (1997). The ideas of the least-squares approach and the intrinsic risk of a claim in the discrete-time setting of Chapter 3 are due to Föllmer and Sondermann (1986). The notion of the minimal martingale measure comes from Föllmer and Schweizer (1990) and Schweizer (1995) (also see Föllmer and Schied (2002)). The definitive treatment of arbitrage theory in a general setting may be found in Delbaen and Schachermayer (2006).

An elementary discussion of Brownian motion is given in the book by Karlin and Taylor (1975). Progressively more advanced treatments of the subject along with stochastic calculus are presented in Karatzas and Shreve (1988), Revuz and Yor (2004) and Rogers and Williams (2000a,b).

As the Black–Scholes model appears in a very high proportion of books on mathematical finance the student new to the subject may be spoiled for choice. The critical factor is the degree of mathematical sophistication required of the reader in the respective treatments. An approach at a comparable mathematical level to that of this work is Shreve (2004b); similar mathematical demands are made by Lamberton and Lapeyre (2008) and Dana and Jeanblanc (2003) while somewhat more background is required by Karatzas and Shreve (1998).

Very good surveys of interest-rate models may be found in the books by James and Webber (2000) and Hunt and Kennedy (2004). Original references for the various named interest-rate models are Vasicek (1977), Hull and White (1990), Cox, Ingersoll and Ross (1985), Ho and Lee (1986) and Heath, Jarrow and Morton (1992). A discussion of affine models is given in the book by Duffie (2001), which also provides useful material on many of the topics included in other chapters. The LIBOR market model was pioneered by Miltersen, Sandmann and Sondermann (1997) and by Brace, Gatarek and Musiela (1997). The treatment of the Gaussian random field model is taken from Kennedy (1994, 1997).

For the mathematical preliminaries in the Appendix, to gain introductions to the language of measure-theoretic probability and to martingales, the first choice is the book by Williams (1991) which sets out many formal ideas in an informal way; another well-regarded treatment is Billingsley (1995). The book by Neveu (1975) is a further good source for martingales in discrete time and it also has a discussion of the notion of the essential supremum of random variables which is relevant to optimal stopping and American options.

References

Billingsley, P. (1995). *Probability and Measure,* Third edition. John Wiley & Sons, Chichester.

Brace, A., Gatarek, D. and M. Musiela (1997). The market model of interest rate dynamics. *Mathematical Finance*, **7**, 127-155.

Cox, J. C., Ingersoll, Jr., J. E. and S. A. Ross (1985). A theory of the term structure of interest rates. *Econometrica*, **53**, 385-407.

Cox, J. C., Ross, S. and M. Rubinstein (1979). Option pricing: a simplified approach. *J. Fin. Econ.*, **7**, 229-263.

Dana, R-A. and M. Jeanblanc (2003). *Financial Markets in Continuous Time.* Springer-Verlag, New York.

Delbaen, F. and W. Schachermayer (2006). *The Mathematics of Arbitrage.* Springer-Verlag, New York.

Duffie, D. (2001). *Dynamic Asset Pricing Theory,* Third edition. Princeton University Press, Princeton.

Fama, E. F. and K. R. French (2004). The capital asset pricing model: theory and evidence. *J. Economic Perspectives*, **18**(3), 25-46.

Föllmer, H. and A. Schied (2002). *Stochastic Finance, An Introduction in Discrete Time.* Walter de Gruyter, Berlin.

Föllmer, H. and M. Schweizer (1990). Hedging of contingent claims under incomplete information, *in* M. H. A. Davis and R. J. Elliott (eds.), *Applied Stochastic Analysis*, Vol. 5 of *Stochastic Monographs*, Gordon and Breach, New York, pp. 389-414.

Föllmer, H. and D. Sondermann (1986). Hedging of non-redundant contingent claims, *in* W. Hildebrand and A. Mas-Colell (eds.), *Contributions to Mathematical Economics in Honor of Gérard Debreu*. North-Holland, Amsterdam, pp. 205-233.

Harrison, J. M. and D. M. Kreps (1979). Martingales and arbitrage in multiperiod securities markets. *J. Econ. Theory*, **20**, 381-408.

Harrison, J. M. and S. R. Pliska (1981). Martingales and stochastic integrals in the theory of continuous trading. *Stochastic Proc. Appl.*, **11**, 215-260.

Harrison, J. M. and S. R. Pliska (1983). A stochastic calculus model of continuous trading: complete markets. *Stochastic Proc. Appl.*, **15**, 313-316.

Heath, D., Jarrow, R. A. and A. J. Morton (1992). Bond pricing and the term structure of interest rates: a new methodology for contingent claims valuation. *Econometrica*, **60**, 77-105.

Ho, T. S. Y. and S-B. Lee (1986). Term structure movements and pricing interest rate contingent claims. *J. Finance*, **41**, 1011-1029.

Hull, J. C. (2008). *Options, Futures and Other Derivatives,* Seventh edition. Prentice-Hall, London.

Hull, J. C. and A. White (1990). Pricing interest rate derivative securities. *Rev. Financial Studies*, **3**, 573-592.

Hunt, P. J. and J. E. Kennedy (2004). *Financial Derivatives in Theory and Practice,* Revised edition. John Wiley & Sons, Chichester.

Ingersoll, Jr., J. E. (1987). *Theory of Financial Decision Making.* Rowman & Littlefield, Savage, MD.

James, J. and N. Webber (2000). *Interest Rate Modelling.* John Wiley & Sons, Chichester.

Jarrow, R. A. (1988). *Finance Theory.* Prentice-Hall, London.

Karatzas, I. and S. E. Shreve (1988). *Brownian Motion and Stochastic Calculus.* Springer-Verlag, New York.

Karatzas, I. and S. E. Shreve (1998). *Methods of Mathematical Finance.* Springer-Verlag, New York.

Karlin, S. and H. M. Taylor (1975). *A First Course in Stochastic Processes,* Second edition. Academic Press, New York.

Kennedy, D. P. (1994). The term structure of interest rates as a Gaussian random field. *Mathematical Finance*, **4**, 247-258.

Kennedy, D. P. (1997). Characterizing Gaussian models of the term structure of interest rates. *Mathematical Finance*, **7**, 107-118.

Lamberton, D. and B. Lapeyre (2008). *Stochastic Calculus Applied to Finance,* Second edition. Chapman & Hall/CRC, Boca Raton, FL.

Markowitz, H. M. (1987). *Mean-Variance Analysis in Portfolio Choice and Capital Markets.* Blackwell, Oxford.

Miltersen, K. R., Sandmann, K. and D. Sondermann (1997). Closed form solutions for term structure derivatives with log-normal interest rates. *J. Finance*, **52**, 409-430.

Neveu, J. (1975). *Discrete-Parameter Martingales.* North-Holland, Amsterdam.

Pliska, S. R. (1997). *Introduction to Mathematical Finance, Discrete Time Models.* Blackwell, Oxford.

Revuz, D. and M. Yor (2004). *Continuous Martingales and Brownian Motion,* Corr. third edition. Springer-Verlag, New York.

L. C. G. Rogers and D. Williams (2000a). *Diffusions, Markov Processes and Martingales. Vol. 1, Foundations,* Second edition. Cambridge University Press, Cambridge.

L. C. G. Rogers and D. Williams (2000b). *Diffusions, Markov Processes and Martingales. Vol. 2, Itô Calculus,* Second edition. Cambridge University Press, Cambridge.

Schweizer, M. (1995). Variance-optimal hedging in discrete time. *Math. Operations Res.*, **20**, 1-32.

Shreve, S. E. (2004a). *Stochastic Calculus for Finance I, The Binomial Asset Pricing Model.* Springer-Verlag, New York.

Shreve, S. E. (2004b). *Stochastic Calculus for Finance II, Continuous-Time Models.* Springer-Verlag, New York.

Vasicek, O. (1977). An equilibrium characterization of the term structure. *J. Fin. Econ.*, **5**, 177-188.

Williams, D. (1991). *Probability with Martingales.* Cambridge University Press, Cambridge.

Index

Printed and bound by CPI Group (UK) Ltd, Croydon, CR0 4YY
21/10/2024
01777085-0008